Insect Neurochemistry and Neurophysiology • 1989

Experimental and Clinical Neuroscience

Insect Neurochemistry and Neurophysiology • 1989,
edited by ***A. B. Borkovec and E. P. Masler,*** 1990

Insect Neurochemistry and Neurophysiology • 1986,
edited by ***A. B. Borkovec and Dale B. Gelman,*** 1986

Molecular Pathology of Nerve and Muscle: *Noxious Agents and Generic Lesions,* edited by ***Antony D. Kidman, John K. Tomkins, Carol A. Morris, and Neil A. Cooper,*** 1983

Neural Membranes, edited by ***Grace Y. Sun, Nicolas Bazan, Jang-Yen Wu, Giuseppe Porcellati, and Albert Y. Sun,*** 1983

Insect Neurochemistry and Neurophysiology • 1989 •

Edited by

A. B. Borkovec and E. P. Masler

USDA, Beltsville, Maryland

Humana Press
Clifton, New Jersey

Library of Congress Cataloging-in-Publication Data

Softcover reprint of the hardcover 1st edition 1990

Crescent Manor
PO Box 2148
Clifton, New Jersey 07015

Insect neurochemistry and neurophysiology, 1989 / edited by A. B. Borkovec and E. P. Masler.
p. cm. — (Experimental and clinical neuroscience)
Includes bibliographical references.
ISBN-13:978-1-4612-8854-1 e-ISBN-13:978-1-4612-4512-4
DOI: 10.1007/978-1-4612-4512-4

1. Insects—Physiology—Congresses. 2. Neurochemistry— Congresses. 3. Neurophysiology—Congresses. 4. Nervous system—Insects—Congresses. I. Borkovec, A. B. (Alexej B.), 1925– II. Masler, Edward P., 1948– . III. Series.
QL495.I4964 1990
595.7'0188—dc20

90-4001
CIP

Preface

This volume contains the proceedings of the Third International Conference on Insect Neurochemistry and Neurophysiology (ICINN'89), which was held at the University of Maryland at College Park, Maryland, on July 10–12, 1989. The ICINN meetings were initiated in 1983 at a time when insect neurobiology was emerging as a vigorous sector of entomology, requiring a degree of attention and autonomy long before granted to its sister field, vertebrate neurobiology. The distinctly medical aspects of the latter were replaced by potentialities for developing new approaches to insect control, and the difference was reflected in the kind of sponsorship that has made the ICINNs possible. It is to the credit of the sponsors of the present conference, i.e., Agricultural Research Service, US Department of Agriculture, American Cyanamid, BASF, Dow Chemical USA, E. I. DuPont de Nemours, Monsanto, Nippon Soda, Rohm and Haas, and Sandoz Crop Protection, that they recognized the need for providing a forum for specialized scientific presentations as well as a medium by which ideas and information could be exchanged directly and cooperation could be planned.

As in the previous conferences in 1983 and 1986, the ICINN'89 program contained state-of-the-art plenary lectures or reviews by invited speakers as well as shorter research communications presented orally or as posters. The latter two are included here in three sections: neuroanatomy, neurochemistry, and neurophysiology. Each of these sections is prefaced by an overview that highlights and summarizes the section's contents. Overview authors were not limited in their assessments of the field to those papers that appear in their specific sections and were encouraged to consider all papers that might have a bearing on their overviews. Plenary speakers were selected by the ICINN'89 Program Chairman, Thomas J. Kelly, with the generous advice of the Organizing Committee, to feature areas of particular current interest to insect neuroscientists. The subject matter of contributed papers, in contrast, was not preselected to highlight any specific area of insect neuroscience.

Alexej B. Borkovec
Edward P. Masler

Contents

v Preface

Reviews

3 Neuropeptides: Unity and Diversity, A Molecular Approach, ***J. Joosse and W. P. M. Geraerts***
39 Isolation Techniques for Insect Neuropeptides, ***D. A. Schooley, H. Kataoka, S. J. Kramer, and A. Toschi***
63 Insect Neurodevelopment: Gene Cascades in *Drosophila*, ***R. Bodmer***
105 Modulation and Repair of the Insect Blood–Brain Barrier, ***J. E. Treherne, E. A. Howes, P. K. Schofield, and P. J. S. Smith***
127 Drug Design, ***T. G. Kempe***
163 Insect Neuroendocrinology: Its Past; Its Present; Future Opportunities, ***L. L. Keeley, T. K. Hayes, and J. Y. Bradfield***

Summaries of Recent Research

Neurochemistry

207 Overview: Progress in Insect Neurochemistry, ***A. K. Raina***
211 *Bombyx* Eclosion Hormone, ***A. Suzuki, H. Nagasawa, T. Kono, B. Sato, T. Kamito, H. Tanaka, Y. Sakagami, A. Mizoguchi, H. Ishizaki, and H. Fugo***
215 Expression of the Eclosion Hormone Gene in the Brain of *Manduca sexta*, ***F. M. Horodyski, L. M. Riddiford, and J. W. Truman***
219 Sequencing and Biological Activities of Neuroparsins Isolated from Corpora Cardiaca of *Locusta migratoria*, ***J. Girardie, J.-C. Huet, J.-C. Pernollet, and A. Girardie***
223 Isolation and Structure of Three Neuropeptides from the Corpora Cardiaca of the American Cockroach, ***J. A. Veenstra***
227 Isolation, Purification, and Characterization of a Diuretic Peptide(AP/I) from the House Cricket, *Acheta domesticus*, ***G. M. Coast, C. H. Wheeler, N. F. Totty, R. J. Philp, and G. J. Goldsworthy***

231 Members of the AKH/RPCH-Peptide Family in Coleoptera, ***G. Gäde***

235 Structures in the AKH Family of Neuropeptides, ***C. H. Wheeler, A. F. Drake, C. M. Wilmot, J. M. Thornton, G. Gäde, and G. J. Goldsworthy***

239 Quaternary Structure and Hydrophobic Interactions of *Drosophila* Acetylcholinesterase in Wild Type Flies and in Several Mutants of the *Ace* Locus, ***J.-P. Toutant and M. Arpagaus***

243 Insect Hypertrehalosemic Hormone: Structure-Activity Studies in *Periplaneta americana*, ***G. Gäde and T. K. Hayes***

247 Insect Hypertrehalosemic Hormone: Structure-Activity Studies in *Blaberus discoidalis*, ***T. K. Hayes, M. M. Ford, and L. L. Keeley***

251 The Role of Mosquito Oostatic Hormone in the Regulation of Midgut Serine Proteases, ***D. Borovsky and D. A. Carlson***

255 Recent Developments in the Use of HPLC for the Purification of House Fly Oostatic Hormone, ***A. B. DeMilo, T. J. Kelly, E. P. Masler, and R. A. Bell***

259 Characterization of Testis Ecdysiotropin(TE) from Brains of Lepidoptera, ***M. J. Loeb, A. B. DeMilo, and C. A. Sheppard***

263 Binding of Synthetic Dipteran Neuropeptide Preparations to Head and Abdominal Proteins of 3-Day-Old Face Flies, ***D. K. Hayes***

267 Neuropeptide Metabolism: Properties of an Endopeptidase from Heads of *Drosophila melanogaster*, ***R. E. Isaac and R. M. Priestly***

271 Degradation of Dipteran Neuropeptides by *Musca autumnalis*, ***R. M. Wagner, T. G. Bird, and J. Kochansky***

275 Metabolism of Inositolphosphates in the Cerebral Ganglia of the American Cockroach, *Periplaneta americana* L., ***P. D. Verhaert, H. R. Walgraeve, and R. G. H. Downer***

281 Tyramine-β-Hydroxylase Activity in the Synganglion of the Cattle Tick *Boophilus microplus* (Acari: Ixodidae), ***L. R. C. Kempton, R. J. Willis, J. B. Pillmoor, and R. E. Isaac***

285 An Ecologically Acceptable Strategy for the Use of Genetically Engineered Baculovirus Pesticides, ***H. A. Wood, P. R. Hughes, N. van Beek, and M. Hamblin***

Neurophysiology

291 Overview: Progress in Insect Neurophysiology, ***C.-M. Yin***

295 Gene Expression During Pupal-Adult Neurogenesis in *Manduca sexta*, ***M.-T. B. Davis, R. S. Hakim, and J. A. Svoboda***

299 Receptivity Terminating Factor in *Heliothis zea* and Its Interaction with PBAN, ***A. K. Raina, T. G. Bird, and J. M. Giebultowicz***

303 Regulation of Sex Pheromone Biosynthesis in *Chrysodeixis chalcites* and *Heliothis peltigera* by PBAN, ***M. Altstein, Y. Gazit, and E. Dunkelblum***

309 Stimulation of *de novo* Pheromone Biosynthesis by In Vitro Pheromone Glands of *Heliothis spp*, ***A. Rafaeli, V. Soroker, J. A. Klun, and A. K. Raina***

313 Regulation of Sex Pheromone Titer in Mated Gypsy Moth Females, ***J. M. Giebultowicz, A. K. Raina, and E. C. Uebel***

317 Changes in Number and Size of Corpus Allatum Cells of *Blattella germanica* During Oocyte Maturation, ***A.-S. Chiang, M. Gadot, and C. Schal***

321 Patterns of Corpora Allata Activity in Ovariectomized *Blattella germanica*, ***M. Gadot, A.-S. Chiang, E. Burns, and C. Schal***

325 Protein Kinase C and Calcium Are Involved in the Action of Juvenile Hormone on Follicle Cells of *Rhodnius prolixus*, ***V. L. Sevala and K. G. Davey***

329 Voltage-Dependent Sodium and Calcium Channels of the Corpora Allata of the Cockroach *Diploptera punctata*, ***A. R. McQuiston, C. S. Thompson, and S. S. Tobe***

333 Juvenile Hormone Production by Wild Type and *Apterous*^{4} Mutant *Drosophila melanogaster* Corpora Allata In Vitro, ***M. Altaratz, D. Segal, D. S. Richard, L. I. Gilbert, and S. W. Applebaum***

337 Developmental Profile of Prothoracicotropic Hormone (PTTH) Activity in the Gypsy Moth, *Lymantria dispar*, ***B. S. Thyagaraja, T. J. Kelly, E. P. Masler, R. A. Bell, R. B. Imberski, and A. B. Borkovec***

341 Endocrinology of Embyrogenesis and Late Embryonic Diapause in the Gypsy Moth, *Lymantria dispar*, ***R. A. Bell, T. J. Kelly, E. P. Masler, B. S. Thyagaraja, A. B. DeMilo, and A. B. Borkovec***

345 Prothoracicotropic Hormone(PTTH) Regulation of the Larval-Pupal Molt in the Gypsy Moth *Lymantria dispar*: Observations on the Development of a Larval In Vivo Bioassay, ***B. S. Thyagaraja, T. J. Kelly, E. P. Masler, R. A. Bell, R. B. Imberski, and A. B. Borkovec***

349 Developmental Profiles for the Regulation of Ecdysteroid Production by the Prothoracic Glands of Female Gypsy Moth, *Lymantria dispar*, Pupae, ***H. W. Fescemyer, E. P. Masler, and T. J. Kelly***

353 Developmental Changes in *Lymantria dispar* (L.) Prothoracicotropic Hormone Activity from Embryo to Adult, ***E. P. Masler, T. J. Kelly, B. S. Thyagaraja, R. A. Bell, D. B. Gelman, R. B. Imberski, and A. B. Borkovec***

357 Prothoracicotropic Hormone and Ecdysteroid Ketoreductase from Pre-Hatch Eggs of the Gypsy Moth, *Lymantria dispar*, ***T. J. Kelly, E. P. Masler, B. S. Thyagaraja, R. A. Bell, D. B. Gelman, R. B. Imberski, and A. B. Borkovec***

361 Ecdysiotropins and Ketoreductase in the Brain of the European Corn Borer, *Ostrinia nubilalis*, Promote the Production of Ecdysone and 3-Dehydroecdysone by Prothoracic Glands of the Gypsy Moth, *Lymantria dispar*, ***D. B. Gelman, B. S. Thyagaraja, T. J. Kelly, E. P. Masler, R. A. Bell, and A. B. Borkovec***

365 Molt Inhibition by Azadirachtin; Effects on Prothoracicotropic Hormone, Prothoracic Glands, and Ecdysteroid Titers in *Manduca sexta* Pupae, ***M. P. Pener, D. B. Rountree, S. T. Bishoff, and L. I. Gilbert***

369 Metabolic Changes in the Nerve Tissue of the Silkworm, *Bombyx mori*, Following Repeated Topical Application of Endosulfan, ***G. K. Jadhav and V. L. Kallapur***

373 Precocene II Suppresses the Biosynthesis of Juvenile Hormone, Ecdysteroids, and Vitellogenin in the Black Blowfly, *Phormia regina* Meigen, ***C.-M. Yin, B.-X. Zou, and J. G. Stoffolano, Jr.***

377 Regulation of Vitellogenesis in Gypsy Moth, *Lymantria dispar* (L.) Fifth-Instar Larvae, ***R. E. Davis, H. W. Fescemyer, T. J. Kelly, E. P. Masler, R. A. Bell, B. S. Thyagaraja, and A. B. Borkovec***

381 6-Hydroxydopamine Retardation of Ovarian Development in the Housefly, *Musca domestica* L., ***J. W. Gerst and T. S. Adams***

385 The Hypolipaemic Hormone of the Locust, ***B. G. Loughton***

389 Demonstration of a Human Atrial Natriuretic Peptide-Like Material in the Stable Fly, *Stomoxys calcitrans*, ***A. C. Chen***

393 Diuresis in Adult *Heliothis zea*: Neural and Humoral Regulation, ***D. W. Bushman and J. O. Nelson***

397 Characterization of FMRF-amide-Like Peptides in the Locust *Shistocerca gregaria*, ***P. D. Evans, S. Robb, and I. Cournil***

401 FMRF-amide in the Corn Earworm (*H. zea*): Immunoreactivity in the Midgut and Cerebral Nervous System, ***J. W. Crim, A. C. Jenkins, M. R. Brown, G. A. Herzog, and A. O. Lea***

405 Uptake and Binding of Ponasterone A in the Brain of Fifth Stadium *Manduca sexta* Larvae, ***H.-J. Bidmon, W. E. Stumpf, and N. A. Granger***

409 The Role of Ecdysone on the Chitinase Enzyme in the Development of *Spodoptera litura*, ***T. Thangaraj, T. J. Kelly, K. Karaiyan, M. Aruchami, and A. B. Borkovec***

413 Prospective Chemical Regulators of Female Reproductive Muscle Function in the Stable Fly *Stomoxys calcitrans*, ***B. J. Cook and R. M. Wagner***

417 Serotonergic Modulation of the Locust Mandibular Closer Muscle, ***R. A. Baines and R. G. H. Downer***
421 Octopaminergic Modulation of Sense Organs Involved in Locust Flight, ***J.-M. Ramirez, I. Orchard, and K. G. Pearson***

Neuroanatomy

427 Overview: Progress in Insect Neuroanatomy and Neurogenesis, ***S. M. Meola***
431 Ultrastructure of a Sensory Receptor in the Abdominal Body Wall of *Rhodnius prolixus*, ***R. G. Chiang, J. A. Chiang, and K. G. Davey***
435 The Association of Proctolin with the Female Reproductive Tissue of *Rhodnius prolixus*, ***A. B. Lange***
439 Serotonin-Like Immunoreactive Dorsal Unpaired Median Neurons in *Rhodnius prolixus*, ***I. Orchard, A. B. Lange, H. Cook, and J.-M. Ramirez***
443 Immunocytochemical Localization of Leucomyosuppressin (LEM-MS) in the CNS of the Cockroach, *Leucophaea maderae* and the Stable Fly, *Stomoxys calcitrans*, ***S. M. Meola, M. S. Wright, G. M. Holman, and J. M. Thompson***
447 A Vasopressin-Like Neurohemal System in *Lymantria dispar* and *Manduca sexta*, ***N. T. Davis and H. Lehman***
451 Histocytochemical Identification of Neurosecretory Cells in the Brain of Short Horned Grasshopper, ***M. U. Patil***

455 List of Participants
463 Author Index
467 Subject Index

Reviews

From *Insect Neurochemistry and Neurophysiology • 1989* Edited by A. B. Borkovec and E. P. Masler

NEUROPEPTIDES: UNITY AND DIVERSITY, A MOLECULAR APPROACH

J. Joosse and W.P.M. Geraerts
Department of Biology, Vrije Universiteit,
De Boelelaan 1087, 1081 HV Amsterdam,
The Netherlands

SUMMARY

Neuropeptides are neuronal messengers which occur throughout the animal kingdom in a great molecular diversity. In recent years the molecular structure and gene coding of a number of neuropeptides of invertebrates has been elucidated. It appears from this data that we can distinguish some families of neuropeptides which occur in various groups of invertebrates as well as in vertebrates. These peptides, therefore, have an extremely long evolutionary history, up to 600 million years. As examples of these peptides we discuss here insulin and insulin-related peptides, the FMRFamide family, the head activator peptide and the vasopressin/vasotocin family.

On the other hand, there are neuropeptides which appear to be restricted to specific animal groups. Possibly these peptides evolved for control of a group specific adaptation. From this category we present here the egg-laying hormones of snails, and the adipokinetic and eclosion hormones of insects.

Besides, the genomically coded peptides, nerve cells use enzymatically produced "classical" transmitters, such as acetylcholine, as messengers. The advantages of this messenger dualism is discussed.

INTRODUCTION

In the history of the study of the nervous system various phases can be distinguished, each of which is characterized by the introduction of new concepts and techniques to unravel the secrets of the most complex organ system of animals and man. In this section an impression of this history is given in the light of the phylogenetic relationships of the animal groups. The first comprehensive anatomical pictures of the human brain and peripheral nervous system were published in 1543 by Andreas Vesalius in his beautifully illustrated "De humane corporis fabrica". The earliest detailed morphological pictures of the nervous systems of invertebrates, i.e. of insects and snails, were published in 1737 by Jan Swammerdam in his "Biblia Naturae". Of crucial importance for the progress in the morphology of the brain was the introduction of excellent histological techniques by Ramón y Cajal (1894, 1911). This allowed the identification of neurones as an entity including axons, collaterals and dendrites. This approach formed the basis of our understanding of behavioural actions as a result of specific and localizable connections between sensory and motor structures. The morphological studies showed that all animals, except sponges, have nervous systems which consist of neurones and supporting glial elements. These neurones have been formed very early in the evolution of animals. The comparative morphology of the nervous system appeared to be particularly useful for the study of phylogenetic relationships of animals.

Although Galvani, already in the 18th century, had demonstrated that nerves innervating muscles show electrical phenomena, the concept of the action potential was not introduced until 1930. At that time also the role of chemical tranmission (neurotransmission) was recognized. These two concepts have stimulated brain research enormously and led to the development of electrophysiology as a discipline. It became generally accepted that neurones produce action potentials and release chemical transmitters. With regard to the neurotransmitters, until recently attention was nearly exclusively focussed on the role of so-called classical transmitters such as acetylcholine, serotonin and amino acids. From the phylogenetic point of view the results in this field demonstrated a high degree of unity in the animal groups. Classical transmitters were found in nearly all animal groups (see below), and in some cases specializations occur, such as the GABA-ergic (instead of cholinergic) innervation of the abdominal motor system of crustaceans (cf. Leake and Walker, 1980).

So far we have considered the nervous system as a system with input from sensory organs and output to muscles with integration centres in between. However, as early as 1928 Ernst Scharrer postulated an endocrine role for the nervous system on the basis of the presence of secretory material in hypothalamic neurones in the brain of a fish. It was not until 1949 that the concept of neurosecretion was generally accepted. This occurred after the demonstration of the hypothalamic neurohypophysial pathway by Bargmann (1949). The products of neurosecretory neurones appeared to have a peptide nature (peptidergic neurones, neuropeptides). Since these products appeared to be released in the blood, neuropeptides were initially considered neurohormones. These new insights led to the conclusion that the nervous system acts not only directly on muscles and glands, but also at distance by endocrine mechanisms. Some of the neurohormones appeared to be involved in the control of endocrine organs: in fact at present all endocrine organs are considered to be controlled by neuroendocrine messengers. From the early histological studies of Berta Scharrer (e.g., 1935) and functional investigations of Sir Vincent B. Wigglesworth (1936) on molluscs and insects, respectively, it was already evident that the phenomenon of neurosecretion occurred in all animal groups at the early stage of the development of the concept of neurosecretion. Since the number of endocrine organs in invertebrates is very low, endocrine control in these groups is mainly exerted by the nervous system.

The discovery of neurohormones and the universal occurrence of this type of neuronal messenger throughout the animal kingdom led to the development of a new discipline, neuroendocrinology, which was mainly in the hands of endocrinologist. The nearly complete absence of neurophysiologists and biochemists in this field has delayed the next and most recent breakthrough in this area of brain research. Attention for chemical transmission dates back as far as 1921. Since then numerous chemical transmitters have been discovered. These are nowadays termed classical transmitters, e.g., acetylcholine, dopamine, serotonin, GABA, various amino acids, etc. By using antibodies raised to neuropeptides, the first signs for a neurotransmitter role of neuropeptides were observed (e.g., Buijs and Swaab, 1979). These findings led to a wealth of publications showing that a great deal of all known peptide messenger molecules, irrespective as to whether these have a neuronal or non-neuronal origin, can be identified in the central and/or peripheral nervous system. I this respect there are also no limitations among the various animal groups: vertebrate peptides are found in the brain of invertebrates and *vice versa* (Schot *et al.*, 1984; Boer and Van Minnen, 1988). Evidence is now overwhelming that they act not only as neurohormones but also as neurotransmitters or neuromodulators, often in cooperation with classical neurotransmitters (e.g., Jan and Jan, 1983). The fact that various neuropeptides are produced by the same neuron and are co-localized with classical transmitters meant the end of Dale's concept (one neuron/one messenger) and led to the introduction of the concept of multimessenger neurones (cf., Hökfelt *et al.*, 1987). These discoveries have changed profoundly our view on the functioning of the nervous system. They have also led to an active participation of electro-physiologists in the studies of the functions of neuropeptides.

It is now becoming clear that the molecular diversity of neuropeptides by far exceeds that of the classical transmitters. Estimates of 200 peptides in one animal species are reported (cf., Joosse, 1986; Eipper *et al.*, 1986).

Moreover, peptides seem to occur throughout the nervous system (cf. Nieuwenhuys, 1985). It seems that the importance of the role of neuropeptides in the functioning of the nervous system can hardly be overestimated.

Peptides consist primarily of amino acids, the sequence of which is encoded by genes. In the early days of the study of peptide structure attention was mainly focussed on the vertebrates. Initially, this led to the conclusion that peptides would be messengers with a moderate diversity and phyletic specificity ("vertebrate peptides"). With the development of microanalytical methods in peptide biochemistry it became possible to determine the amino acid sequence of invertebrate peptides which usually occur in minute amounts. Recently, molecular biological techniques have been introduced in this area which allow the determination of the structure of mRNAs and genes coding for the neuropeptides and also for their receptors. This molecular biological approach in the study of the nervous system is highly promising and is a rapidly expanding field. The first results show that knowledge about the genetic coding and the biosynthetic pathways of neuropeptides is extremely informative for understanding the way in which peptidergic neurones function. It is also clear that there is unity as well as diversity of peptide messengers in the various animal groups. It is of these recent developments that this paper will present an impression. The emphasis will be on results obtained with invertebrates.

First, attention will be given to the question why in all animal groups (except *Hydra*) both classical transmitters and neuropeptides have evolved as neuronal messengers. Does this mean that there are really basic differences between these categories of messengers? This question, which is ignored in the literature, will be discussed and some arguments in favour of the maintenance of this "dualism" of messengers will be presented.

Second, attention is focussed on the molecular structure of neuropeptides. It will be demonstrated how similarity at the peptide and gene level can be used to draw conclusions about the evolution of peptide messengers.

WHAT IS SPECIAL ABOUT PEPTIDES AND CLASSICAL TRANSMITTERS AS NEURONAL MESSENGERS?

Neuropeptides are produced and used as chemical messengers by all animals, even by the most simply organized such as the coelenterate freshwater polyp *Hydra* (Schaller *et al.*, 1984). In all animals with blood vascular systems some neuropeptides are transported via the blood and are known to act as neurohormones. Besides this role, neuropeptides may act as typical synaptic neurotransmitters and as paracrine or neuroendocrine regulators after release at non-synaptic sites (Buijs and Swaab, 1979; Buma and Roubos, 1985; Nieuwenhuys, 1985). The latter mode of action is particularly well developed inside the central nervous system and is considered to be a mechanism that allows peptides to diffuse in the intercellular space to more widely distributed receptors. In this way peptides may recruit numerous neurones in a nucleus. This mechanism of action of peptides has been shown to be important in the situation of co-release with classical transmitters, as was first demonstrated by Jan and Jan (1983).

Classical neurotransmitters are also found throughout the animal kingdom, possibly with the exception of coelenterates (Martin and Spencer, 1983). The almost universal presence of both classical transmitters and neuropeptides as messengers of the nervous system, raises the question of what is special about them. This question is discussed below.

Previously, little attention has been given to the eco-physiological aspects of neuronal messengers. For example, what are the energy costs involved in the production of various messengers? In this respect, studies should be made of the energy budget of peptide transport from the site of synthesis (cell body) to the release point. Classical transmitters are often synthesized near their site of release, which reduces transport costs. Recycling of materials is also important. Little is known about reprocessing of peptides and their amino acid components. (Probably the single amino acid transmitters are the most economic signal molecules.) Acetylcholine reabsorption from the synaptic cleft is a well-known recycling phenomenon: after exocytosis, the membranes of the elementary granules are reprocessed locally. In peptidergic neurones, the multilamellar bodies which contain similar membrane material need to be transported retrogressively to the cell bodies at considerable energy costs (cf., Golding and Pow, 1988).

Any consideration of the selective value of neurotransmitter variety must include their genetics. Classical neurotransmitters are synthesized by one enzyme or a series of enzymes, from precursors in the blood or in the cells, as components of typical metabolic pathways. Thus, it is not the structures of the neurotransmitters, but that of the synthesizing enzymes that are encoded by the genome. On the other hand, with neuropeptides the primary molecular structures of the signal molecules themselves are encoded by the genome: triplets of nucleotides code for each amino acid of the peptide molecules. Here we perceive a basic difference between categories of neuronal messenger (cf., Joosse, 1986). Classical transmitters have an indirect relation with the genome, whereas for peptides this relation is direct. The mechanisms which operate in changing the genome are not restricted to point mutations, but show great diversity. All such changes may affect the nucleotide sequence in genes coding for peptides, and consequently the composition and sequence of the amino acids comprising the peptide messengers. Clearly, genomic changes will also affect genes coding for enzymes involved in the synthesis of classical transmitters. The effects of these changes are difficult to predict, but it may be that the quantity of messenger produced changes, or that a particular enzyme in the synthetic pathway of the messenger is unable to function. However, the chances are extremely low that the structure of the messenger concerned is changed. Moreover, classical transmitters are small molecules, and a change in their structure will often be lethal, since their functioning is not only dependent on receptor binding, but also on subsequent enzymatic breakdown and/or reabsorption. On the other hand, changes in the amino acid composition of peptides have the advantage of being gradual: a change from one hydrophobic amino acid to another may alter the characteristics of the complete molecule only slightly, whereas a change from any amino acid into proline could alter the shape of the entire molecule with severe consequences for receptor binding. This graduality of structural alterations makes peptide messengers suitable candidates for a crucial role in the evolutionary development of regulatory systems. Small changes in molecular structure may, for instance, alter the rate or duration of the process controlled and therefore may be more easily successful when exposed to selection pressure. Another positive aspect of the direct genomic coding of neuropeptide structure is the occurrence of families of genes coding for the same neuropeptide precursor molecule, but with slightly different molecular structures of the final product (Hakanson and Thorell, 1985; cf., Geraerts *et al.*, 1989). (Besides point mutations also changes of larger fragments of the genome are important in the generation of new peptide messengers; see Geraerts *et al.*, 1982). From the viewpoint of population genetics this is a highly advantageous situation for the successful control of adaptive changes.

In conclusion, it appears that peptides are a special category of messenger. These molecules are suitable candidates for the control of the great number of processes related to animal adaptation. This may explain why peptides are involved in the control of a great diversity of physiological processes and why these molecules have persisted as messengers throughout animal evolution, despite their apparently higher energy budget compared with classical transmitters. On the other hand, the molecular structure of the enzymatically synthesized classical transmitters is subdue to a much stronger selection pressure than that of peptides. This may be considered to be an adaptation to their role in the morphologically specialized synaptic contacts, which function in extremely sensitive well-defined time schedules. In fact, in these contacts the specialization of the morphology and the messenger molecules are so accurately intertwined that they form an entity which will hardly allow any change, neither in the morphology, nor in the messenger structure.

UNITY OF NEUROPEPTIDES

Until recently there were conflicting views on the phylogenetic relationships of neuronal messengers in the various animal groups. On the one hand, it was accepted as self-evident that animals use a variety of classical transmitters, the molecular structure of which is identical in all animal groups. On the other hand, there was a great reservedness to accept that vertebrate peptides would also occur in invertebrates. This hesitation concerned in particular peptide hormones such as insulin which is produced by the islets of Langerhans, that would certainly not occur in non-vertebrate animals. However, for some peptides there is now a definitive proof for their presence in vertebrates as well as in invertebrates. The most convincing of these is insulin, since for this messenger not only data on sequencing of the amino acids of the peptide are known, but there is also information about the gene structure available. The other example of a "vertebrate" peptide identified in invertebrates discussed here is vasotocin. The FMRFamide family and the head activator are used as examples of "invertebrate" peptides identified in vertebrates.

Insulin and insulin-related peptides

In vertebrates, carbohydrate metabolism is controlled by a number of hormones. Among these insulin plays a central role. The biochemical processes involved in glucose and glycogen metabolism are considered to be uniform in the various animal groups. This does not necessarily imply that the factors involved in the control of carbohydrate metabolism would also be similar in vertebrates and invertebrates. This is the more so, since in vertebrates, hormones like insulin and glucagon are produced in tissues, the homologues of which are not present in invertebrates.

The earliest indication about the occurrence of a substance similar to insulin in an invertebrate dates back as far as 1923 when Collip demonstrated that an extract of the tissues of the marine bivalve *Mya arrenaria* caused the typical convulsions of a hypoglycemic condition in a rabbit in which this extract was injected. This effect could be counteracted by an injection of dextrose.

With the development of immunocytochemical and biochemical techniques for the identification of small amounts of peptides in sections or in extracts, from 1970 and onwards numerous publications appeared about the presence of insulin-like peptides in invertebrates, particularly in insects and molluscs (for references, see Thorpe and Duve, 1984; Joosse and Geraerts, 1988; Conlon *et al.*, 1988). This material was mainly found in the central nervous system of these animals. Since the nervous system is the source of nearly all known peptide messengers in invertebrates, the nervous orgin of the insulin-like peptides, deviating from that of the vertebrates, seemed acceptable. However, in recent years numerous observations have indicated that in vertebrates the majority of gastro-intestinal peptides are also produced by nerve cells in the brain (Nieuwenhuys, 1985). (For insulin the production in the mammalian brain is still not proved beyond doubt: Baskin *et al.*, 1987). Therefore, the question is warranted whether in invertebrates the centrally identified insulin-like peptides are also found in peripheral tissues.

For some molluscan species there is good evidence for the presence of insulin-like peptides in the digestive tract. Plisetskaya and her associates saw a loss of secretion product from insulin immunoreactive cells in the small intestine of the bivalves *Unio pictorum* and *Anodonta cygnea* after injections of glucose. At the same time, the insulin titer in the haemolymph, determined with a RIA for mammalian insulin, increased. Extracts of the insulin-like material of both species were more potent in the species of origin than mammalian insulin. In the marine bivalve *Mytilus galloprovincialis* the extract appeared to stimulate the activity of glycogen synthase, whereas mammalian insulin had little effect. Moreover, application of an antiserum against molluscan insulin revealed a ten times higher insulin titer in the RIA than for mammalian insulin (Plisetskaya *et al.*, 1978a,b; Rusacov and Kazakov, 1979; cf., Plisetskaya and Joosse, 1985). These results clearly show that in bivalves the gut produces and releases an insulin-like peptide that differs from that of the mammals.

Insulin-like peptides have also been extracted from the intestinal wall of the freshwater gastropod *Lymnaea stagnalis* (Ebberink and Joosse, 1985). Their molecular weight is similar to that of mammalian insulin, but the molecules are slightly more hydrophobic. Further studies resulted in the purification of two pure insulin-like substances, isolated from the midgut. Efforts to sequence the A and B chains after cleavage failed, since both chains have a blocked N-terminal (Ebberink *et al.*, 1989). These intestinal snail insulin-like peptides appear to compete with ^{125}I-labeled bovine insulin for the insulin receptors of isolated rat fat cells. Thus, in conclusion, among molluscs the occurrence of biologically active intestinal insulin-like peptides is beyond doubt, although the full structure of these molecules is not yet elucidated.

The first records about the molecular structure of invertebrate insulin-like peptides came as a surprise, even for the investigators concerned. The first observation was made by Ishizaki in cooperation with Nagasawa and Suzuki. After many years of trial (Ishizaki, 1986) they succeeded in purifying and sequencing the prothoracicotropic hormone (PTTH) of the silk-worm *Bombyx mori* (Nagasawa *et al.*, 1986). This peptide is now called Bombyxin II. Bombyxin is clearly an insulin-related peptide and is produced in neurones which are involved in control of ecdyson secretion by the prothoracic glands shortly before and during ecdysis. (Recently, a

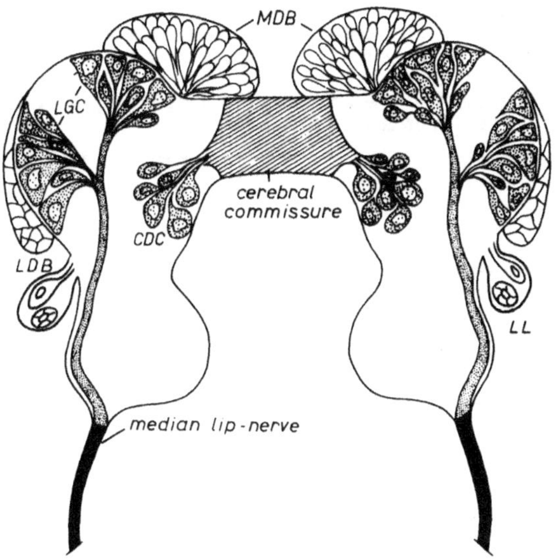

Fig. 1. Diagrammatic transverse section through the cerebral ganglia of *Lymnaea stagnalis*. LGC, light green cells; CDC, caudo-dorsal cells; LL, lateral lobe; MDB and LDB, medio- and latero-dorsal body. The periphery of the median lip nerves is the neurohaemal area of the LGCs. The periphery of the intercerebral commissure is the neurohaemal area of the CDCs.

cDNA-clone coding for an insulin-related peptide has been identified in *Locusta migratoria* by Lagueux and Hoffmann (personal commun.; cf., Table I).)

The second observation was based on the nucleotide sequence of a neuron-specific cDNA clone by Smit *et al.* (1988). The neurones concerned are the Light Green Cells (LGC) in the cerebral ganglia (Fig. 1) of the pond snail *Lymnaea stagnalis*. The total number of these giant neurosecretory neurones is about 200. From classical endocrinological experiments it is known that the LGC control body and shell growth, mobilization of glycogen stores, regulation of the blood glucose concentration and neurite outgrowth from neuronal cell bodies *in vitro* (Geraerts, 1976; Ebberink and Joosse, 1985; cf., Joosse and Geraerts, 1983; Joosse, 1988; Ebberink *et al.*, 1989). The *Lymnaea* insulin-like peptide was called molluscan insulin-related peptide 1 (MIP-1). The term "related" was chosen to distinguish it from "-like", the latter being reserved for peptides of which the molecular structure is not yet known.

The physiological roles of both these invertebrate insulin-related peptides are at first sight different from that of vertebrate insulins. However, the effect of insulin on protein synthesis and growth in verte-

Table 1. Comparison of the amino acid sequence of the A and B chain of LIRP (Locust insulin related peptide) of *Locusta migratoria*[*], Bombyxin II (PTTH) of *Bombyx mori*[+], one of the molluscan insulin-related peptides (MIP-1) of *Lymnaea stagnalis*[†] and human insulin.

A-chains

	1 5 10 15 20
LIRP :	T R G V F D E C C R K T C S I S E L Q T Y C G (R R)
Bombyxin II :	G I V D E C C L R P C S V D V L L S Y C
MIP-1 :	Q G T T N I V C E C C M K P C T L S E L R Q Y C P
Human Insulin:	G I V E Q C C T S I C S L Y Q L E N Y C N

B-chains

	-10 -5 1 5 10 15 20 25
LIRP :	S G A P Q P V A R Y C G E K L S N A L K L V C R G N Y N T M F
Bombyxin II :	pQ Q P Q A V H T Y C G R H L A R T L A D L C W E A G V D
MIP-1 :	Q F S A C N I N D R P H R R G V C G S A L A D L V D F A C S S S N Q P A M V
Human Insulin:	F V N Q H L C G S H L V E A L Y L V C G E R G F F Y T P K T

[*] M. Lagueux and J. Hoffmann, personal communication; [+] After Nagasawa *et al.*, 1986; [†] After Smit *et al.*, 1988

brates is also important. Moreover, moulting in insects is strongly related to body growth, whereas the spectrum of PTTH effects may be wider than hitherto known. Possibly the invertebrate insulin-related peptides are primarily involved in the control of body growth and secondarily in the control of energy metabolism, whereas in vertebrates the control of metabolism prevails.

The molecular structure of the invertebrate insulin-related peptides is remarkably similar to that of vertebrate insulins (Table 1). They consist of two chains and have the 6 cysteines, important for the formation of disulphide bridges at identical positions. MIP-1 has an additional 2 cysteines which almost certainly will form a fourth disulphide bridge (Fig. 2). From the cDNA data of MIP-1 it is clear that this molecule is derived from a precursor that is remarkably similar to preproinsulin in its organization, i.e., it has a B and A chain and a typical C-peptide and 2 dibasic processing sites for conversion to a 2 chain insulin-related peptide (Smit *et al.*, 1988). All these data suggest that these insulin-related peptides of invertebrates will have the typical 3-dimensional structure known from vertebrate insulins.

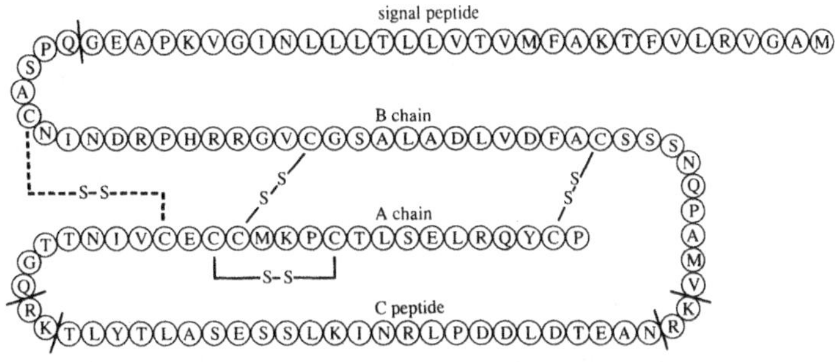

Fig. 2. Amino acid sequence of prepro-molluscan insulin-related peptide (preproMIP-1). The putative proteolytic processing sites are indicated (bold lines). The three disulphide bridges and the putative molluscan specific fourth bridge (dashed line) are indicated. (With permission of Dr. A.B. Smit.)

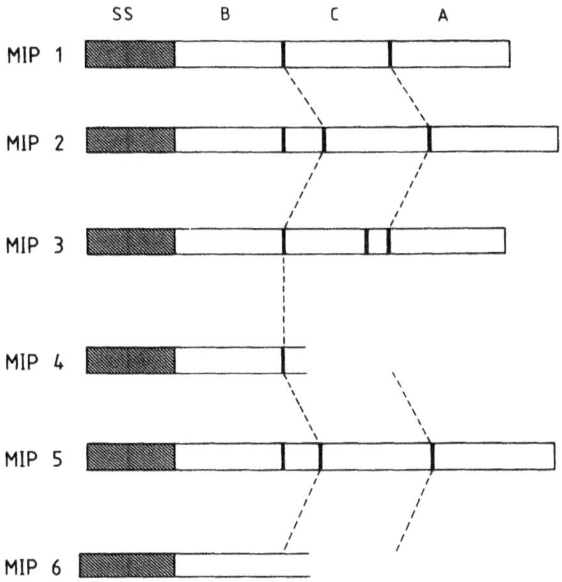

Fig. 3. Schematic representation of the organization of the prepro MIPs in *Lymnaea stagnalis*. The data are based on the analysis of cDNA and genomic clones. The structure of the MIP 4 and 6 precursors is only partially elucidated. Indicated are the putative proteolytic processing sites, the signal sequences (SS), the A and B chains and the C-peptides (C). The amino-acid sequence similarity of the A and B chains is about 50%. The C-peptides of the MIP 1, 2 and 5 precursors are identical, whereas the C-peptide of the MIP 3 precursor shows only about 40% sequence homology. The C-peptide of the MIP 3 precursor contains, in addition, an extra proteolytic cleavage site, and the C-peptide domain of the MIP 2 and 5 precursors contains an extra partial C-peptide sequence. (With permission of Dr. A.B. Smit.)

There are also important differences. Amino acid substitutions occur at many positions involved in insulin receptor binding, including the sequence B23-B26, which plays an important role in receptor binding enhancement. Due to these substitutions it is likely that bombyxin and MIP will bind to receptors which differ from those of mammalian insulin.

Recently, Iwami *et al.* (1989) have succeeded in elucidating the structure of one of the genes coding for a bombyxin in *Bombyx mori*. This gene appears to have no introns. In contrast, in *Lymnaea stagnalis* the organization of the genes coding for the various MIPs is very similar to that of the human insulin gene, i.e., the genes have three exons separated by two introns (A.B. Smit, personal commun.).

An interesting and invertebrate specific phenomenon is the occurrence of a relatively large number of genes coding for the insulin-related peptides in both *Bombyx* and *Lymnaea*. In *Bombyx* multiple microheterogeneous molecular forms of bombyxins occur (Nagasawa *et al.*, 1987b). In *Lymnaea* 4 MIP cDNAs have now been structurally characterized and the MIPs derived from them have only about 50% homology (Fig. 3). Of two other MIP precursors the structure is as yet only partially elucidated. One of the MIPs, MIP5, appears to be a complex hybrid molecule of MIPs 1 and 2. These findings indicate a relatively high rate of divergence of the MIPs during evolution, involving gene duplications, genetic rearrangements of preexisting MIP genes and point mutations (cf. Geraerts *et al.*, 1989). Interestingly, the C-peptides of MIPs 1, 2 and 5 are almost 100% conserved. Therefore, a vital physiological function for the molluscan C-peptide is to be expected.

The above data allow the conclusion that insulin and insulin-related peptides have an extremely long evolutionary history. In view of the early divergence of the various animal groups and their polyphyletic origin (Fig. 4), the common ancestors of insects, molluscs and vertebrates date back as far as 600 million years. The basic structure of insulin (2 peptide chains connected by disulphide bridges) is retained as the back bone of the molecule. The resulting nuclear insulin structure (Steiner and Chan, 1988) makes these peptide messenger molecules relatively large, compact and therefore resistent to proteolytic degradation. Nucleotide changes have taken place frequently and these persisted as long as they did not interfere with the molecular frame, even when the C-peptide is not removed as is the case with the vertebrate insulin-like growth factors (IGFs). The latter may represent more recently evolved adaptations rather than more primitive evolutionary antecedents of insulin (Steiner and Chan, 1988).

The discovery of the invertebrate insulin-related peptides has extended the insulin superfamily. It is likely that further research about the presence of insulins in invertebrates will yield a great number of new insulin-related

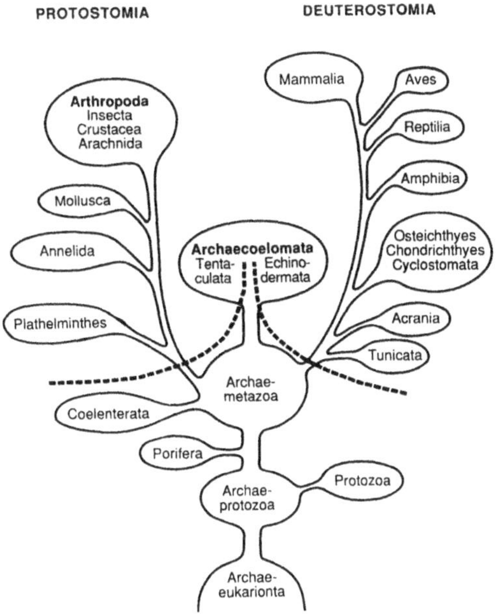

Fig. 4. Phylogenetic relationships in the animal kingdom. The figure shows the polyphyletic origin of the phyla of the animal kingdom. The evolution of the chordates and the vertebrates is thought to have occurred independently of that of the echinoderms and tentaculates, as well as of that of the coelenterates, molluscs, worms and arthropods. The Porifera (sponges) are a special case; they do not seem to have evolved a nervous communication system.

molecules. There is now good evidence that insulins will be found in all insects, molluscs and worms. It will be of interest to know whether the the great diversity of insulin-related peptides, as found in the presently studied species, will appear to be a specific feature of invertebrates. For this phenomenon there is at the moment no reasonable explanation. However, it can not be excluded that further research on insulins in the vertebrate brain might reveal the existence of a greater variety of insulins in the vertebrates as well.

The definitive and convincing demonstration of insulin-related peptides in invertebrates has attracted great attention (Thorpe and Duve, 1988). It has a unifying effect as it brings vertebrate and invertebrate endocrinologists together. Most probably soon other vertebrate gastrointestinal peptides will be identified in invertebrates, as is already the case with cholecystokinin and gastrin homologues (Nachman *et al.*, 1986).

Vasopressin/oxytocin homologues

Invertebrate homologues of the vertebrate vasopressin/oxytocin peptides were purified and sequenced from an unexpected source, the venom of the fish-hunting cone snails *Conus geographus* and *C. striatus* (Cruz *et al., 1987*, see Table 2). The venoms of these molluscs (conotoxins) are a rich source of biologically active peptides, e.g., peptides which inhibit highly specialized calcium channels and sodium channels, block the acetylcholine receptor, etc. The vasopressin/oxytocin homologues were given the generic term conopressins. In fact, these peptides are more similar to vasotocin, but the authors refused to introduce the term conotocin to avoid confusion with conotoxin. An interesting observation is that the biological activity ("scratching") observed when the conopressins are injected intracerebrally into mice is very similar to that elicited by the vertebrate vasopressins. This suggests that these peptides act on a common receptor.

Table 2. Primary structures of vasopressin/oxytocin-related peptides, and their distribution in the animal kingdom

Vertebrates	
Arg-VP	Cys-Tyr-Phe-Gln-Asn-Cys-Pro-Arg-Gly-NH_2
Vasotocin	Cys-Tyr-Ile-Gln-Asn-Cys-Pro-Arg-Gly-NH_2
Lys-VP	Cys-Tyr-Phe-Gln-Asn-Cys-Pro-Lys-Gly-NH_2
Molluscs (*Conus*)*	
Arg-conopressin	Cys-Ile-Ile-Arg-Asn-Cys-Pro-Arg-Gly-NH_2
Lys-conopressin	Cys-Phe-Ile-Arg-Asn-Cys-Pro-Lys-Gly-NH_2
Insects (*Locusta*)†	
monomer	Cys-Leu-Ile-Thr-Asn-Cys-Pro-Arg-Gly-NH_2
dimer	Cys-Leu-Ile-Thr-Asn-Cys-Pro-Arg-Gly-NH_2 Cys-Leu-Ile-Thr-Asn-Cys-Pro-Arg-Gly-NH_2

*After Cruz et al. (1987); †Proux *et al.* (1987). Divergent amino acid residues are underlined.

In the insects (*Locusta*) two neuropeptides, a monomer and its antiparallel dimer, with high sequence homology to the vertebrate vasopressin/oxytocin peptides, have been identified (Proux *et al.*, 1987). The monomer does not have a known biological function but the dimer acts as a diuretic messenger. Both peptides occur in the same cell type and it is suggested, but not proved, that neurophysin(-related) proteins play a role in the dimerization process.

Both molluscan and insect peptides are clearly members of the vasopressin/oxytocin family, yet they have distinct features. The major differences are at positions 2, 3 and 4. Whether this contributes unique biological properties will be of great interest. Important similarities are the conservation of the exact positions of the cysteine residues and the conservation of N-terminal amino acid residues. All peptides have a basic residue at position 8, which is strongly correlated with pressor activity in the vertebrate peptides.

No invertebrate gene or cDNA sequences coding for the vasopressin/oxytocin homologues have been reported as yet, so we have no information concerning the organization of the invertebrate prohormones and do not know, for instance, whether neurophysin(-related) proteins are generated from invertebrate prohormones. With regard to the conopressins it would be of great interest to know whether the genes coding for these peptides are also expressed in the central nervous system. Indications for the presence of vasopressin/oxytocin-like peptides in the central nervous system of a snail (*Lymnaea*) have been presented by Ebberink and Joosse (1985).

The FMRFamide family

The neuropeptide FMRFamide was isolated in 1977 originally from the clam *Macrocallista nimbosa* (cf., Greenberg *et al.*, 1988). In addition to FMRFamide, various related peptides have subsequently been isolated from the brain of a number of species belonging to different classes of the molluscs (Table 3). These peptides all act as neurotransmitters/neuromodulators and/or neurohormones (see e.g., Brussaard *et al.*, 1988). The gene encoding the FMRFamide precursor in the opisthobranch *Aplysia californica* contains 28 copies of FMRFamide and one copy of FLRFamide (Taussig and Scheller, 1986; Fig. 5). In the gastropod pulmonate *Lymnaea stagnalis*, the precursor is much smaller and contains only 9 copies of FMRFamide and 2 copies of FLRFamide (J. Burke and P. Benjamin, personal communication). In *Lymnaea* also related FMRFamides, SDPFLRFamide and GDPFLRFamide, have been isolated (Ebberink *et al.*, 1987), which must be encoded by (a) separate gene(s), since they are not found on the FMRFamide gene. This might indicate that duplications of this FMRFamide gene have occurred during molluscan

Table 3. Amino acid sequences for FMRFamide and some related molluscan and vertebrate peptides*

Molluscs	Phe-Met-Arg-Phe-NH_2
	Phe-Leu-Arg-Phe-NH_2
	Ser-Asp-Pro-Phe-Leu-Arg-Phe-NH_2
Coelenterates	pGlu-Gly-Arg-Phe-NH_2
Arthropods	Ser-Asp-Arg-Asn-Phe-Leu-Arg-Phe-NH_2
	Asp-Pro-Lys-Gln-Asp-Phe-Met-Arg-Phe-NH_2+
Vertebrates	
Chicken brain	Leu-Pro-Leu-Arg-Phe-NH_2
Bovine brain	Phe-Leu-Phe-Glu-Pro-Gln-Arg-Phe-NH_2

*After Greenberg *et al.* *(1988)*; +After Nambu *et al.* (1988)

evolution, followed by mutational events leading to a diversification of the gene product.

The possibility that the molluscan FMRFamide and related peptides might be related to peptides or peptide families in other phyla has been and continues to be investigated. Several new peptides in other phyla were isolated and appear to be related to the molluscan FMRFamide family on the basis of immunological and biological re-activity. Although no genetic evidence for kinship has been demonstrated as yet, certain biological activities of FMRF-related peptides seem to be conserved throughout the evolution (e.g., cardioexcitatory in invertebrates versus blood pressure effect in mammals and opioid antagonism both in *Cepaea* as well as in mammals). The list includes peptides isolated from coelenterates, arthropods and vertebrates (Table 3). In addition, a number of known vertebrate peptides appear to have C-terminal sequences sufficiently FMRFamide-like to convey the potential for functional and immunological cross-reactivity. Most of these substances, however, reacted very weakly, if at all, in an RIA and the standard molluscan bioassays (Greenberg *et al.*, 1988). The exceptions are the pancreatic polypeptide-related peptides and γ-MSH which are functionally cross-reactive. The opioid peptide Met-enkephalin-Arg^6-Phe^7, though it shares amino acid residues with FMRFamide, has no functional and immunological cross-reactivity with FMRFamide. Furthermore, since the genes encoding the precursors of both peptides are completely dissimilar, we can only conclude that the

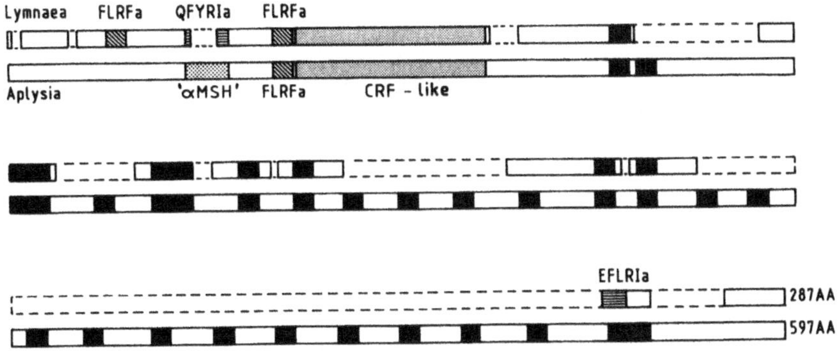

Fig. 5. Comparison of the structure of the FMRFamide precursors of *Lymnaea stagnalis* (with permission of Drs. J. Burke and P. Benjamin) and *Aplysia californica* (after Taussig and Scheller, 1986).

similarity between these peptides is an example of convergence. Nevertheless, there is some compelling evidence of a (functional) FMRFamide-opioid connection (Geraerts *et al.*, 1988; Greenberg *et al.*, 1988). In mammals but also in the snail *Cepaea* FMRFamide acts as an endogenous modulator of opioid activity. It has been argued that these effects of FMRFamides are mediated through specific receptors different from opioid receptors, thus indicating a physiological antagonism of FMRF-related peptides and opioids (Brussaard *et al.*, 1989b).

In conclusion, the ancestral form of FMRFamide and related peptides has appeared early in evolution. The C-terminal "core" sequence -Arg-Phe-NH_2 is often conserved. In our laboratory it has recently been shown that this C-terminal sequence is sufficient for receptor activation in the CNS of *Lymnaea* (Brussaard *et al.*, 1989a). Extensions at the N-terminal side of this core sequence show a large variety, probably reflecting the evolution of other biological functions or modifications thereof. However, in *Lymnaea* N-terminal extension did not affect the bioactivity on central targets and similarly it was shown that FMRFamide-like receptors in rat brain are also activated by the C-terminal Arg-Phe-NH_2 (Brussaard *et al.*, 1989a,b). Possibly, therefore, extensions may primarily be developed to vary the degradation rates of the messengers.

It is difficult to delineate precisely the extended family of FMRFamide and related peptides. It is entirely possible that the C-terminal -Arg-Phe-NH_2 represents general characteristics of associations between peptides and proteins (e.g., receptor molecules); thus, peptides (e.g., many known vertebrate peptides) that comprise this sequence may belong to different peptide families.

Head-activator: from coelenterates to mammals

One of the highlights of neuropeptide unity is demonstrated by the head activator peptide, originally identified in the small (1 cm) freshwater polyp *Hydra* (Schaller and Bodenmüller, 1981). The head activator is one of the four messengers involved in control of morphogenesis of *Hydra*. This animal reproduces asexually under favourable conditions by budding, with a 3-day doubling time. Under less favourable conditions (food scarcity and low temperatures) the animal enters a sexual cycle in which instead of buds, gonads are formed (cf., Lenhoff, 1982).

It was found that two sets of substances regulate head and foot specific differentiation in *Hydra*, an activator and inhibitor of head formation and a similar set for foot formation. The two activators are small peptides with a molecular weight of about 1000 daltons, the two inhibitors are not of a peptide nature and have a smaller molecular mass (<500). As yet the amino acid sequence of the head activator only is known (Schaller and Bodenmüller, 1981; Table 4). This decapeptide is active at extremely low concentrations in the picomolar range. One *Hydra* contains less than one femtomole of head activator. For the identification and molecular characterization, therefore, the authors made use of the fact that the head activator occurs also in sea anemones.

In *Hydra* the head activator is produced in nerve cells in the head. It is not freely diffusing since this small peptide is bound to a large molecular mass (600-800 kilodaltons) carrier (Schaller *et al.*, 1986). Possibly the

Table 4. Molecular structure of the head activator of *Hydra*. Homology with bradykinin, and occurrence in the animal kingdom

Head activator

pGlu-<u>Pro</u>-<u>Pro</u>-<u>Gly</u>-Gly-<u>Ser</u>-Lys-Val-Ile-Phe

Bradykinin

Arg-<u>Pro</u>-<u>Pro</u>-<u>Gly</u>-Phe-<u>Ser</u>-Pro-Phe-Arg

Identified in:

- Coelenterates
- Insects and crustaceans
- Mammals, including man
  hypothalamus, intestine and blood
  developing tissues
  tumours

binding activity of the head activator to its receptor is higher than that to the carrier, thus allowing the messenger to exert its biological action. The importance of the complex formation is apparent from the fate of the free head activator peptide. The molecule either inactivates itself rapidly by dimerazation or is degraded by proteolytic enzymes. The dimer is not biologically active (Bodenmüller *et al.*, 1986).

The head activator has a number of biological effects. It acts as a true mitogen on interstitial cells (Schaller, 1976). Furthermore, it induces head-specific epithelial cell differentiation and the differentiation of interstitial stem cells to nerve cells.

Attempts to identify the head activator in the mammalian brain were astonishingly successful (Bodenmüller and Schaller, 1981). The more so since the molecular structure of the mammalian peptide appeared to be completely identical to that of *Hydra*. In mammals, the peptide is produced by nerve and/or neuroendocrine cells in the hypothalamus and the intestine and it is present in the circulating blood, suggesting that it may act as a hormonal messenger. One of its functions early in mammalian development is a growth-promoting action on nerve precursor cells (as it has in *Hydra*). In line with this effect is the finding that the head activator is found in tumours of neural or endocrine origin, in cell lines derived from such tumours, and in blood of patients with such tumours. Attempts are undertaken to use serum levels of the head activator as a marker for tumour formation (Schaller *et al.*, 1988). Besides in coelenterates and mammals, the head activator has also been identified in neural tissues of amphibia, birds, insects and crustaceans.

In conclusion, the head activator appears to have passed through the evolutionary history of all animal groups without any change in its molecular structure. This stability equals that of the classical transmitters. It means that the molecule must be under extremely strong selection pressure and that it will have a vital role in all animals. As yet little is known about messengers structurally related to the head activator. It shares a number of amino acid residues with bradykinins (Table 4), which are encoded by a separate gene.

NEUROPEPTIDE DIVERSITY

For some neuropeptides there are at the moment no indications that they occur in identical or related molecular forms in other animals groups than in those in which they have been originally discovered. Three examples will be presented. They have been selected from invertebrate groups. Each of these peptides is produced in a neuroendocrine system that controls a complex physiological activity which has passed through a specific phylogenetic history within the phylum in which it occurs. Therefore, these systems may have developed within these phyla.

The organization of each of the neuroendocrine systems presented here is extremely different, ranging from one of the most sophisticated systems, the egg-laying controlling system in snails, to the most simply organized system known at the moment, the eclosion hormone system of insects.

The egg-laying controlling system of snails

The neuroendocrine control of egg laying and egg-laying behaviour has been studied in great detail in the marine snail *Aplysia californica* and the freshwater snail *Lymnaea stagnalis* (for recent reviews, see Roubos, 1984; Geraerts *et al.*, 1988). These snails are simultaneous hermaphrodites with a highly specialized female accessory sex organ system in which numerous egg cells are packaged into eggs and fused to an egg mass. Egg laying is characterized by internal events (ovulation of numerous oocytes, fertilization, egg and egg-mass formation and oviposition) and simultaneous behaviours (alterations of locomotion and feeding movements). The main purpose of this behaviour is to ensure that the egg mass is deposited in a habitat with conditions which are favourable for the development of the embryos and hatchlings. It is therefore interesting to note that in *Lymnaea* the neuroendocrine system that induces egg laying is activated by the same factors which define the optimum conditions for embryonic development.

The egg-laying controlling system of *Lymnaea* consists of the neuroendocrine Caudo-Dorsal Cells (CDC). A total number of 100 CDCs are located in two symmetrically arranged clusters in the cerebral ganglia (Fig. 1). These neurones have the periphery of the intercerebral commissure as their neurohaemal area. Each cluster of CDCs contains a small subset of specialized large neurones (the ventral cells) that receive various inputs, and relay these inputs electrotonically to the other CDCs. The ventral cells each have a second axon that passes through the commissure to contact the other cluster of CDCs. In the commissure these axons form a collateral system (Schmidt and Roubos, 1987). From these collaterals non-synaptic release of CDC peptides inside the nervous system occurs, which most probably affects circuits of interneurones and motor neurones (Jansen and Bos, 1984; Jansen and Ter Maat, 1985; Fig. 6) that are involved in the control of the egg-laying behaviours.

In the giant snail *A. californica* the system homologous to the CDC is the Bag Cell (BC) system. This consists of about 800 neurones, located in paired distinct clusters of electrotonically coupled cells on the pleuro-abdominal connectives adjacent to the abdominal ganglion. The neurohaemal area is located in the connective tissue sheath surrounding the connectives and the abdominal ganglion.

BCs and CDCs are normally electrically silent, but prolonged after

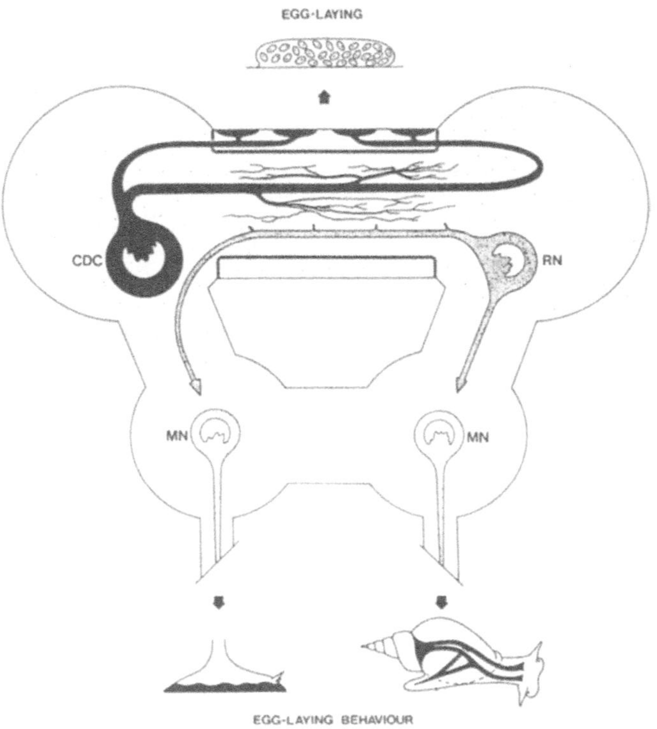

Fig. 6. Tentative scheme of the role of nonsynaptic release of peptides from the CDC collateral system in control of egg-laying behaviour in *Lymnaea stagnalis*. The peptides are released in the commissure where one of the axons of the ring neuron (RN) passes. This ring neuron modulates the activity of numerous motor neurones which control foot (below left) and shell retractor (right) muscle systems which play an important role in egg-laying behaviour. (Based on Schmidt and Roubos, 1987, and Jansen and ter Maat, 1985; Courtesy Dr. E.W. Roubos.)

discharges (burst of synchronous spiking activity) can be evoked in all cells of the network by electrical stimulation of a single neuron *in vitro*. Discharges with a pattern of action potentials similar to the afterdischarges recorded in isolated brains can be recorded from CDCs and BCs by fine wire electrodes in freely moving animals. These discharges always precede spontaneous egg laying. Discharges followed by egg laying can be induced in *Lymnaea* by different environmental conditions or a combination of these, such as a raise in temperature, food supply, long

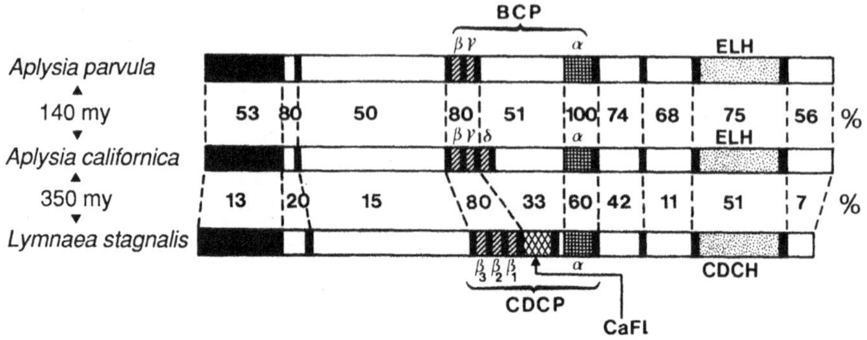

Fig. 7. Comparison of the ELH precursors and signal sequences (SIG) from the BCs of *Aplysia parvula* and *A. californica*, and the CDCH precursor from the CDCs of *Lymnaea stagnalis*. Percentage similarities between regions are indicated. Further are indicated the positions of identified peptides (hatched). Known or potential cleavage sites are indicated with a solid bar.

day photoperiods, a change to clean water, oxygen supply, etc.

The most interestings aspect of the CDC and BC systems is the molecular organization of their messengers. The nominative peptides, CDCH and egg-laying hormone (ELH), respectively, are amidated peptides of 36 amino acid residues, with a 50% similarity. However, the primary structure and organization of the egg-laying peptide precursors is of an extreme complexity (Fig. 7). Generation from these prohormones of a set of 11 peptides is predicted. None of these peptides has significant similarity to other known peptide messengers. Therefore, we consider these systems as group specific. For *Lymnaea* the functions of some of these peptides is known. CDCH induces ovulation, egg and egg-mass formation and part of the egg-laying behaviours. Calfluxin acts on the albumen gland cells (Dictus and Ebberink, 1988). It induces calcium movements in these cells which are probably related to cellular release of the secretion products during the formation of eggs. In the BC and CDC systems the discharge is maintained by an autoexcitatory mechanism. In *Lymnaea* this mechanism is now fully elucidated. A mixture of peptides (CDCH, α-CDCP$_{1\text{-}11}$ and the from the latter derived peptides, α-CDCP$_{3\text{-}11}$ and α-CDCP$_{3\text{-}10}$) is necessary and sufficient to induce and maintain a discharge of the CDC (Brussaard *et al.*, 1989). Thus, also peptides modified after their liberation from the precursor are necessary to keep the CDC firing. It is hypothesized that the three β-CDCPs will act as transmitters in inducing part of the egg-laying behaviours. The β-peptides have a high degree of homology, both among each other and in comparison to the homologous peptides in *Aplysia*.

Recent findings concerning the processing and packaging of the egg-laying peptide precursors have shown that there is differential packaging of the various peptides in distinct granule classes (Fisher *et al.*, 1988; cf.,

Sossin *et al.*, 1989). The segregation occurs via a rapid initial proteolytic cleavage of the prohormone followed by sorting at the trans Golgi. C-terminally located peptides of the prohormones, including the egg-laying inducing peptides, are loaded into neurosecretory granules and transported and released at the axon terminals in the neurohaemal area. N-terminal peptides are largely included in granules that stay in the cell body and only small amounts are included in neurosecretory granules that are transported towards the neurohaemal area. Thus, intracellular packaging and routing of the peptides cleaved from the prohormones regulate their localization and levels in the CDCs and BCs.

The CDCH and ELH genes are members of small prohormone gene families. The CDCH gene family consists of two types of genes: the CDCH-I and -II genes (Vreugdenhil, 1988; Fig. 8). These genes display over 90% similarity. The most striking difference is a small deletion near the C-terminal region of CDCH, including the processing site. This

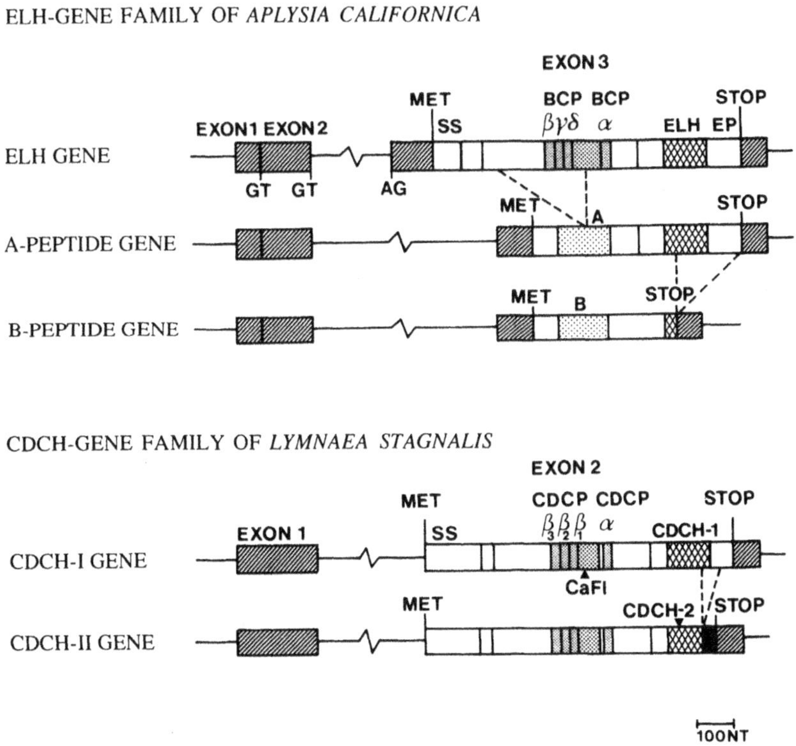

Fig. 8. Organization of the ELH- and CDCH-gene families. For details, see text. Based on Scheller *et al.* (1983), Mahon *et al* (1985) and Vreugdenhil (1988).

deletion gives rise to a CDCH-like peptide with a completely different C-terminal sequence. In addition, several amino acid substitutions occur in other peptide domains. As a result, each of the two CDCH genes encodes a different, but overlapping set of neuropeptides which are thought to control various aspects of the egg-laying action pattern.

In *Aplysia*, the ELH gene family comprises 5 distinct genes, of which 3 have been fully characterized (Fig. 8). The peptide A and B genes are considerably smaller than the ELH gene, due to the deletion of a 240 base pair DNA segment which encodes 80 amino acids.

The biological significance of the existence of multigene families can be examplified by the CDCH system (van Minnen *et al.*, 1989; Fig. 9). Both CDCH genes are expressed in all CDC and in paired groups of ectopic CDC-like neurones in the pleural ganglia, while a group of small

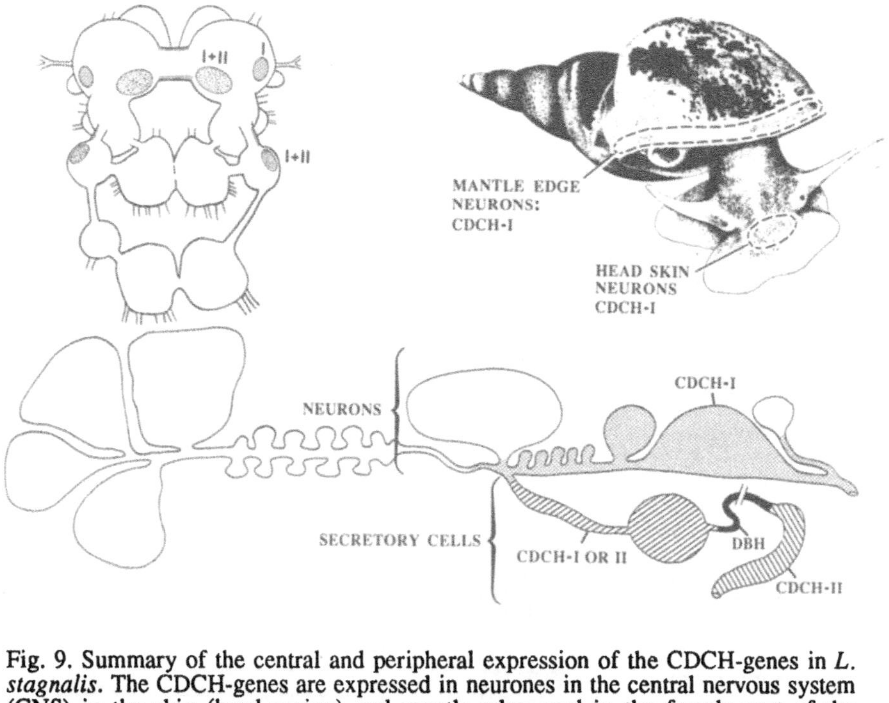

Fig. 9. Summary of the central and peripheral expression of the CDCH-genes in *L. stagnalis*. The CDCH-genes are expressed in neurones in the central nervous system (CNS) in the skin (head region) and mantle edge, and in the female part of the reproductive tract (dotted areas), while expression in epithelial (exocrine) cells is found in the male part of the reproductive tract (hatched areas). A, albumen gland, B, bursa copulatrix, C, carrefour area, Ce, cerebral ganglion, G, gonad, M, muciparous gland, O, oothecal gland, P, pars contorta, Pe, penis, Pg, prostate gland, Pl, pleural ganglion, Sd, sperm duct, Sv, seminal vesicles, V, vagina, Vd, vas deferens. (Courtesy Dr. J. van Minnen.)

neurones in the cerebral ganglia expresses the CDCH-I gene only. In addition, a widespread expression of the CDCH genes has been demonstrated in peripheral tissues. Neurones in the female part of the reproductive tract express the CDCH-I gene. This gene is also expressed in sensory neurones in the head skin and mantle edge. The CDCH-I and -II genes are expressed, each gene in a different cell type, in the exocrine male accessory sex organs. This central and peripheral pattern of CDCH gene expression suggests that in the signal transduction from the environment to the actual egg formation all sensory cells and neurones involved might use these genes as messengers. The question then remains why these genes are also expressed in the male system. It appears that the peptides derived from these genes are secreted into the lumen and transferred to the female copulant (these snails are simultaneous hermaphrodites). One possible pheromone-like function of these peptides might be the acceleration and synchronization of the onset of egg laying in virgin snails. It has been shown that transfer of semen is necessary to induce the start of egg laying in virgin snails which are kept individually (van Duivenboden, 1983). A similar action is described for peptides produced in the male accessory sex gland in *Drosophila melanogaster* (Chen *et al.*, 1988).

As regards the evolution of the egg-laying controlling systems of snails and of the occurrence of multigene families, recently some suggestions have been presented by the authors (Joosse, 1986, 1987a,b; Geraerts *et al.*, 1989). The ancestors of *Aplysia* and *Lymnaea* were primitive gonochoristic marine snails which released their eggs and sperm in the water where fertilization occurred. The synchronization and seasonal timing of this activity might have been controlled by one peptide messenger which might have been the ancestral molecule of CDCH and ELH. Simultaneous with the development of accessory sex organs, hermaphroditism, internal fertilization and the formation of egg masses, the number of processes that had to be synchronized increased enormously. The major changes in the genomic organization of the egg laying inducing peptides will have been the intragenic duplication and subsequent modification of the code for the pentapeptides and α-peptide, and gene duplication. The latter event opened the possibility that the peptides in the redundant gene were free to mutate, because the unchanged peptide encoded by the other gene remained functional. In this way not only point mutations but also deletions in one of the genes could persist. In view of the fact that the organization of the egg laying peptide precursors is very similar in *Aplysia* and *Lymnaea* (belonging to different classes), we must assume that this organization was already present in their common ancestor. It is estimated that this common ancestor lived about 350 million years ago (Fig. 7; Moore and Pitrat, 1960).

Adipokinetic hormones of insects

Adipokinetic hormone (AKH) was first discovered in the locust, but is now known to occur in many insects. The hormone is present in two molecular forms, the originally discovered decapeptide (Stone *et al* , 1976), now indicated as AKH I, and an octapeptide (Siegert *et al.*, 1985) now referred to as AKH II. These peptides have a moderate similarity with the red pigment concentrating hormone (RPCP) of crustaceans, reasons why they are believed to belong to an AKH/RPCP family of arthropod-specific peptides. The two AKH peptides are produced in the paired corpora cardiaca and released during flight (cf., Orchard, 1987, Wheeler *et al.*, 1988). Their action is to guarantee the availability of fuel for long-term (migratory) flight, although the hormones are also present in larvae. The main action of these peptides is on metabolic processes: mobilization of glycogen stores followed by mobilization of lipid stores, stimulation of uptake and use of lipids by flight muscles and arrangement of the proper lipoprotein composition of the haemolymph for the transport of the diacylglycerols to the muscles. However, in flight there is of course also a change of behaviour involved. So, again the question is how metabolism and behaviour are integrated and whether this occurs by the two adipokinetic hormones or whether the precursors of these peptides will contain other messengers which might be co-released with these hormones and might be involved in the control of (part of) the actions during flight.

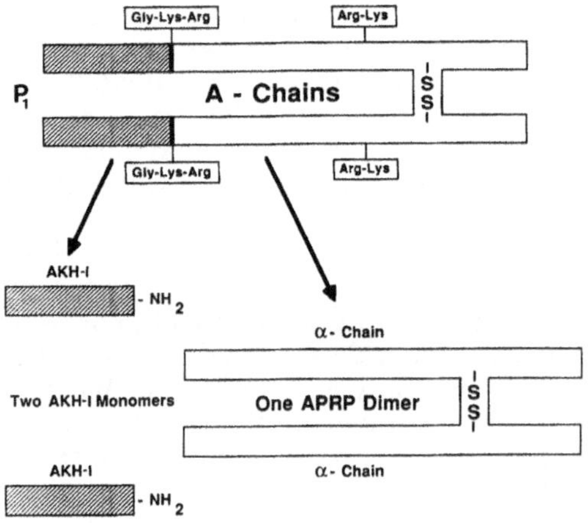

Fig. 10. Biosynthetic pathway of adipokinetic hormone (AKH) and the AKH-precursor related peptide (APRP 1) from their common precursor (P1) in *Schistocerca gregaria* (derived from Hekimi *et al.*, 1989).

By using peptide chemistry (Hekimi *et al.*, 1989) and molecular biological techniques (Schulz-Aellen, 1989), O'Shea and his co-workers succeeded in elucidating the structure of the precursors of AKH I and -II (Fig. 10). It appeared that each of the AKH peptides is encoded by a different mRNA. These give rise to two proteins, which are called the pre-A-chain and pre-B-chain, respectively. Interestingly, at this stage these proteins associate with one another to form three dimer precursors (P_1, P_2 and P_3). From these AKH I and AKH II are derived as well as three other peptides called adipokinetic hormone precursor related peptides (APRPs). P_1 is a homodimer of two A-chains. Each A-chain of this precursor consists of AKH I, a processing site and a 28 residue peptide called the α-chain. The B-chain consists of AKH II, a processing site and a similar 28 residue peptide called the β-chain. Processing of the two homo- and one heterodimer of these peptides results in the generation of AKH I and II peptides and three different APRPs (O'Shea et *al.*, 1989).

It is stated by the authors that biosynthesis of the AKHs and APRPs is unprecedented (Schulz-Aellen *et al.*, 1989). Instead of having repeated copies of one and the same peptide in the mRNA, in this case multiple copies of the AKH molecules are derived from dimer molecules that are formed before the processing of the precursor.

APRPs are stored in the glandular cells of the corpora cardiaca and are co-released when these cells are depolarized (Hekimi and O'Shea, 1989). This suggests that the functions of AKH and APRP may be related. However, as yet there is no function of the APRPs known.

Eclosion hormone of insects

Growth and metamorphosis in insects is characterized by a species specific number of molts during which a new cuticle is produced. Ecdysteroids and juvenile hormones play an important role in the preparation of the skin and cuticle for the complex phenomenon of molting. However, for the circadian timing of ecdysis, various physiological activities directly related to ecdysis (e.g., cuticle plasticization and the programmed degeneration of various muscle groups) and in particular for the behavioural programme of the shedding of the old cuticle, a specific neuropeptide, the eclosion hormone, is responsible (Truman, 1983). The receptors for eclosion hormone are induced by ecdysteroids, shortly before ecdysis is to take place (Morton and Truman, 1989).

Recently, the eclosion hormone has been purified and sequenced from the silkworm *Bombyx mori* (Kono *et al.*, 1987) and the tobacco hornworm *Manduca sexta* (Marti *et al.*, 1987; Kataoka *et al.*, 1987). This revealed a 62 amino acid peptide showing 80% homology between the two species (Table 5). In *Bombyx* eclosion hormone shows

Table.5 Comparison of the amino acid sequence of the eclosion hormones of *Manduca sexta* (Marti et al., 1987; Kataoka et al., 1987) and *Bombyx mori* (Kono et al., 1987).

										10
Manduca	H-Asn	Pro	Ala	Ile	Ala	Thr	Gly	Tyr	Asp	Pro
Bombyx	H-Ser	–	–	–	–	Ser	Ser	–	–	Ala
										20
	Met	Glu	Ile	Cys	Ile	Glu	Asn	Cys	Ala	Gln
	–	–	–	–	–	–	–	–	–	–
										30
	Cys	Lys	Lys	Met	Leu	Gly	Ala	Trp	Phe	Glu
	–	–	–	–	Phe	–	Pro	–	–	–
										40
	Gly	Pro	Leu	Cys	Ala	Glu	Ser	Cys	Ile	Cys
	–	Ser	–	–	–	–	–	–	–	–
										50
	Phe	Lys	Gly	Lys	Leu	Ile	Pro	Glu	Cys	Glu
	Ala	Arg	–	–	Asp	–	–	–	–	–
										60
	Asp	Phe	Ala	Ser	Ile	Ala	Pro	Phe	Leu	Asn
	Ser	–	–	–	–	Ser	–	–	–	–
	Lys	Leu-OH								
	–	–								

a considerable molecular heterogeneity (Nagasawa *et al.*, 1987a). The eclosion hormones have no amino acid identity with any other known peptide messenger. Therefore, it is a good example of an intraphyletic development of a neuropeptide messenger.

The eclosion hormone has a mixture of physiological and behavioural actions. In this respect this hormone acts very similar to the egg-laying hormones of *Aplysia* and *Lymnaea*. Therefore, from a comparative endocrinological point of view it would be very interesting to compare the precursor organization of the eclosion hormones with that of the egg-laying hormones. To our great surprise the eclosion hormone precursor has the most extreme simplicity as it consists of a signal peptide and only the full sequence of the eclosion hormone, without any additional other peptides (Horodyski *et al.*, 1989). Moreover, there is only one gene that encodes this precursor, which is expressed in 2 pairs of ventromedial neurosecretory cells only, in larval, diapausing pupal and developing adult brains. So there are no signs of peripheral gene expression as is the case with the egg-laying hormone genes in snails. This makes the molecular organization of the eclosion prohormone the simplest system known in neuroendocrinology.

There is an additional question raised by this monofunctional structure of the eclosion prohormone. The behavioural components of the ecdysis must be induced by this 62 amino acid peptide. How does this peptide get in contact with the motor neurones involved in ecdysis behaviour? For a direct action this peptide has to pass the blood-brain barrier of the insect nervous system. So far there is no reasonable explanation for a direct action. Possibly specific mechanisms will have developed to transfer the blood brain barrier.

ACKNOWLEDGEMENTS

We thank Dr. A.B. Smit for allowing us to use preliminary information on the MIPs, Drs. J. Burke and P. Benjamin for the use of data on the FMRFamide precursor of *Lymnaea*, Dr. A.B. Brussaard for his assistance in the FMRF section and Dr. F.M. Horodyski for making available preliminary information on the eclosion hormone precursor. We also thank Ms. Thea Laan for tying and composing this manuscript.

REFERENCES

Bargmann W (1949) Über die neurosekretorische Verknüpfung von Hypothalamus und Neurohypophyse. *Z. Zellforsch.* **34**, 610-634.

Baskin D.G., Figlewisz D.P., Woods S.C., Porte D. and Dorsa D.M. (1987) Insulin the the brain. *Ann. Rev. Physiol.* **49**, 335-347.

Bodenmüller H. and Schaller H.C. (1981) Conserved amino acid sequence of a neuropeptide, the head activator, from coelenterate to humans. *Nature* **293**, 579-580.

Bodenmüller H., Schilling E., Zachmann B. and Schaller H.C. (1986) The neuropeptide head activator loses its biological activity by dimerisation. *EMBO J.* **5**, 1825-1829.

Boer H.H. and Van Minnen J. (1988) Immunocytochemistry of hormonal peptides in molluscs: optical and electron microscopy and the use of monoclonal antibodies. In *Neurohormones in invertebrates (Seminar series Soc. exp. Biol. 33*) (Thorndyke M.C. and Goldsworthy G.J., eds.) pp. 19-42. Univ. Cambridge Press, Cambridge.

Brussaard A.B., Kits K.S., ter Maat A., van Minnen J. and Moed P.J. (1988) Dual inhibitory action of FMRFamide on neurosecretory cells controlling egg laying behaviour in the pond snail. *Brain Res.* **447**, 35-51.

Brussaard A.B., Kits K.S. and ter Maat A. (1989a) One receptor type mediates two independent effects of FMRFamide on neurosecretory cells of *Lymnaea. Peptides* **10-2**, 289-297.

Brussaard A.B., Kits K.S., ter Maat A., Mulder A. and Schoffelmeer A.N. (1989b) Peripheral injection of DNS-RFamide, a FMRFamide

agonist, suppresses morphine induced analgesia in rats. *Peptides* **10-4**, In press.
Buma P. and Roubos E.W. (1985) Ultrastructural demonstration of nonsynaptic release sites in the central nervous system of the snail *Lymnaea stagnalis*, the insect *Periplaneta americana*, and the rat. *Neuroscience* **17**, 867-879.
Buijs R.M. and Swaab D.F. (1979) Immunoelectron microscopical demonstration of vasopressin and oxytocin synapses in the rat limbic system. *Cell Tiss. Res.* **204**, 355-365.
Cajal S.R. (1894) La fine structure des centres nerveux. *Proc. R. Soc. London (B)* **55**, 444-468.
Cajal S.R. (1911) *Histologie du système nerveux de l'homme et des Vertébrés*, Vol. 2. Maloine, Paris.
Chen P.S., Stumm-Zollinger E., Aigaki T., Balmer J., Benz M. and Böhlen D. (1988) A male accessory gland peptide that regulates reproductive behavior of female *D. melanogaster*. *Cell* **54**, 291-298.
Collip J.B. (1923) The demonstration of an insulin-like substance in the tissue of the clam. *J. Biol. Chem.* **55**, 39.
Conlon J.M., Reinecke M., Thorndyke M.C. and Falkmer S. (1988) Insulin and other islet hormones (somatostatin, glucagon and PP) in the neuroendocrine system of some lower vertebrates and that of invertebrates – a minireview. *Horm. metabol.* **20**, 406-410.
Cruz L.J., de Santos V, Zafaralla GC, Ramilo CA, Zeikus R, Gray WR and Olivera BM (1987) Invertebrate vasopressin/oxytocin homologs *J. Biol. Chem.* **262**, 15821-15824.
Dictus W.J.A.G. and Ebberink R.H.M. (1988) Structure of one of the neuropeptides of the egg-laying hormone precursor of *Lymnaea*. *Mol. & Cell. Endocrinol.* **60**, 23-29.
Ebberink R.H.M. and Joosse J. (1985) Molecular properties of various snail peptides from brain and gut. *Peptides* **6**, suppl. 3, 451-457.
Ebberink R.H.M., Price D.A., Van Loenhout H, Doble K.E., Rhiem J.P., Geraerts W.P.M. and Greenberg M.J. (1987) The brain of *Lymnaea* contains a family of FMRFamide-like peptides. *Peptides* **8**, 515-522.
Ebberink R.H.M., Smit A.B. and Van Minnen J. (1989) The insulin family: evoluton of structure and function in vertebrates and invertebrates. *Biol. Bull.* **177**, In press.
Eipper B.A., Mains R.E. and Herbert E. (1986) Peptides in the nervous system. *Trends Neurosci.* **9**, 463-468.
Fisher J.M., Sossin W., Newcomb R. and Scheller R.H. (1988) Multiple neuropeptides derived from a common precursor are differentially packaged and transported. *Cell* **54**, 813-822.
Geraerts W.P.M. (1976) Control of growth by the neurosecretory hormone of the light green cells in the freshwater snail *Lymnaea stagnalis*. *Gen. Comp. Endocrinol.* **29**, 61-71.

Geraerts W.P.M., Vreugdenhil E. and ter Maat A. (1988). The peptidergic neuroendocirne control of egg-laying behavior in *Aplysia* and *Lymnaea*. In *Invertebrate Endocrinology, Vol. 2, Endocrinology of Selected Invertebrate Types* (Laufer H. and Downer R.G.H., eds.), pp. 141-231. Alan R. Liss, Inc., New York.

Geraerts W.P.M., Smit A.B. and Joosse J. (1989) Peptide messenger evolution: the invertebrate contribution. In *Procs. 4th ENEA Conference Santiago de Compostella, Spain*. In press.

Golding D.W. and Pow D.V. (1988) The new neurobiology – ultrastructural aspects of peptide release as revealed by studies of invertebrate nervous systems. In *Neurohormones in invertebrates* (Thorndyke M.C. and Goldsworthy G.J., eds.), *Soc. f. exp. Biol. Seminar Series* **33**, 8-18. University Cambridge Press, Cambridge.

Greenberg M.J., Payzee K., Nachman R.J., Holman G.M. and Price D.A. (1988) Relationships between the FMRFamide-related peptides and other peptide families. *Peptides* **9**, 125-135.

Hakanson R. and Thorell J. (eds.) (1985) *Biogenetics of neurohormonal peptides*. Academic Press, pp. 303.

Hekimi S. and O'Shea M. (1989) Biosynthesis of adipokinetic hormones (AKHs): further characterization of precursors and identification of novel products of processing. *J. Neurosci.* In press.

Hekimi S., Burkhart W., Moyer M., Fowler E. and O'Shea M. (1989) Dimer structure of a neuropeptide precursor established: consequences for processing. *Neuron* **2**, 1363-1368.

Hökfelt T. *et al.* (1987) Chemical neuroanatomy of the hypothalamo-pituitary axis: focus on multimessenger systems. In *Integrative neuroendocrinology: molecular, cellular and clinical aspects* (McCann S.M.and Weiner R.I., eds.) pp. 1-34. Karger, Basel.

Horodyski F.M., Riddiford L.M. and Truman J.W. (1989) Isolation and expression of the eclosion hormone gene from the tobacco hornworm, *Manduca sexta*. *Proc. Natl. Acad. Sci.* In press.

Ishizaki H. (1986) A PTTH pilgrimage from *Luchdorfia* to *Bombyx*. *Zool. Sci.* **3**, 921-929.

Iwami M, Kawakami A., Ishizaki H., Takahashi S.Y., Adachi T., Suzuki Y., Nagasawa H. and Suzuki A. (1989) Cloning of a gene encoding bombyxin, an insulin-like brain secretory peptide of the silkmoth *Bombyx mori* with prothoracicotropic activity. *Develop. Growth and Differ.* **31**, 31-37.

Jan Y.N. and Jan L.Y. (1983) A LHRH-like peptidergic neurotransmitter capable of 'action at a distance' in autonomic ganglia. *TINS* **6**, 320-325.

Jansen R.F. and Bos N.P.A. (1984) An identified neuron modulating the ovulation hormone producing Caudodorsal Cells in the pond snail. *J. Neurobiol.* **15**, 161-167.

Jansen R.F. and ter Maat A. (1985) Ring neuron control of columellar motor neurons during egg-laying behavior in the pond snail *Lymnaea stagnalis*. *J. Neurobiol.* **16**, 1-14.

Joosse J. (1986) Neuropeptides: peripheral and central messengers of the brain. In *Comparative endocrinology, developments and directions Progr. Clin. biol. Res.*, vol. 205 (Ralph C.L., ed.) pp. 12-32. Alan R. Liss, New York.

Joosse J. (1987a). Functional and evolutionary perspectives of neuropeptides and their precursors. In *Integrative neuroendocrinology: molecular, cellular and clinical aspects* (McCann S.M.and Weiner R.I., eds.) pp. 176-190. Karger, Basel.

Joosse J. (1987b) Evolutionary aspects of neuropeptides. In *Neuropeptides and Brain Function* (de Kloet E.R., Wiegant V.M. and de Wied D., eds.) *Progr. Brain Res.* **72**, 35-45.

Joosse J. (1988) The Hormones of Molluscs. In *Invertebrate Endocrinology, Vol. 2, Endocrinology of Selected Invertebrate Types* (Laufer H. and Downer R.G.H., eds.), pp. 89-140. Alan R. Liss, Inc., New York.

Joosse J. and Geraerts W.P.M. (1983) Endocrinology. In *The Mollusca, Vol. 4, Physiology, Part 1* (Wilbur K.M., ed.) pp. 317-406. Academic Press, New York.

Kataoka H., Troetschler R.G., Kramer S.J., Cesarin B.J. and Schooley D.A. (1987) Isolation and primary structure of the eclosion hormone of the tobacco hornworm, *Manduca sexta. Biochem. Biophys. Res. Commun.* **146**, 746-750.

Kono T., Nagasawa H., Isogai A., Fugo H. and Suzuki A. (1987) Amino acid sequence of eclosion hormone of the silkworm *Bombyx mori. Agric. Biol. Chem.* **51**, 2307-2308.

Leake L.D. and Walker R.J. (1980) *Invertebrate Neuropharmacology.* Blackie, Glasgow.

Lenhoff H.M. (1982) *Hydra: Research Methods.* Plenum Press, New York and London.

Mahon A.C., Nambu J.R., Taussig R., Shyamala M., Roach A. and Scheller R.H. (1985) Structure and expression of the egg-laying hormone gene family in *Aplysia. J. Neurosci.* **5**, 1872-1880.

Marti T., Tahio K., Walsh K.S., Terzi G. and Truman J.W. (1987) Microanalysis of the amino acid sequence of the eclosion hormone from the tobacco hornworm *Manduca sexta. FEBS Lett.* **219**, 415-418.

Martin S.M. and Spencer A.N. (1983) Neurotransmitters in coelenterates. *Comp. Biochem. Physiol. C* **74**, 1-14.

Moore E.C. and Pitrat C.W. (1960) *Feature on invertebrate paleontology,* part 1, Mollusca 1. Geological Society of America, Boulder/ University of Kansas Press, Lawrence.

Morton D.B. and Truman J.W. (1989) The molecular basis for steroid regulation of eclosion hormone action in the tobacco hornworm,

Manduca sexta. Proc. Int. Symp. Comp. Pharmacol. of Neuropeptides. In press.

Nachman R.J., Holman G.M., Cook B.J., Haddon W.F., Ling N. (1986) Leucosulfakinin-II, a blocked sulfated insect neuropeptide with homology to cholecystokinin and gastrin. *Biochem. Biophys. Res. Comm.* **140**, 357-364.

Nagasawa H., Kataoka H., Isogai A., Suzuki S., Suzuki A., Mizoguchi H., Fujiwara Y., Susuki At., Takahashi S.Y. and Ishizaki H. (1986) Amino acid sequence of the prothoracicotropic hormone of the silkworm *Bombyx mori. Proc. Natl. Acad. Sci USA* **83**, 5640-5843.

Nagasawa H, Mikogami T., Koni T., Fugo H. and Suzuki A. (1987a) Molecular heterogeneity of eclosion hormone in adult heads of the silkworm, *Bombyx mori. Agric. Biol. Chem.* **51**, 1741-1743.

Nagasawa H. *et al.* (1987b) Structure and synthesis of bombyxin from the silkworm, *Bombyx mori*. In *Peptide Chemistry* (Shiba S. and Sakahibara S., eds.) pp. 123-126. Protein Res. Foundation, Osaka.

Nambu J.R., Murphy-Erdosh C., Andrews P.C., Feistner G.J. and Scheller R.H. (1988) Isolation and characterization of a *Drosophila* neuropeptide gene. *Neuron* **1**, 55-61.

Nieuwenhuys R. (1985) *Chemoarchitecture of the brain*, pp. 1-246. Springer, Berlin.

Orchard I. (1987) Review: adipokinetic hormones: an update. *J. Insect Physiol.* **33**, 451-463.

O'Shea M, Hekimi S and Schulz M.-F. (1989) Adipokinetic hormone biosynthesis: molecular biological and biochemical analysis of precursors and processing. *Abstracts XIth Int. Symp. Comp. Endocr. Malaga*, L-69.

Plisetskaya E. and Joosse J. (1985) Endocrine control of metabolism in molluscs. In *Current Trends in Comparative Endocrinology* (Lofts B. and Holmes W.N., eds.), pp. 1077-1079. Hong Kong University Press, Hong Kong.

Plisetskaya E., Soltitskaya L. and Rusacov Y.I. (1978a) Production and role of insulin in the regulation of carbohydrate metabolism in freshwater and marine bivalve molluscs. In *Comparative Endocrinology* (Gaillard P.J. and Boer H.H., eds.), pp. 449-453. Elsevier/North Holland Biomedical Press, Amsterdam.

Plisetskaya E., Kazakov V.K., Soltitskaya L. and Leibson L.G. (1978b) Insulin-producing cells in the gut of freshwater bivalve molluscs *Anodonta cygnea* and *Unio pictorum* and the role of insulin in the regulation of their carbohydrate metabolism. *Gen. Comp. Endocrinol.* **35**, 133-145.

Proux J., Miller C.A., Li J.P., Carney R.L., Girardie A., Delange M. and Schooley D.A. (1987) Identification of an arginine vasopressin-like diuretic hormone from *Locusta migratoria. Biochem. Biophys. Res. Comm.* **149**, 180-186.

Roubos E.W. (1984) Cytobiology of the ovulation-neurohormone producing Caudo-Dorsal cells of the snail *Lymnaea stagnalis*. *Int. Rev. Cytol.* **89**, 295-346.

Rusacov Y.I. and Kazakov V.K. (1979) Preparation of insulin-like substance from molluscs and antiserum against it. *Zh. Evol. Biokhim Fiziol.* **15**, 617-619 (in Russian).

Schaller H.C. (1976) Head regeneration is initiated by the release of head activator and head inhibitor. *Wilhelm Roux's Arch.* **180**, 287-295.

Schaller H.C. and Bodenmüller H. (1981) Isolation and amino acid sequence of a morphogenetic peptide from *Hydra*. *Proc. Natl. Acad. Sci.* **78**, 7000-7004.

Schaller H.C., Hoffmeister S. and Bodenmüller, H. (1984) Hormonal control of regeneration in *Hydra*. In *Biosynthesis, metabolism and mode of action of invertebrate hormones* (Hoffmann J. and Porchet M., eds.) pp. 5-9. Springer, Berlin.

Schaller H.C., Roberge M., Zachmann B., Hoffmeister S., Schilling E. and Bodenmüller H. (1986) The head activator is released from regenerating hydra bound to a carrier molecule. *EMBO J.* **5**, 1821-1824.

Schaller H.C., Schilling E., Theilmann L., Bodenmüller H. and Sachsenheimer W. (1988) Elevated levels of head activator in human brain tumors and in serum of patients with brain and other neurally derived tumors. *J. Neurooncology*. In press.

Scharrer E. (1928) Die Lichtempfindlichheit blinder Elritzen (Untersuchungen über das Zwischenhirn der Fische. I). *Z. Vergl. Physiol.* **7**, 1-38.

Scharrer B. (1935) Ueber das Hanströmsche Organ X bei Opisthobranchien. *Publ. Staz. Zool. Napoli* **15**, 132-142.

Scheller R.H., Jackson J.F., McAllister L.B., Rothman B.S., Mayeri E. and Axel R. (1983) A single gene encodes multiple neuropeptides mediating a stereotyped behavior. *Cell* **32**, 7-22.

Schmidt E.D. and Roubos E.W. (1987) Morphological basis for nonsynaptic communication within the central nervous system by exocytotic release of secretory material from the egg-laying stimulating neuroendocrine caudodorsal cells of *Lymnaea stagnalis*. *Neuroscience* **20**, 247-257.

Schot L.P.C., Boer H.H. and Montagne-Wajer C. (1984) Characterization of multiple immunoreactive neurons in the central nervous system of the pond snail *Lymnaea stagnalis*. *Histochem.* **81**, 373-378.

Schulz-Aellen M.-F., Roulet E., Fischer-Lougheed J. and O'Shea M. (1989) Synthesis of a homodimer neurohormone precursor of *Locust* adipokinetic hormone studied by *in vitro* translation and cDNA cloning. *Neuron* **2**, 1369-1373.

Siegert K.J., Morgan D. and Mordue W. (1985) Primary structures of locust adipokinetic hormones II. *Hoppe-Seylers Z. Physiol. Chem.* **366**, 723-727.

Smit A.B., Vreugdenhil E., Ebberink R.H.M., Geraerts W.P.M., Klootwijk K. and Joosse J. (1988) Growth-controlling neurons produce the precursor of an insulin-related peptide. *Nature* **331**, 535-538.

Sossin W.S., Fisher J.M. and Scheller R.H. (1989) Cellular and molecular biology of neuropeptide processing and packaging. *Neuron* **2**, 1407-1417.

Steiner D.F. and Chan S.J. (1988) Perspective: an overview of insulin evolution. *Horm. metabolic. Res.* **20**, 443-444.

Stone J.V., Mordue W., Batley K.E. and Morris H.R. (1976) Structure of locust adipokinetic hormone that regulates lipid utilisation during flight. *Nature* **263**, 207-211.

Swammerdam Jan (1737) *Biblia Naturae*. Severinus, van der Aa & van der Aa, Leiden.

Taussig R. and Scheller R.H. (1986) The *Aplysia* FMRFamide gene encodes sequences related to mammalian brain peptides. *DNA* **5**, 453-461.

Thorpe A. and Duve H. (1984) Insulin- and glucagon-like peptides in insects and molluscs. *Mol. Physiol.* **5**, 235-260.

Thorpe A. and Duve H. (1988) Insulin found at last. *Nature* **331**, 483-484.

Truman J.W. (1983) Hormonal control of ecdysis. In *Comprehensive Insect Physiol. Biochem. Pharmacol.* (Kerkut G.A. and Gilbert L.I., eds) **13**, pp. 413-440.

Van Duivenboden Y. (1983) Transfer of semen accelerates onset of egg-laying in female copulants of the hermaphrodite freshwater snail, *Lymnaea stagnalis*. *Int. J. Invert. Reprod.* **7**, 249-257.

Van Minnen J., Smit A.B. and Joosse J. (1989) Central and peripheral expression of genes coding for egg-laying inducing and insulin-related peptides in a snail. *Arch. Histol. Cytol.* **52**, suppl. In press.

Vesalius Andreas (1543) De humani corporis fabrica (reprinted by de Forel b.v., Nieuwendijk, The Netherlands, 1975).

Vreugdenhil E. (1988) The molecular basis of egg laying and egg-laying behaviour in the freshwater snail *Lymnaea stagnalis*. Thesis, Vrije Universiteit, Amsterdam.

Wheeler C.H., Gäde G. and Goldsworthy G.J. (1988) Humoral functions of insect neuropeptides. In *Neurohormones in invertebrates. Seminar Series Soc. exp. Biol. 33* (Thorndyke M.C. and Goldsworthy G.J., eds.) pp. 19-42. Univ. Cambridge Press, Cambridge.

Wigglesworth V.B. (1936) The function of the corpus allatum in the growth and reproduction of *Rhodnius prolixus* (Hemiptera). *Q. Jl. micros . Sci.* **79**, 91-121.

From *Insect Neurochemistry and Neurophysiology • 1989* Edited by A. B. Borkovec and E. P. Masler

Isolation Techniques for Insect Neuropeptides

David A. Schooley*[1], Hiroshi Kataoka[2], Steven J. Kramer[3], and Anne Toschi[3]

[1]Department of Biochemistry, University of Nevada, Reno, Nevada, 89557, USA; [2]Dept. of Agricultural Chemistry, University of Tokyo, Tokyo 113, Japan; [3]Sandoz Crop Protection, Palo Alto, CA, 94303, USA

INTRODUCTION

In this paper we review certain key isolation schemes for insect neurohormones, with an attempt to gain a perspective on the methods previously used, on the relative abundance of various neuropeptides, and how parameters such as *in vitro* metabolism of neuropeptides impact on planning their isolation. In addition, we briefly review key principles for isolation of neuropeptides first developed for vertebrate neuropeptides, and how experiences gained by researchers in these systems can improve efficiency of isolation of insect neuropeptides. Our review of the literature is selective rather than comprehensive.

ISOLATION OF PROCTOLIN, AKH, AND 4 K PTTH

Proctolin was the first insect neuropeptide to be isolated (Brown and Starratt, 1975) and identified (Starratt and Brown, 1975). The amount of tissue used, 125 kg, was of necessity quite large because of the minute content of this peptide and the relative insensitivity of peptide sequencing procedures then available. The 11-step isolation procedure used resulted in an estimated recovery of 12% (Fig. 1). This landmark work was carried out half a decade before the routine use of modern liquid

Proctolin Isolation Scheme

~125,000 whole cockroaches (mainly adults, 125 kg)

↓ 1) Homogenize with cold 7% $HClO_4$, filter

↓ 2) Neutralize with KOH, filter off $KClO_4$, concentrate

↓ 3) Dowex 50W-X8, H^+ form, eluted w. 4 M NH_3 in 50% EtOH

↓ 4) Dowex 50W-X8, NH_4^+ form, eluted w. 0.05 M NH_3

↓ 5) Acidic alumina, eluted w. 0-70% H_2O in MeOH/0.1% BHT gradient

↓ 6) Rexyn 101, NH_4^+ form, eluted w. 0-0.04 M NH_3 gradient

↓ 7) CCD partition, BuOH:CH_3CO_2H:H_2O 4:1:5

↓ 8) partition chromatography on paper, same solvent as 7)

↓ 9) HVPE pyridine:CH_3CO_2H:H_2O 25:1:350, pH 6.4

↓ 10) HVPE pyridine:CH_3CO_2H:H_2O 1:10:445, pH 3.5

↓ 11) Sephadex G-15 0.02 M NH_4CO_2H

~180 μg pure proctolin (~270 nmol); ~12% recovery, corrected purification factor 83 x 10^6

2.2 pmol proctolin per cockroach

Fig. 1. Isolation scheme for proctolin, from data of Brown and Starratt (1975). Abbreviations: BHT, butylated hydroxytoluene; CCD, counter-current distribution; HVPE, high-voltage paper electrophoresis.

chromatographic (LC) procedures for peptide isolation. One of the noteworthy features of this isolation is the use of 7% $HClO_4$ for extraction; very vigorous conditions which would denature proteases. Similarly, the high capacity of ion exchange chromatography was exploited (steps 3 and 4) when the mass of impurities was high. The use of adsorption chromatography on alumina (Step 5) is most unusual for peptides, and may be feasible only for small, hydrophilic neuropeptides. Further purification of proctolin was achieved by counter-current distribution (CCD), both by partition and on paper; this established and powerful technique is little used currently. In subsequent steps, high voltage paper electrophoresis was used to purify proctolin. With the exception of ion exchange chromatography, these methods have been largely supplanted by LC methods. The amount of proctolin isolated, ~180 μg, is very large by current standards. Based on the estimated recovery of biological activity, the amount of proctolin per cockroach would correspond to a little over 2 pmol per animal.

Only one year later the structure of adipokinetic hormone was determined from two species of locust (Stone *et al.*, 1976). The isolation scheme for AKH was much simpler than for proctolin, due to the use of dissected lobes of the corpora cardiaca (CC) as a tissue source and due to the high content of AKH per locust (100-250 pmol, about two orders of magnitude higher than the amount of proctolin in a single roach). Only two chromatographic steps had to be utilized (Fig. 2), even though the investigators did not have the advantage of access to modern LC methods.

Many peptides of the AKH family have been isolated subsequently from a host of insect species, using this approach but with LC separation. Frequently these peptides have been isolated to homogeneity in a single LC step following extraction of the CC. Unfortunately, this ease of isolation is undoubtedly the exception rather than

AKH Isolation Scheme

3,000 glandular lobes of locust corpora cardiaca

↓ 1) Homogenize in MeOH, spin, evaporate supernatant

↓ 2) Size exclusion; controlled-pore glass beads, H_2O

↓ 3) Silica gel TLC, 2-propanol:H_2O:CH_3CO_2H, 25:10:1

300-750 nmol from *L.migratoria*,
600-750 nmol from *Schistocerca gregaria*

100-250 pmol AKH per locust

Fig. 2. Scheme for isolation of adipokinetic hormone from two species of locusts (Stone *et al.*, 1976).

the rule, and many non-specialists in peptide isolation may be misled from these studies, to anticipate that isolation of neuropeptides with different types of activity will be as facile as for AKH, rather than as difficult as for proctolin.

The isolation of *Bombyx mori* 4K prothoracicotropic hormone (PTTH, the first morphogenetic insect neurohormone to be isolated and identified)(Nagasawa *et al.*, 1984) is an even more graphic demonstration of the difficulty of isolating some insect neuropeptides than is the isolation of proctolin. While the latter is a small, hydrophilic peptide, PTTH is a large, relatively hydrophobic peptide. Perhaps because of the properties of 4K PTTH, its 15-step isolation from 648,000 *B. mori* heads (4.86 kg) resembles that of a protein. After extracting delipidated heads with 2% NaCl and heat-treating the extract to denature proteins, the next four purification steps involve precipitation with $(NH_4)_2SO_4$, acetone, or picric acid-steps that are facile and suited to handling large quantities. Next the Suzuki group utilized six column chromatography steps, including three separations by molecular size (Sephadex G-50) and three by ion exchange (DEAE-Sepharose and SP-Sephadex), followed by a final purification by reversed-phase LC. Correcting for an ~8% recovery of biological activity, the 11 nmol of hormone recovered amounts to a content of only 0.2 pmol of PTTH per head, an order of magnitude lower than the content of proctolin per insect!

PEPTIDE METABOLISM

One of the chief problems in isolating neuropeptides is the possibility of degradation by proteases during isolation. Examples of steps taken to minimize degradation by proteases can be seen in the above isolations: for proctolin (Fig. 1), tissue was extracted with 7% $HClO_4$, an extremely acidic and denaturing solvent for peptides, while for PTTH, extracts were boiled to denature proteins. There is currently little knowledge on metabolism of insect neuropeptides. However, proctolin has been shown to have a half-life in *P. americana* tissue homogenates of only <1 to ~2 min at 5 nM *in vitro* (Quistad *et al.*, 1984) or <5 min at the much higher, pharmacological concentration of 20,000 nM *in vivo* in *P. americana* (Starratt and Steele, 1984). Such rapid inactivation would be consistent with its putative role as a neurotransmitter. On the other hand, metabolism studies on a cardioacceleratory factor from *P. americana*, Pea CAH-II (also known as periplanetin CC-1, M 1, and neurohormone D), have shown a half-life *in vivo* of ~60 min at the low concentration of 1-2 nM (Skinner *et al.*, 1987). Thus, from the limited metabolic data available, there appears to be a difference in rate of metabolism of about two orders of magnitude between proctolin and Pea CAH-II, a member of the AKH family. This is significant, as the lack of strongly denaturing conditions in the isolation of AKH-like neuropeptides may not be a liability. However, use of conditions that denature proteases must ordinarily be assumed to be extremely important.

A PERSPECTIVE: ISOLATION OF VERTEBRATE NEUROPEPTIDES

It is widely acknowledged that research in insect neurochemistry and neuroendocrinology has substantially lagged behind the rapid advances made in the last decade for the corresponding research in mammalian systems. The fundamental advances made in peptide isolation and sequencing have been applied first in mammalian systems. A number of reviews of isolation methodology have been published; see Esch *et al.* (1983) and Wilson (1988). As there are no apparent reasons why isolation of neuropeptides from insects should be methodologically different than their isolation from vertebrates, interested readers are urged to consider such references. We will illustrate how careful consideration of some of these principles, especially those outlined by Esch *et al.* (1983), led to a successful scheme used for the efficient isolation of eclosion hormone, a diuretic hormone, and an allatotropin from heads of *Manduca sexta*.

Tissue Source for Isolation

While neuropeptides can be isolated successfully from extracts of whole animals (Brown and Starratt, 1975), purifications are simplified if a tissue with a high content of the neuropeptide is used for the isolation. With this strategy, surgical procedures are used as a mechanical purification, as exemplified by the isolation of AKH, or the earlier isolation from dissected vertebrate hypothalami of releasing factors which act on the pituitary (see Esch *et al.*, 1983). However, an interesting point is raised by the well-documented fact that levels of biogenic amines and cyclic nucleotides are higher in brains of vertebrates removed from animals sacrificed by microwave treatment *vs.* decapitation (see, for example, (Mir *et al.*, 1983)). Heating of the nervous tissue induced by microwave treatment denatures enzymes contained in the tissue whose purpose is to inactivate these chemical messengers.

While there do not appear to be published reports of similar differences in neuropeptide content in tissues removed from animals without microwave treatment, it seems reasonable to speculate that considerable degradation of neuropeptides may occur in tissues during dissection. A usual step to mitigate neuropeptide proteolysis subsequent to dissection is to immediately freeze tissues in dry ice or liquid nitrogen. A number of researchers have preferred to dissect brains from insects, rather than using whole heads. However, it is obviously slower to remove the brain from an insect than to slice off its head directly into a container of dry ice. Although the latter technique will require more chemical isolation steps, it would seem a prudent choice if the neuropeptide of interest is of unknown metabolic stability.

Choice of Purification Steps

The stages of a neuropeptide isolation can be divided into the extraction step, the defatting step, the pre-LC purifications, and the LC purification steps (Esch *et al.*, 1983). This strategy is a testament to the power and simplicity of purification of peptides by LC techniques, in particular reversed-phase LC. This technique has truly revolutionized neuropeptide isolation.

Esch *et al.* (1983) advocate extraction of tissues with an acidic solvent, followed by partitioning of this extract with an organic solvent such as hexane or chloroform to remove lipids. In the authors' hands, during such partitions with organic solvents, emulsions frequently form that make phase separation difficult and incomplete. The reverse order of these steps was used for isolation of *B. mori* PTTH (Nagasawa *et al.*, 1984), namely, to first extract the tissue with acetone to remove fats. Then the defatted tissue residue is extracted with an appropriate solvent to give neuropeptides. This order of steps is more rapid and may give higher recoveries.

The majority of vertebrate neuropeptides are somewhat basic peptides, explaining the success and popularity of extraction with acidic solvents. Typical of these procedures are extraction with 2 N acetic acid containing 10 mg/l of PMSF (a serine protease inhibitor) and 10 mg/l of pepstatin A (an acidic protease inhibitor) (Esch *et al.*, 1983). The activity of most enzymes is much lower at low pH, and addition of protease inhibitors further reduces proteolysis during extraction. An interesting example of attempts to denature proteases by a combination of low pH and high salt content is the "Bennett solvent": 5% formic acid, 1% trifluoroacetic acid (TFA), and 1% NaCl in 1 M HCl (Bennett *et al.*, 1981). It should be borne in mind that HCl is corrosive to stainless steel, and that Bennett *et al.* used a glass homogenizer with this solvent.

It is possible that acidic neuropeptides may be poorly extracted with these solvents, in which case use of neutral pH solvents could be more appropriate. Several commercial suppliers (Boehringer Mannheim, Pharmacia, and USB Biochemicals) recommend use of "cocktails" of various protease inhibitors, which they sell, for stabilizing proteins during extraction. It should be borne in mind that buffers for isolating proteins are intended to preserve enzymatic activity, and that for neuropeptide isolation modifications to decrease enzymatic activity (such as use of high salt (Bennett *et al.*, 1981), (Nagasawa *et al.*, 1984)), in addition to use of protease inhibitors, may be advisable.

Pre-LC Purifications

The crude extract of tissue will now contain a host of water-soluble materials, including salts, sugars, pigments, and other low-molecular weight biochemicals, as well as peptides and proteins. Many of the latter may be enzymes in varying states of denaturation, and it is conceivable that if steps are performed where the pH is raised, or salts and inhibitors removed chromatographically, that some proteases may again become active. It is therefore desireable that these initial steps be performed using conditions that are denaturing to enzymes. Due to the exceptional resolving power of modern LC techniques, it is advisable to use as few "classical" purifications as possible before separating the sample on LC. In designing a separation, it is also advisable to avoid repeating use of a technique with the same mechanism of separation for sequential purification steps, as maximum efficiency is obtained by varying the mechanism of separation. Precipitations with $(NH_4)_2SO_4$, acetone, or picric acid have been used to purify neuropeptides from massive amounts of *B. mori* tissues. While these methods are simple, relatively fast, and well suited for large quantities, they usually result in relatively low purification factors compared with chromatographic methods. Moreover, these procedures are frequently used for isolating enzymes and do not satisfy the need of utilizing denaturing conditions.

Size-exclusion, or gel filtration, chromatography has been widely utilized as a first step for purification of crude extracts containing peptides. With a support such as Sephadex G-50 or G-75 eluted at 4° C with 3-5 M acetic acid, proteins will be largely denatured and hydrophobic interactions of the support and the peptide suppressed (Esch *et al.*, 1983). The peptide will be recovered in a fraction containing materials of similar molecular weight, usually freed of salts and pigments, as well as most proteins. While this technique is very appealing, the resolving power of size-exclusion chromatography is lower than ion exchange chromatography or reversed-phase chromatography.

Ion exchange chromatography supports generally have a very high capacity for sample, making them well suited for preparative isolations. In addition, the number of peaks which can be resolved by ion exchange supports using gradient elution is substantially higher than the number of peaks resolvable by size-exclusion chromatography. The pH of the extract must be in an appropriate range for the peptide of interest to bind to the column (see section, Developing an Ion Exchange Separation on LC). Acidic extracts of tissue can be loaded directly on a strong cation exchange resin (such as a sulfopropyl resin) with confidence that all but very acidic peptides will bind.

Reversed-phase chromatography on alkyl-bonded silica packings under low resolution conditions also lends itself well to purification of crude extracts. Supports for this method are of the same type as utilized in high-pressure applications, but should be obtained with a larger particle diameter to minimize the pressure required to elute the sample (large diameter supports are also significantly cheaper). The early bonded-phase packings were usually C_{18} functionalized silicas, but shorter chain lengths (such as C_3 or C_4) have been shown to be preferable for hydrophobic peptides and proteins. The solution of peptide for loading should be acidic. Even with the largest diameter packings currently commercially available, the sample will have to be loaded on the packing using a pump (Esch *et al.*, 1983) or pulled through using vacuum (Kataoka *et al.*, 1987). Thus, the packing must be contained in a suitable apparatus, generally different from columns used for open column chromatography. Salts and other hydrophilic molecules will pass through without absorption. Elution of the column with 60% acetonitrile or 30% 1-propanol, containing 0.1 % TFA, will elute all but extremely hydrophobic peptides. If a higher degree of purification is desired, step gradients of gradually increasing organic concentration can be used.

Affinity chromatography is a powerful method for separating peptides from crude extracts due to its exceptional specificity. This method is limited in scope, as the requisite ligand will usually not be available unless the experimenter plans to isolate a peptide homologous to another to which appropriate antisera are available. The specificity is obtained at the cost of preparing an immunoaffinity support. Antibodies must be purified from sera, and the pure antibody then coupled to a suitably activated packing, such as an activated Sepharose. For isolation of a number of peptides with homology to the antigen used for raising the antibody, this considerable investment of preparing the support will be worthwhile.

FACTORS CONTRIBUTING TO LOSSES DURING ISOLATION

Peptides and proteins tend to bind to vessels used in common laboratory operations. The tendency to bind is greatest for hydrophobic peptides and is dependent on the nature of the surface. Polypropylene seems to show the least affinity for peptides. Glass must be used with caution, and other materials (polystyrene in particular) avoided. Adsorptive losses are usually greatest when the sample is highly purified, as adsorption is a surface phenomenon: once the surface is saturated with sample, the former should show reduced adsorptive activity. A commonly used precaution to avoid loosing sample during purification steps is to add to every fraction a pure and inexpensive protein, such as bovine serum albumin (BSA), in quantity greater than the peptide of interest. Clearly, this precaution must be avoided on the last purification prior to sequence and amino acid analysis, so deciding on when to add BSA will depend on the experimenter's judgment as to whether or not the chromatographic step has yielded a pure product.

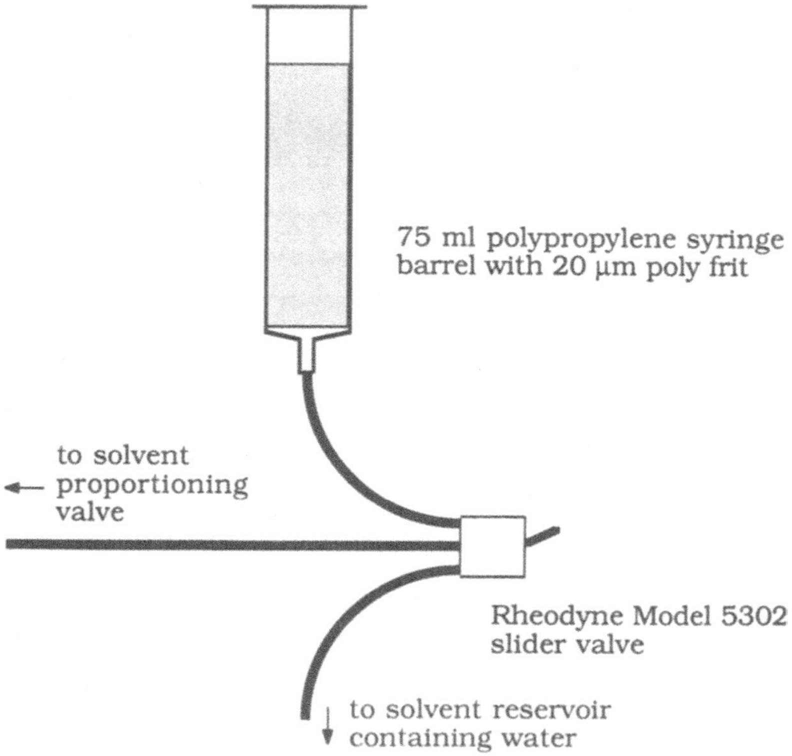

Fig. 3. Apparatus used by the authors for loading water-diluted fractions from RPLC onto a column for subsequent RPLC separation. A low pressure slider valve is used before the pump in the liquid chromatograph to select flow (normally from the water reservoir) to a polypropylene reservoir holding the diluted solution. It is reasonable to load hundreds of milliliters onto a 10 mm i.d. column with this device.

A related, but more serious, problem is that the propensity of peptides to adsorb is aggravated greatly by concentrating the solvent. Lyophilizing even semi--purified extracts is to be avoided, and removing organic solvent from LC fractions in a vacuum centrifuge or with a nitrogen stream can cause severe losses. In fact, Esch *et al.* (1983) recommend loading fractions from one purification step onto LC columns, without concentration, by pumping the solution onto the LC column. If the fraction contains organic solvent from reversed-phase chromatography, it must be diluted with at least two to three times its volume of water, or the peptide of interest may not bind to the reversed-phase column. With certain makes of LC pumps, the solution can be loaded readily by means of a valve built into the pump. We have devised an apparatus (Fig. 3) which allows loading solutions onto a column with virtually any pumping system. Our device utilizes a switching valve to select a suitable reservoir instead of the aqueous buffer reservoir. It is desirable to slightly pressurize the reservoir with helium (~ 0.5 atm) so that the solution is fed to the pump under slight pressure, as it is difficult to avoid introducing air bubbles into the pistons of the pump using this technique.

Several amino acids, in particular methionine, are susceptible to oxidation by atmospheric oxygen. The extent of oxidation will depend on the purity of the sample, the number of purification steps, and the amount of time the solution stands in exposure to oxygen. Other parameters affect the rate of oxidation; low pH has been shown to promote oxidation (Wilson, 1988). When an extract is not highly purified, proteinaceous impurities will suffer the majority of the oxidation. As the degree of purity increases, the peptide of interest will be more likely to oxidize. The most effective way of dealing with this problem is to add an antioxidant in large excess which does not interfere with the bioassay. Thiodiglycol (Hayes and Keeley, 1984) has been used for this purpose, although we prefer 2-(methylthio)ethanol. These closely related substances are completely soluble in water and are both sulfides, like methionine. The advantage of the latter reagent is that it is volatile and can be removed from fractions dried down for bioassay, unlike the former.

While it is clearly preferable to keep peptides in solution as opposed to dried down, peptides are susceptible to hydrolysis on prolonged storage, particularly in acidic solution. Aspartic acid residues in particular are sites for preferential acid cleavage at the alpha-carboxyl group. Solutions of peptides should be stored at 4° C to minimize hydrolysis. Freezing solutions of peptides slowly by placing them in a deep freeze is likely to concentrate solutes in the last portion of fluid to solidify, which will have the same consequences as concentrating solvent. Quick freezing solutions using liquid nitrogen is safer (the vessel must be turned rapidly while freezing the solution). It is clearly important to use the purified peptide for amino acid and sequence analysis as expeditiously as possible in light of uncertainties regarding optimal storage conditions for small amounts of bioactive peptides.

BIOASSAY OF ALIQUOTS DURING PURIFICATION

It is necessary during many purification steps to utilize buffer salts or acids, whose presence in fractions will be incompatible with the majority of bioassays. It is obviously desireable to use volatile buffers if possible, such as ammonium or triethylammonium acetate. Similarly, volatile ion-pairing reagents, such as TFA or heptafluorobutyric acid (HFBA) are almost universally used in reversed-phase LC so that aliquots may be completely dried down in a vacuum centrifuge. However, most salts used in ion exchange separations are non-volatile and may be present in rather high concentration. In such cases, aliquots of fractions may be desalted easily prior to bioassay by adsorbing an aliquot to a reversed-phase solid phase extraction cartridge (such as a C_{18} Sep Pak or C_4 Bond Elut) using aqueous buffer, and then eluting the cartridge with 60% CH_3CN in 0.1% aqueous TFA. The organic-containing effluent from the solid phase cartridge may then be evaporated in the vacuum centrifuge. One difficulty in this methodology is that hydrophobic peptides are likely to be lost on dry-down prior to assay, but adding ~100-250 μg of BSA prior to dry-down will usually mitigate this problem. Unfortunately, solid phase extraction cartridges packed with wide-pore supports (optimal for proteins and peptides) are not currently available, and performing this technique on a large number of fractions is time-consuming. Because of these problems, we have utilized commercially available empty reservoirs for solid-phase extraction cartridges, filled with 100 mg of bulk Vydac C_4 packing (20-30 μm particle diameter). Each cartridge can generally be used for a number of samples, provided it is properly re-equilibrated with aqueous buffer after eluting the prior sample with 60% acidic acetonitrile.

LC PURIFICATION

Reversed-phase LC techniques (RPLC) have proven extremely powerful for separation of a wide variety of molecules. In performing gradient separations, it is possible to resolve more than a hundred peaks with an average column. With judicious choice of mobile phase to move desired components away from major impurities, the purification factors possible in a single HPLC run are extremely high. Columns for peptide separation on analytical scale instruments are currently available with inner diameters ranging from 2.1 to 22 mm. While most analytical instruments are designed to operate with 4.6 to 10 mm i.d. columns, many can function adequately with the "microbore" (2.1 mm) columns and/or preparative columns (22mm). There is a hundred-fold difference in sample capacity between the former and latter columns. Accordingly, relatively crude extracts can be applied to preparative or semi-preparative columns, fractionated, and enriched fractions chromatographed on successively smaller columns. In isolating small quantities of peptides for sequence analysis, it is beneficial to use a microbore column for the final purification step in order to obtain the peptide in the smallest volume possible. With current peptide sequencers, the usual maximum volume of solvent for a single application to the peptide support is 15-30 µl. This limit places substantial restrictions on the percentage of a single RPLC fraction that can be sequenced. While multiple applications of solvent to the disk are possible, it is a decided advantage to have the sample in the smallest volume feasible without an evaporation step.

The popularity of RPLC has eclipsed other separation modes which can be performed using a liquid chromatograph. This is unfortunate as techniques such as ion exchange and size-exclusion chromatography offer different mechanisms of separation that may provide the ability to separate peptides which happen to coelute on several RPLC systems. Ion exchange LC can be an especially powerful tool, and seems to be growing in popularity. Because peptides are usually eluted from an ion exchange column with salt gradients in buffers, this technique is most profitably employed at a stage that is not expected to be the last purification step. RPLC is an ideal final purification step as the sample can be obtained in an entirely volatile solvent.

Manipulating Selectivity in RPLC

The power of RPLC for peptide isolation lies not only in the inherent resolution of the technique, but also in the variety of parameters which can be adjusted to affect the selectivity of separation. Manipulation of a variety of parameters allows the experienced practitioner to resolve easily peptides which coelute in a single fraction on a system with a different mobile phase. There is a tremendous volume of chromatographic literature comparing effects of different column packings, organic modifiers, ion pairing reagents, and other mobile phase additives.

The choice of column packing is quite important. Early reversed-phase packings consisted generally of silica with 60-100Å pore sizes to which had been bonded linear C_{18} hydrocarbon chains (such as µBondapak C_{18}). Attachment of the alkyl groups further decreases the effective size of these pores, which contain a large percentage of the surface area of the support. Thus, peptides of even moderate molecular weight may be excluded from the pores. This results in decreased sample capacity and non-optimal separations. It is now generally agreed that silica-based reversed-phase supports should have a pore size of at least 150-300Å for chromatography of peptides.

Hydrophobic peptides, and especially proteins, were found to be retained excessively on C_{18}-bonded phase columns. This led to the synthesis of RPLC supports with shorter carbon chains, including linear C_8, C_4, and even C_3 functionality. Currently the C_4 supports appear to have gained the widest acceptance for RPLC of hydrophobic peptides. Not only are peptides eluted with a lower concentration of organic modifier from these supports, but recoveries are usually improved. For small and/or hydrophilic peptides (such as those from a tryptic digest), use of C_{18} columns will generally be advantageous to enhance retention. If the peptide is so hydrophilic that it is eluted by only a few percent of organic modifier, it may be poorly retained on C_4 columns when aqueous extracts, or enzymatic digests, are loaded.

In addition, other stationary phases are available, such as phenyl-bonded silicas and polymer-based columns. The mechanism of interaction of phenyl columns will differ from the n-alkane phases, which may be of use for difficult separations. A recent study (Burton *et al.*, 1988) reports that separation of proteins on at least one polymeric reversed-phase column gave results comparable to that achieved with the best C_4 silica packings tested.

Usually the most fruitful approach to varying the selectivity of separation is to alter the mobile phase rather than to change columns. While changing the rate of the gradient from fast to slow to isocratic (constant composition) will improve the separation, it is generally far more useful to vary the organic modifier or the ion suppression/ion pairing reagent. That is, a sample that has been purified on an acetonitrile/0.1% TFA system will likely benefit little upon being rechromatographed using the same mobile phase with a different gradient, as compared with changing the mobile phase to 1- or 2-propanol/0.1% TFA. The mechanism of interaction of organic solvents with peptides differs: acetonitrile is rich in pi-electrons and will interact with aromatic regions of the peptide and perhaps the amide bond, while the propanols will hydrogen bond to nitrogens. These different mechanisms of interaction will account for differences in selectivity. It is worth noting that methanol is a weaker eluent than acetonitrile on reversed-phase LC. Methanol must be used at such high concentration for hydrophobic peptides that it may precipitate them, and is accordingly little used. Ethanol is also rarely used, probably because of the difficulty of obtaining high purity grades of tax-exempt ethanol. Propanol is reported to have little tendency to precipitate proteins at the concentrations used in reversed-phase LC. Peptides usually elute at about one-half the concentration of propanol as compared to acetonitrile.

Ionic peptides or proteins do not chromatograph well on reversed-phase in the absence of a buffer or ion pairing reagent. Phosphoric acid is an example of an ion suppression reagent, whose low pH suppresses ionization of the carboxyl moieties and protonates basic functionalities, leading to symmetrical chromatographic peaks. TFA similarly suppresses ionization of the carboxyls, but will also form ion pairs with positively charged groups, making the peptide somewhat more hydrophobic. HFBA forms substantially more hydrophobic ion pairs than TFA, so that peptides with a greater content of positively charged residues will be substantially more strongly retained. Thus, use of HFBA in the mobile phase allows the experimenter to combine some of the attributes of cation exchange chromatography and RPLC. It should be borne in mind that the greater hydrophobicity imparted by HFBA will also lead to a greater tendency for peptides which interact with it to adsorb to containers, especially if solvent is removed!

A number of buffer salts, organic and inorganic, have been used successfully for peptide separation. These buffers also improve peak shape by controlling the ionization of the peptide. Certain buffers may also ion pair, such as triethylammonium phosphate or acetate, where the buffer cation will interact with carboxyl functionalities. However, literature reports of their use seem to have declined in recent years, and similarly with phosphoric acid. The advantages of using totally volatile solvents (e.g., those containing TFA or HFBA) are very readily perceived for those occasions when solvent must be removed, such as for bioassay or analysis of fractions for amino acid composition or sequence. In addition, by using sequential separations with all combinations of TFA, HFBA, and the organic modifiers available, there are few peptides which cannot be resolved.

Developing an Ion Exchange Separation on LC

There is far less literature on principles of altering selectivity using ion exchange LC than in RPLC. While relatively acidic peptides are better separated on anion exchange, such as a diethylaminoethyl column, and those with high pI should be separated on cation exchange, such as a sulfopropyl or carboxymethyl column, there appears to be no well understood reason for choosing cation over anion exchange columns for peptides of intermediate pI values. As a pure ion exchange mechanism should depend only on the pI values of various constituents, this may not be surprising. As a rule of thumb, retention of peptides will be increased by lowering the buffer pH in cation exchange, or raising it in anion exchange. Separations are usually done at a constant pH, and the samples eluted using a salt gradient. While it is also possible to perform separations with a pH gradient, these are experimentally very difficult to control and to reproduce. A substantial inconvenience is that fractions for bioassay must generally be desalted prior to testing (see section on bioassay of aliquots). It is possible to affect separation not only by altering pH, but by changing the salt used for the eluting gradient. In cation exchange, for example, sodium ion is a stronger eluent than lithium. The advantages of changing salts do not appear to be as important, or as well understood, as the advantages of changing mobile phase or ion pairing reagent in RPLC. Nevertheless, inclusion of this technique in an isolation protocol will provide selectivity different from that achieved by changing from a TFA containing system to one containing HFBA.

***M. sexta* Allatotropin (AT) Isolation Scheme**

10,000 trimmed heads from pharate adults (420 g)

↓ 1) Homogenize with acetone, filter (delipidation)

↓ 2) Extract with strongly acidic buffer containing protease inhibitors

↓ 3) SP Sephadex chromatography; AT in 0.1&0.2 M NH_4OAc, pH 7.0

↓ 4) Vydac C_4 cartridge; AT in 60% CH_3CN/0.1% TFA fraction

↓ 5) Vydac C_4 semiprep LC, 0-60% CH_3CN/0.1% TFA gradient

↓ 6) " " " , 10-30% CH_3CN/0.1% HFBA gradient

↓ 7) TSK SP-5PW LC, 0-0.5 M NaCl/0.02 M $H_2PO_4^-$, pH 6.25 gradient

↓ 8)Vydac C_{18} analytical LC, 10-40% CH_3CN/0.1% TFA gradient

~1.5 nmol pure AT (~2.2 μg); purification factor 190 x 10^6 from heads

0.15 pmol/head

Fig. 4. Scheme for isolation of an allatotropin from heads of *Manduca sexta*, based on data from Kataoka *et al.* (1989a).

EXAMPLE OF AN ISOLATION SCHEME BASED ON THESE PRINCIPLES

We recently isolated the eclosion hormone (Kataoka *et al.*, 1987), an allatotropin (Kataoka *et al.*, 1989a), and the diuretic hormone (Kataoka *et al.*, 1989b) of *Manduca sexta* using slight variants of a very similar procedure. This procedure requires only eight or nine steps from whole heads. As this isolation scheme is relatively simple and possibly of more general utility, we describe its crucial features using the isolation of allatotropin as an example (Fig. 4).

The starting material for isolating these peptides was 10,000 heads of pharate adult *Manduca sexta*, which were cut off directly into dry ice. A 5 mm cork borer was used to "trim" a tissue portion containing the brain and corpus allatum/corpus cardiacum complex away from the eyes, proboscis, and other parts of the head. The trimmed heads (0.42 kg) were homogenized in acetone using a Brinkmann polytron to remove lipids. The extract was filtered, the acetone discarded, and the residue re-extracted with the polytron using an acidic solvent with protease inhibitors: 1 M acetic acid containing 20 mM HCl, 0.1 mM phenylmethylsulfonyl fluoride (PMSF), and 0.01 mM pepstatin A. This solvent must be made up just before use, as PMSF is hydrolytically unstable. Both protease inhibitors are poorly soluble in water, so they are dissolved in the acetic acid required for the final concentration, which is then diluted into water. The extract was centrifuged, the supernatant decanted, and the pellet re-extracted. All steps were performed at 0-4° C. The combined supernatants were loaded onto an SP-Sephadex C-25 column in the cold room, which was eluted sequentially with 1 liter each of 1 M acetic acid, 50 mM NH_4OAc (pH 4), and then 50 mM, 100 mM, 200 mM, 400 mM, and 800 mM NH_4OAc (pH 7). Aliquots from this step elution were desalted as described in the Bioassay section. Eclosion hormone activity was found to elute in the 50 mM NH_4OAc fraction, allatotropin in the 100 and 200 mM NH_4OAc fractions, and diuretic hormone activity in the 400 and 800 mM NH_4OAc. (While these isolations were not performed simultaneously, it is convenient to discuss them as if they were). These results showed that eclosion hormone was relatively acidic, allatotropin less so, and diuretic hormone probably somewhat basic.

The relatively large volumes of buffer were concentrated and desalted by applying them to a large cartridge column containing 10 g of Vydac C_4 bulk packing contained in a 75 ml polypropylene syringe barrel (a scale-up of the desalting procedure described in the Bioassay section). The solution was pulled through by vacuum, and the column washed with 100 ml of 0.1% TFA. All fractions from SP-Sephadex contained varying amounts of pigment, most of which was not retained on the Vydac C_4. The desalted peptides could then be eluted using step gradients of acetonitrile/water containing 0.1% TFA to provide some enrichment, or the column could simply be eluted with 60% acetonitrile/0.1% aqueous TFA to elute all peptides together. We used 100 ml each of 20%, 35%, and 60% acetonitrile/0.1% TFA to elute EH and DH. Both hormones were recovered in the 35% acetonitrile/0.1% TFA fraction, freeing them of both very hydrophilic and very hydrophobic

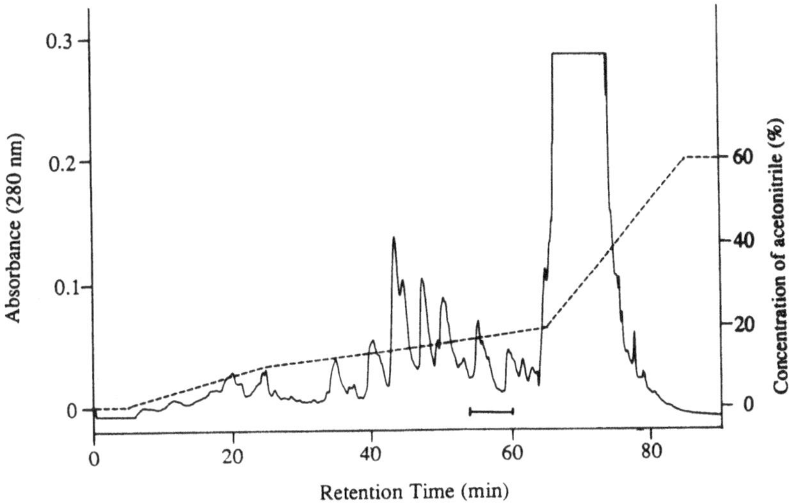

Fig. 5. First semi-preparative isolation of *M. sexta* allatotropin on a 25 x 1.0 cm Vydac C_4 column. The solid line indicates absorbance at 220 nm, while the dashed line indicates the composition of the acetonitrile/0.1% TFA gradient. The biologically active fractions eluting from 54-60 min are marked with a bar. Note that a very slow gradient was used in the area where the biological activity eluted, with a subsequent fast gradient used to elute massive amounts of UV-absorbing impurities. Fine-tuning a gradient in this way requires several pilot experiments, but is generally most worthwhile at the early stages of isolation (Kataoka *et al.*, 1989a).

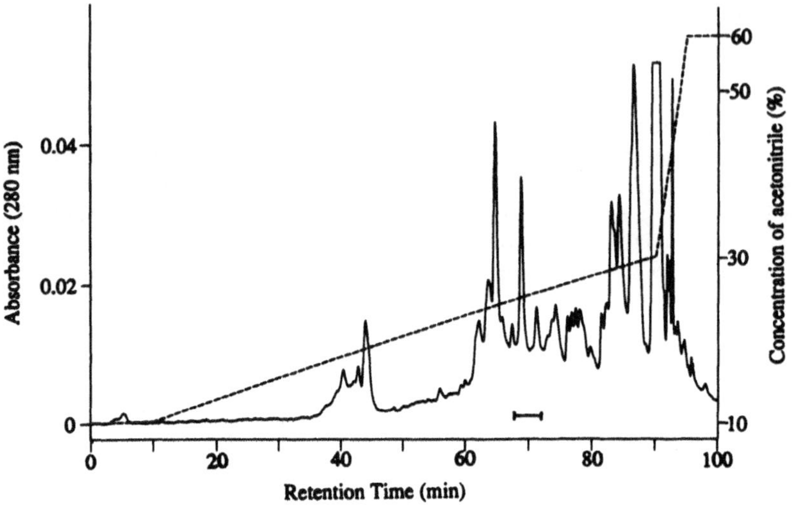

Fig. 6. Second semi-preparative isolation of *M. sexta* allatotropin. The sample was the active fractions from Fig. 5. The same column was used as in Fig. 5, but HFBA was used instead of TFA as ion-pairing reagent. The solid line indicates absorbance at 220 nm, while the dashed line shows the percentage of acetonitrile. The biologically active fractions elute from 68-72 min (solid bar with tics), in a region of the chromatogram with relatively few constituents.

peptides. However, we eluted allatotropin together with other peptides using 60% acetonitrile.

For each hormone, the active organic fraction was then diluted with twice its volume of water and pumped onto a 25 x 1.0 cm Vydac C_4 semipreparative column using the pump, as shown in Fig. 4. (It is advisable that the pump and column be equilibrated with 0.1% TFA prior to loading). The choice of solvent for the first semi-preparative separation can be crucial, to avoid coelution with massive amounts of protein, perhaps structural proteins from the cuticle (see very large peak eluting just after allatotropin, Fig. 5). For EH we used a gradient of 1-propanol/0.1% TFA, to provide different selectivity from the C_4 cartridge step. For both allatotropin and DH, we used an acetonitrile/0.1% TFA gradient.

In the case of allatotropin, biological activity was recovered in a zone which corresponded to about 17-19% acetonitrile on the gradient (see Fig. 5).

For each hormone, active fractions from the first semipreparative RPLC purification were pooled, diluted with twice the volume of water, and again pumped onto the same column. In each case the second semipreparative step used a different mobile phase from the prior step. For allatotropin, this was a 10-30% acetonitrile/0.1% HFBA gradient; biological activity was recovered in fractions corresponding to 24.5-25.5% acetonitrile, 6-7% higher than when TFA was the ion pairing reagent.

The subsequent purification (step 5, Fig. 4) was an ion exchange LC separation. For allatotropin and DH, we used a TSK SP-5PW analytical column, equilibrated with 20 mM sodium phosphate buffer, pH 6.25, eluted with a gradient of sodium chloride from 0-0.5 M. For the more

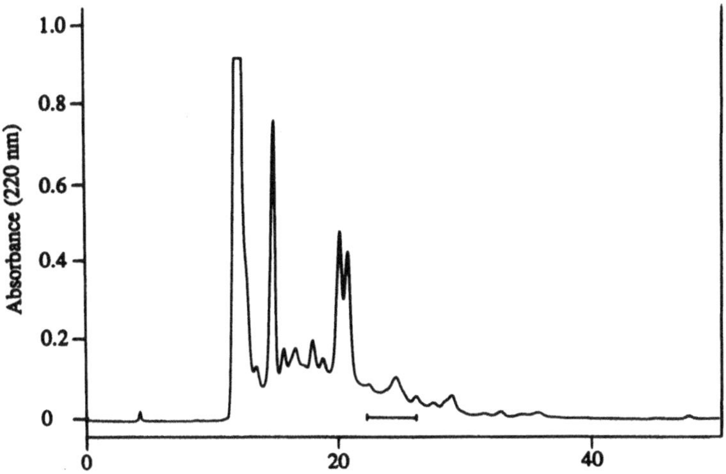

Fig. 7. Separation of active fraction from Fig. 6 on a hydrophilic ion exchange LC column (Toyo Soda TSK SP-5PW), eluted with a gradient of 0 to 0.5 M NaCl in a 20 mM sodium phosphate buffer, pH 6.25. The active fraction eluted from 22-26 min (shown by solid bar with tics), which corresponded to about 0.09-0.10 M NaCl in the gradient. Buffers contained 10% acetonitrile to retard microbial growth.

acidic eclosion hormone, we used anion exchange, a TSK DEAE-5PW column and pH 8.0 Tris-HCl buffer. These useful columns are packed with hydrophilic polymers having low adsorptive properties (unlike polystyrene polymers).

The active fraction from ion exchange chromatography was loaded directly onto a Vydac C_{18} analytical column (10 x 0.46 cm) eluted with a gradient of 10-40% acetonitrile/ 0.1% TFA. A number of peaks eluted rather close together, with one biologically active peak (arrow in Fig. 8) chromatographing in the tail of several other peaks at a retention time of 34.3 min, corresponding to ~17% acetonitrile. This is an example of changing the column packing, but using a mobile phase already employed earlier (in step 5, Fig. 5). As can be seen, the impurities are rather close eluting. In this case, use of 1-propanol (not tried earlier) with TFA or HFBA might have provided better selectivity. However, the peak indicated was sufficiently pure for complete sequence analysis of the 13-residue peptide, H-Gly-Phe-Lys-Asn-Val-Glu-Met-Met-Thr--Ala-Arg-Gly-Phe.

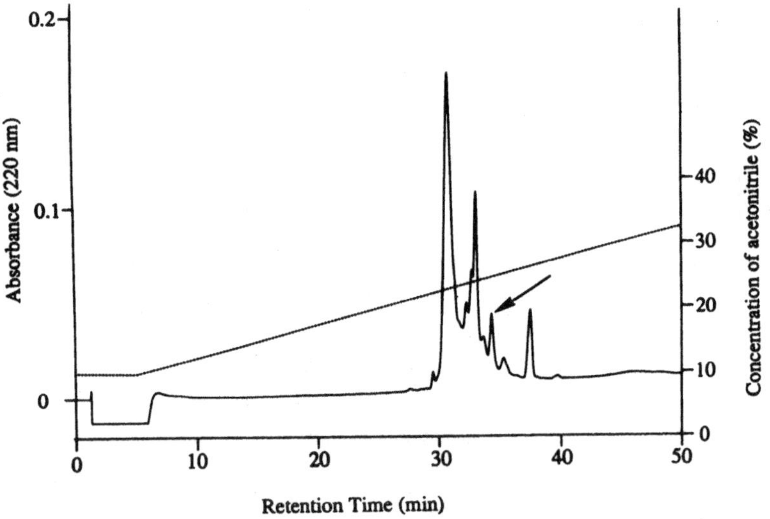

Fig. 8. Separation of active fraction from Fig. 7 on analytical RPLC (10 x 0.46 cm Vydac C_{18} column). A solid line indicates absorbance at 220 nm; the dashed line shows the percentage of CH_3CN in the gradient. The ion pairing reagent was TFA as in Fig. 5. The biologically active peak used for sequence analysis is shown with an arrow.

About 1.5 nmol of pure allatotropin (~2.2 μg) was recovered. Assuming a high recovery (for which we have no evidence, as the bioassay is not quantitative) this would correspond to about 0.15 pmol/head. This is similar to the amount of 4K PTTH isolated from a head of *Bombyx*. The purification factor of allatotropin from heads would correspond to about 190 x 10^6 , averaging about 20 million-fold per step. This is remarkable considering that steps 1 and 4 (Fig. 4) probably have a rather small purification factor, and is a testament to the power of LC techniques for peptide purification.

REFERENCES

Bennett H. P. J., Browne C. A. and Solomon S. (1981) Purification of the two major forms of rat pituitary corticotropin using only reversed-phase liquid chromatography. *Biochemistry* **20**, 4530-4538.

Brown B. E. and Starratt A. N. (1975) Isolation of proctolin, a myotropic peptide, from *Periplaneta americana*. *J. Insect Physiol.* **21**, 1879-1881.

Burton W. G., Nugent K. D., Slattery T. K., Summers B. R. and Snyder L. R. (1988) Separation of proteins by reversed-phase high-performance liquid chromatography. I. Optimizing the column. *J. Chromatogr.* **443**, 363-379.

Esch F. S., Ling N. C. and Böhlen P. (1983) Microisolation of neuropeptides. *Methods Enzymol.* **103**, 72-89.

Hayes T. K. and Keeley L. L. (1984) Isolation of insect neuropeptides using reversed-phase high performance liquid chromatography. In *Insect Neurochemistry and Neurophysiology* (Ed. by Borkovec A. B. and Kelly T. J.), pp. Plenum, New York.

Kataoka H., Toschi A., Li J. P., Carney R. L., Schooley D. A. and Kramer S. J. (1989a) Identification of an allatotropin from adult *Manduca sexta*. *Science* **243**, 1481-1483.

Kataoka H., Troetschler R. G., Kramer S. J., Cesarin B. J. and Schooley D. A. (1987) Isolation and primary structure of the eclosion hormone of the tobacco hornworm, *Manduca sexta*. *Biochem. Biophys. Res. Commun.* **146**, 746-750.

Kataoka H., Troetschler R. G., Li J. P., Kramer S. J., Carney R. L. and Schooley D. A. (1989b) Isolation and identification of a diuretic hormone from the tobacco hornworm, *Manduca sexta*. *Proc. Natl. Acad. Sci. USA* **86**, 2976-2980.

Mir A. K., Pallot D. J. and Nahorski S. R. (1983) Biogenic amine-stimulated cyclic adenosine-3',5'-monophosphate formation in the rat carotid body. *J. Neurochem.* **41**, 663-669.

Nagasawa H., Kataoka H., Hori Y., Isogai A., Tamura S., Suzuki A., Guo F., Zhong X., Mizoguchi A., Fujishita M., Takahashi S. Y., Ohnishi E. and Ishizaki H. (1984) Isolation and some characterization of the prothoracicotropic hormone from *Bombyx mori*. *Gen. Comp. Endocrinol.* **53**, 143-152.

Quistad G. B., Adams M. E., Scarborough R. M., Carney R. L. and Schooley D. A. (1984) Metabolism of proctolin, a pentapeptide neurotransmitter in Insects. *Life Sci.* **34**, 569-576.

Skinner W. S., Quistad G. B., Adams M. E. and Schooley D. A. (1987) Metabolic degradation of the peptide periplanetin CC-2 in the cockroach, *Periplaneta americana*. *Insect Biochem.* **17**, 433-437.

Starratt A. N. and Brown B. E. (1975) Structure of the pentapeptide proctolin, a proposed neurotransmitter in insects. *Life Sci.* **17**, 1253-1256.

Starratt A. N. and Steele R. W. (1984) In vivo inactivation of the insect neuropeptide proctolin in *Periplaneta americana*. *Insect Biochem.* **14**, 97-102.

Stone J. V., Mordue W., Batley K. E. and Morris H. R. (1976) Structure of locust adipokinetic hormone, a neurohormone that regulates lipid utilisation during flight. *Nature* **263**, 207-211.

Wilson K. J. (1988) Purification of proteins/peptides for structural studies. In *Protein/Peptide Sequence Analysis: Current Methodologies* (Ed. by Bhown A. S.), pp. CRC Press, Boca Raton, Florida.

From *Insect Neurochemistry and Neurophysiology • 1989* Edited by A. B. Borkovec and E. P. Masler

INSECT NEURODEVELOPMENT:

GENE CASCADES IN DROSOPHILA.

Rolf Bodmer

Howard Hughes Medical Institute, UCSF

Box 0724, San Francisco, CA 94143

ABSTRACT

How is an ectodermal cell specified to become a neural precursor? How do its progeny acquire their correct cellular identities? These questions are of central importance in neural development. Many genes have been identified which act at particular stages in a cascade of events during neurogenesis. A network of genes are involved in determining the number of neural precursors. Other genes have been found to act at the level of specifying cellular identities. One of these, cut, is required to specify the correct identity of a particular class of sensory organs. Another gene, numb, is needed for the progeny of sensory precursors to develop as neurons.

The nature of the products encoded by these genes suggests molecular mechanisms by which these genes exert their function. Some are DNA binding and may regulate transcription of other genes, while others are membrane proteins suggesting a function in cell-cell communication. Understanding the interactions of such genes will elucidate the mechanisms of neural development.

INTRODUCTION

Important questions in neuronal development are: (1) How does an undifferentiated cell become a neuron and how does is acquire its correct identity? (2) How do the neuronal processes like axons and dendrites find their ways to the correct targets and form appropriate connections? The first question addresses the early events of neuronal specification and the second addresses the later events of pathfinding and synaptogenesis. In this article I will concentrate on the first problem and discuss what genes are involved in the creation of a neuron.

In many insect systems a large amount of information is available on neuronal structure and pathways. Insects have been a particularly favorable system, because individual neurons can be identified and pioneering studies have elucidated some of the principles of neuronal connectivity and pathfinding (see for example: Murphy et al., 1985; Blair and Palka, 1985; Goodman et al., 1982; 1984). However, relatively little is known about the actual molecules which contribute to these processes. If one considers development as a continuous interplay between molecules, of which the emerging cells and tissue structures are the macroscopic manifestations, then it becomes clear that the principles of the individual molecular interactions must lay the foundation for the higher form of structural order found in differentiated tissues.

To study neurodevelopment two distinctly different experimental approaches can be envisioned:

(a) Surgical manipulation of a minimal part of the organism.

(b) Genetic manipulation of only a single molecular component.

Surgical manipulations can be very focused on a particular part of the system, e.g. the transsection of a nerve, the transplantation or ablation of a cell or a developing body part. This yields information about macroscopic functional relationships, for example what kinds of spatial and temporal arrangements are important for a certain developmental process to occur. However, the molecular factors that have been changed in such procedures are very numerous and to identify the relevant ones is not always an easy task.

Genetic manipulations can be used to introduce a change of only one molecular component. However, the alterations may affect more than just the desired tissue and the effect on the developing organism be created indirectly. If a mutation is specific in its range of actions and affects only a limited set of tissues, it may be possible to interpret the phenotypic changes as a direct result of the altered single molecular component.

Therefore, mutational analysis constitutes a complementary approach to the analysis of surgical manipulations, allowing investigation of developmental processes from different and independent angles. Unfortunately, most animal systems genetic manipulations are severely limited.

Drosophila melanogaster, because of its short generation time and the relative simplicity of its nervous system, lends itself favorably to a genetic analysis of neurodevelopment. A wealth of genetic information is available concerning the makeup of this organism as well as a large number of well defined mutations. Drosophila has been used successfully to identify the key genes that are involved in early embryonic development that transform the morphogenic gradients in the egg into the first subdivisions and ultimately into the correct segmental arrangements (e.g. Nuesslein-Volhard and Wieschaus, 1980; Scott and Carroll, 1987; Ingham, 1988).

Many aspects of neurodevelopment are quite similar among different insect and those parallels were useful in establishing the phenomenological description of Drosophila's neuronal development. For example, the sensilla of the PNS have similar morphologies (Zacharuk, 1985), and similar division patterns during their genesis were observed (Bate, 1978). In addition, the development of the CNS was found to be remarkably similar to the grasshopper, including cellular identities and axonal trajectories (Thomas et al., 1984).

Specification of neurons, in many aspects, most likely precedes the formation of specific connections. A reasonable approach to understanding neuronal development may therefore be first to study the earlier events of neurogenesis, such as neuronal specification, and then capitalize from the gained insights to understand the later events of pathfinding and synaptogenesis. However, if one is to use such an approach, it is necessary to identify the components of the developing nervous system before differentiation has begun.

In this article, I will review some of the recent results using Drosophila as a model insect system for a genetic analysis of neuronal development and speculate on the hierarchy of genes which might govern neurogenesis.

RESULTS and DISCUSSION

In insects, neurogenesis in the embryo starts out in a seemingly uniform layer of ectodermal cells by singling out individual cells. These cells usually enlarge and move out of the ectodermal sheath to a position just underneath the ectoderm. At this point these cells are becoming neuronal precursor cells or neuroblasts and begin to divide in either a stem cell fashion (Doe and Goodman, 1985a, Bodmer el al.,1989) or by symmetrical divisions (Lawrence, 1966; Bate, 1978; Bodmer et al., 1989). The progeny eventually assume very individual identities including a variety of different types of neurons and glial cells.

In Drosophila, the pattern of PNS neurons and sensory organs is very precise and reproducible, which makes it an ideal model system to investigate the genetic requirements that lead to pattern formation in the nervous system. The embryonic PNS in the body segments is strictly segmentally arranged as is illustrated with a neuron specific antibody in Fig. 1.

The pattern is very stereotyped from one embryo to another and the neurons are arranged in characteristic clusters. There are five different classes of peripheral neurons that have distinct morphological and antigenic characteristics (Bodmer and Jan, 1987). Two of these neuronal classes have single dendrites and innervate different types of sensory organs: The external sensory (es) organs have sensory structures in the cuticle like bristles, etc., and the chordotonal (ch) organs are internal stretch receptor types. The other classes have multiple dendrites with free endings (md neurons). Thoracic and abdominal segments have characteristic distributions of PNS cells (Fig. 2). A variety of cell-type specific antibodies allows investigation of different components of the PNS individually (Jan and Jan, 1982; Ghysen et al.; 1986, Bodmer et al., 1987; Bier et al., 1988). The structure and development of the PNS of Drosophila embryos has been described in considerable detail (Campos-Ortega and Hartenstein, 1985; Ghysen et al., 1986, Dambly-Chaudiere and Ghysen, 1986; Bodmer and Jan, 1987; Hartenstein, 1988; Bodmer et al., 1989).

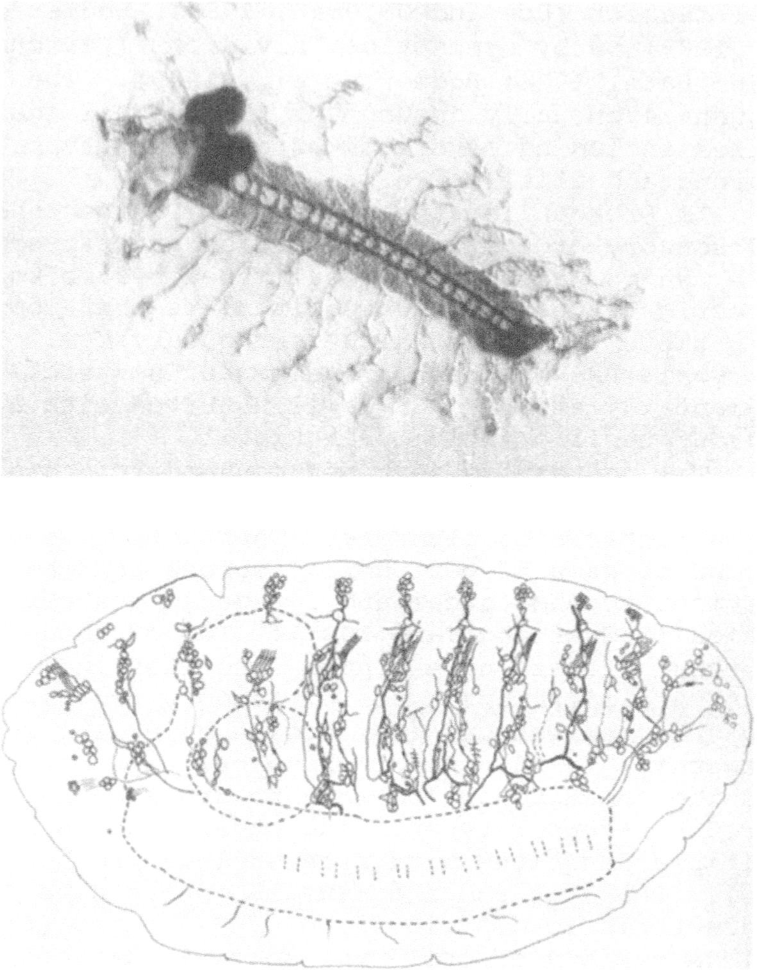

Fig. 1. Drosophila embryo at about 12 hr. of development stained with an antibody specific for neuronal surfaces (Jan and Jan, 1982). T he top embryo is cut open dorsally in a filet preparation and the peripheral epidermis is stuck to the glass surface to expose the PNS. Camera lucida drawing (bottom) of a same stage whole-mount embryo emphasizing the cells in the PNS. CNS is indicated in dashed lines. Anterior is left and dorsal up in this and all subsequent micrographs.

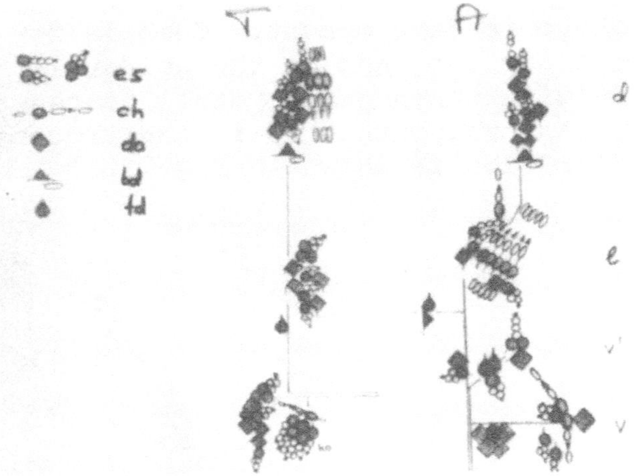

Fig. 2. Arrangement of the five different classes of sensory organs and neurons in a typical thoracic (T) and abdominal (A) segment. The neurons are the filled symbols and the sensory organ support cells are blank. es external sensory organs. ch chordotonal organs. da, td and bd are multiple dendrite (md) neurons.

It is possible to follow the DNA replication pattern of the precursor cells that give rise to the PNS with a newly developed technique using a thymidine analog (BrdU) which can be localized immunocytochemically (Gratzner et al., 1982; Truman and Bate, 1988). This method has revealed that the precursor divisions are as stereotyped as is the differentiated structure of the PNS (Bodmer el al., 1989). The development can therefore be monitored from the time when precursors start dividing to their final differentiation. The PNS divisions happen between 5 and 9 hr. of embryonic development. This is for the most part after the rest of the epidermis has ceased division which greatly facilitates the analysis. An embryo that is in the process of PNS proliferation is shown in Fig. 3.

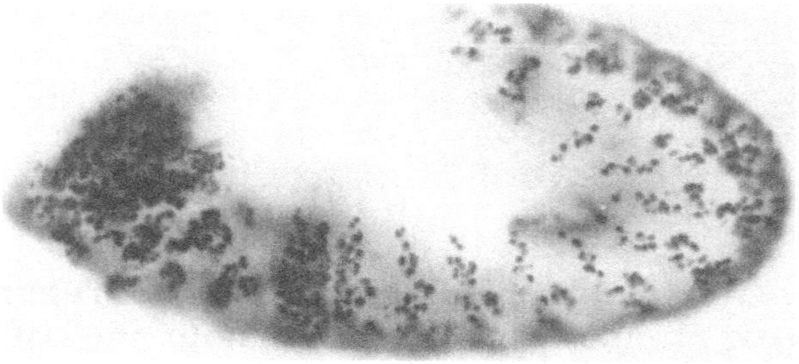

Fig. 3. 7-71/2 hr. embryo (in the process of germ band retraction) that has been incubated for 1/2 hr. with the thymidine analog BrdU, showing the PNS precursors in the process of replication at that time.

Analysis of the division pattern in the PNS also allowed investigation of the lineage relationships between the PNS cells. The cells within each sensory organs are most likely strictly related by lineage (Bodmer et al., 1989). es organ precursors divide symmetrically giving rise to usually four cells which will differentiate into the appropriate cell types that form the sensory organ including the neuron. ch organs, on the other hand, are created from a stem cell type division pattern. In both cases the neuron is produced from the last division.

The detailed knowledge of the stereotyped and invariant structure and development of virtually every cell in the PNS makes it a very useful and sensitive system to investigate the effect of mutations in genes that are playing a role in neurogenesis. A number of new mutations have been discovered in the past few years that are involved in different aspects of the developing nervous system, particularly the PNS. Their analysis has provoked speculations about how these genes might work together with known genes which have already been suggested to be crucial for neuronal development. The model I will discuss below is an attempt to order the action of these genes in a tentative hierarchical sequence of events, focussing mostly on the steps that mediate the development of sensory organs. Many aspects of the model are probably also applicable to the CNS. In fact, some aspects of the model depend on genetic and surgical experiments that have been done in the CNS. Some genes I will describe also affect the development of the compound eye in the fly, and some of the principles found in the development of the embryonic CNS and PNS may also be applicable there. However, many aspects of the eye development show distinct differences, especially in that lineage relationships seem to be much less important, rather eye specific forms of cell-cell interactions play crucial roles (Ready, 1989).

Many of the genes that are involved in neuronal development have been or are being cloned. The protein sequence of some of these genes reveal similarities to proteins whose mode of action is well understood and suggest a similar function for the neuronal genes. Also, the spatial and temporal localization of expression suggests in some cases the range of action for these genes. These results yield additional and complementary information to the mutant phenotype how these neuronal genes act.

A Model for Neurogenesis

A hypothesis for the sequence of events that lead to the specification of the neuronal precursors and their progeny has recently been formulated by Ghysen and Dambly-Chaudiere (1989b). The following model of neurogenesis is largely based on their proposition with some modifications:

Fig. 4 (next page). Model of the consecutive steps that lead from an ectodermal sheath of cells to a defined number of neuronal precursor cells and finally to the mature set of identified neurons. The temporal sequence of hypothetical events in the model are numbered and the genes discussed in the text are indicated in italics.

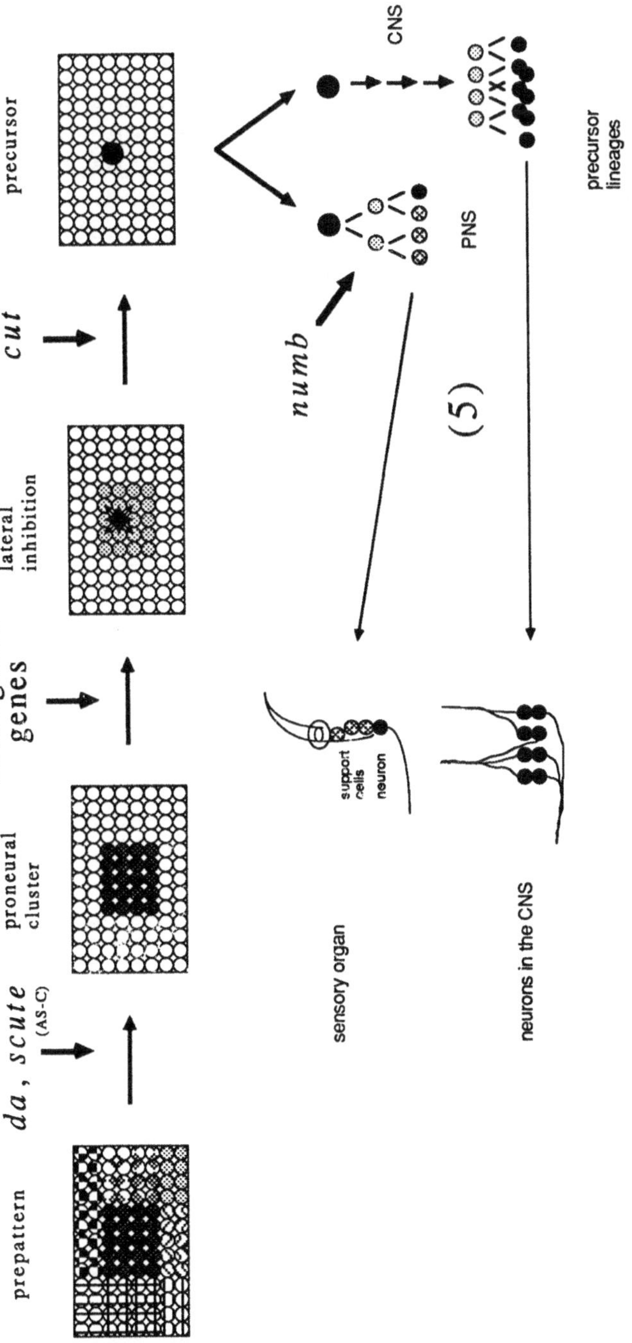
(1)
prepattern
proneural
genes
da, scute
(AS-C)
(2)
proneural
cluster
neurogenic
genes
(3)
lateral
inhibition
selector
genes
cut
(4)
precursor
numb
PNS
CNS
precursor
lineages
(5)
sensory organ
support
cells
neuron
neurons in the CNS

(1) Prepattern Genes

Many genes that govern the early events of Drosophila development have been identified and their hierarchical relationships have been worked out in considerable detail (e.g. Nuesslein-Volhard and Wieschaus, 1980; Scott and Carroll, 1987; Ingham, 1988). The patterns of their expression show extensive overlap in that a given cell expresses more than one of those early position- or region-specific genes like the segmentation, the segment polarity, or the dorso-ventral genes. The idea is that those genes - in addition to or as a consequence of their early function - also define in a concerted fashion the approximate regions where the neuronal precursor cells will eventually be formed. This prepattern is thought to directly or indirectly activate (or repress) other genes which are more closely related to the genes than are involved in neurogenesis. These subregions of the ectodermal sheath become competent to produce neuronal precursor cells. For the ventral nerve cord (CNS) this region encompasses the whole ventral ectoderm (Campos-Ortega, 1988), whereas for the PNS it is thought that certain peripheral patches produce individual PNS precursor cells. The nature of the process by which these early genes interact and initiate such a pattern is unknown and remains speculative.

(2) Proneural Genes

In the subregions, which are predetermined to become the regions of future neuronal precursor cells one or several genes are activated (or repressed) to provide the competent factor(s) for promoting the neuronal pathway. They were termed "proneural" genes (Ghysen and Dambly-Chaudiere, 1989b). They are expressed not only in the ectodermal cell(s) which will eventually become the neuronal precursor cell(s), but also in a patch of neighboring cells. The idea is that all the cells in a proneural cluster have an initial capacity to become the precursor, which will be refined later.

In Drosophila the genes in the achaete-scute complex (AS-C) were found to be essential for the development of a class of sensory organs, because mutations in AS-C eliminate these sensory organs, but have little effect on the rest of the epidermis (Garcia-Bellido, 1979; Ghysen and Dambly-Chaudiere, 1987). There is recent evidence that they are expressed in patches of cells in the region where later precursor cells will develop (Romani et al., 1989). These regions of AS-C expression may correspond to the proposed proneural clusters.

Another gene, daughterless (da), has been recently found to be necessary for all the PNS (Caudy et al., 1988a). The expression of that gene has not been localized yet. The phenotype of a da mutant embryo is shown on Fig. 5, where no neurons are present in the PNS. An AS-C embryo would look the same, except that the ch neurons and some md neurons are not deleted (see Fig. 2). In order to determine whether the precursor cells of the deleted PNS cells were also absent or whether the differentiation of their progeny was impaired in da or AS-C mutants, I investigated mutant embryos for the presence of proliferating PNS precursors early during embryogenesis (Fig. 6). I found that in AS-C as well as in da mutant embryos the precursors for the deleted PNS cells are not replicating their DNA (Fig. 3). Also in the CNS many fewer replicating neuroblasts can be detected in these mutants. It is likely that the precursors do not form because they lack sufficient positional information to specify neuronal development.

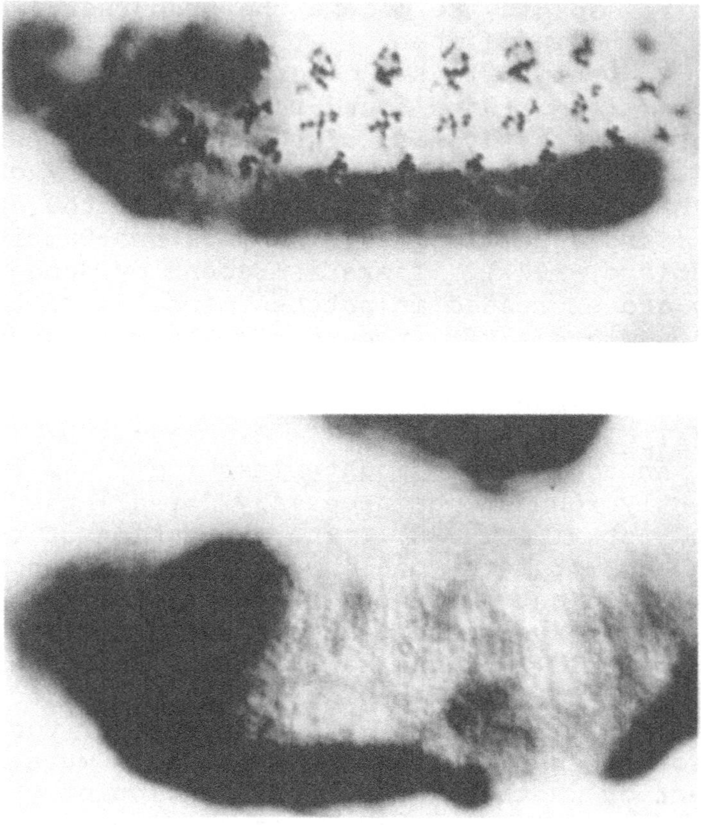

Fig. 5. 14 hr. wild-type embryo (top) that has been stained with an antibody that is specific for all neuronal nuclei (44C11, Bier et al., 1988). Note the segmentally arranged clusters of PNS neurons. Similar stage embryo (bottom) mutant for da, where no peripheral neurons present. da also affects the CNS. Micrographs were kindly provided by Y.N. Jan and M. Caudy.

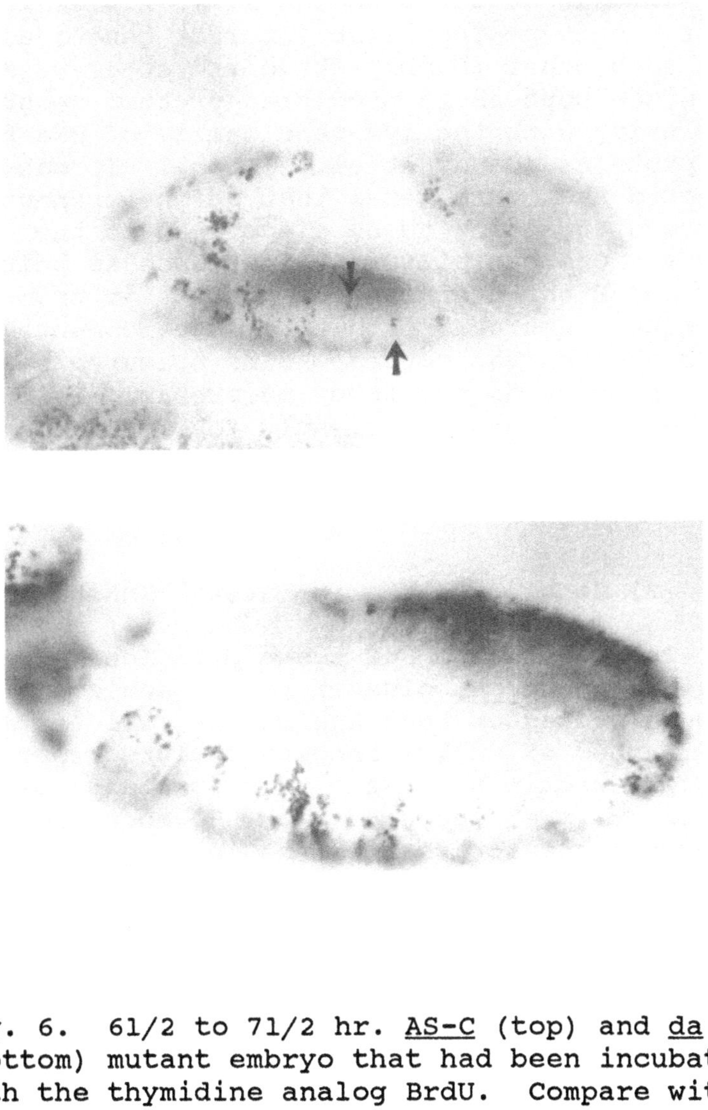

Fig. 6. 61/2 to 71/2 hr. AS-C (top) and da (bottom) mutant embryo that had been incubated with the thymidine analog BrdU. Compare with the wild-type pattern in Fig. 3. The top micrograph shows replicating precursors only of ch precursors (arrows) which are unaffected by AS-C, whereas for da (bottom) no replicating PNS precursors can be detected. Labelled cells are CNS cells or occasional epidermal cells.

AS-C and da may act at a similar level (Fig. 4), since in addition to the similarities in their phenotype they also interact genetically with each other (Dambly-Chaudiere et al., 1988), and since both share predicted protein sequence similarity with the myc-gene family of DNA binding proteins (Caudy et al., 1988b). It can be expected that more genes that define proneural clusters are going to be discovered in the future. One can picture these genes as being at the top of the hierarchy of genes that promote the actual neuronal pathway of development. In their absence the corresponding precursors and their progeny do not form and probably fail to become specified sufficiently for initiation of neurogenesis (Ghysen and O'Kane, 1989; Ghysen and Dambly-Chaudiere, 1989b; Bodmer et al., 1989).

(3) Genes Involved in Lateral Inhibition

The acquisition of acute positional values within a proneural cluster may be achieved through processes that ensure one cell (or a defined number of cells) becomes the precursor cell. The cell within a cluster that starts to become a precursor first (perhaps with the most pronounced precursor characteristics) begins to inhibit its neighboring cells from also becoming a precursor cell. Alternatively, the emerging precursor might also actively promote the formation of ectodermal cells around it. The consequence is formally the same: Neighboring cells are prevented to assume a similar precursor fate. Such a mechanism has been termed "lateral inhibition" (Campos-Ortega, 1988). Evidence that such a mechanism is very likely to be acting in the sharpening of positional information comes from ablation experiments and observations of mutant phenotypes:

a) If a neuroblast in the grasshopper CNS is

ablated during neurogenesis, an ectodermal cell in the neighborhood of the removed neuroblast will develop with the appropriate developmental fate (Doe and Goodman, 1985b). This is true even when the neuroblast started to divide to give rise to ganglion mother cells. These results indicate that several ectodermal cells in a particular area have the potential to develop into the neuroblast specific for that area. Once a neuroblast enlarges and segregates it will inhibit neighboring cells from becoming a neuroblasts, and it retains this capacity over an extended period of time. This process of lateral inhibition is likely to be a general phenomenon in sharpening the positional information to a single cell or a defined number of cells as genetic experiments in Drosophila suggest.

b) Mutations in a set of genes, called the "neurogenic" genes (Campos-Ortega, 1988), apparently disrupt the normal mechanism of lateral inhibition. Strong alleles of these genes cause all the cells in the so called neurogenic region to differentiate as neuronal precursors (Campos-Ortega, 1988). In the region where peripheral sensory organs develop milder alleles of these genes generate an overproduction of the existing organs (Hartenstein and Campos-Ortega, 1986), suggesting the location in which these organs develop and is normal hence independent of the neurogenic genes, but more cells than normal will develop as sensory precursors.

In experiments that used a marker for sensory precursors it was found that the number and pattern of the <u>first</u> precursors that segregate is normal in mutations of the neurogenic genes that were tested. Soon afterwards, however, many more precursors are observed, suggesting that lateral inhibition of the surrounding cells is defective, but the initial specification for a certain type of precursors cell is normal (Ghysen and Dambly-Chaudire, 1989a). This overproduction of precursors could also be shown as a hyperplasia of proliferation in the PNS in neurogenic mutants (R. Bodmer, unpublished observations) compared with wild-type (Fig. 3).

Another set of experiments combined mutations in AS-C and one of the neurogenic genes. It was found that only those sensory organs which are not affected by AS-C mutations are increased in number, and the remaining organs and neurons are absent (Brand and Campos-Ortega, 1988). This again is in congruence with the idea that AS-C determines the competence of proneural clusters, whereas the neurogenic genes this pattern into the exact number of precursors in that area.

(4) Selector Genes

One could imagine that the interaction of genes that specify proneural clusters with genes that ensure a defined number precursors through lateral inhibition would be sufficient for the specification of given neuronal precursor cell. How is this positional information translated into the formation of differentiated neurons? In the early embryo, for a particular segment to develop with the correct identity depends on the activity of the appropriate so called "selector" gene (Garcia-Bellido, 1975). In the absence of these genes whole segmental identities are transformed into those of other segments. These genes seem to be required as the key to translate positional information into a final developmental path. Are there also such genes which are essential in translating the positional information of a newly specified neuronal precursor cell to differentiate with the appropriate identity?

Recently, one such gene, cut, has been described to be required for a set of sensory cells (es organs, see Fig. 2) to differentiate into the sensory organs with the correct identity

(Bodmer et al., 1987; Blochlinger et al., 1988). In the absence of cut gene function sensory organs form nevertheless, but they are transformed into another type (ch organs). Fig. 7 and 8 show this transformation of sensory organs using antibodies that label different aspects of ch organs. In all cases es organs, their neurons as well as support cells, acquire ch organ characteristics. This demonstrates that es organs as a whole are transformed. It appears as if ch organs represent a ground state and that es organs require cut activity to acquire their identity. Therefore, it is likely that cut acts at the precursor level. This is also strongly suggested by the staining pattern of antibodies against the cut gene product (Blochlinger et al., 1988). cut is expressed in all cells of es organs as well as in their precursors (Fig. 9).

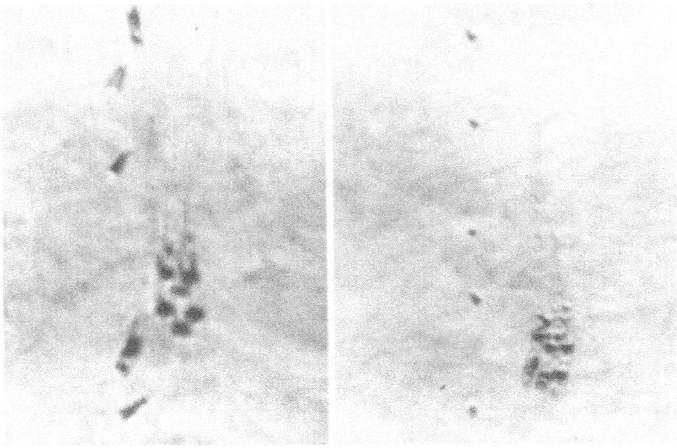

Fig. 7. Dorsal thoracic cluster (see Fig. 2) of a wild-type (right) and a cut mutant (left) whole-mount embryo that have been stained with an antibody which stains exclusively a structure around the tip of es and ch dendrites. In ch organs the structure appears elongated like an arrowhead (they are called scolopales) whereas in es organs it looks like a dot. Note that in the

cut mutant the dots are transformed into ch-like scolopales (see also Bodmer et al., 1987).

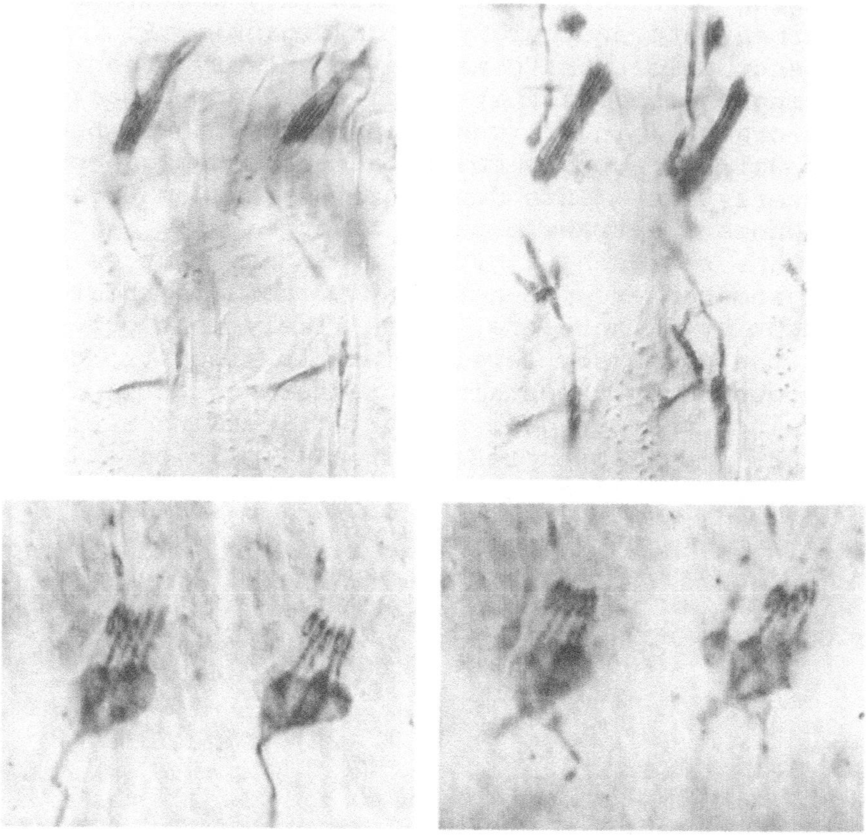

Fig. 8. Top: Two hemisegments of a wild-type (left) and a cut mutant embryo (right) stained with a ch support cell specific antibody. Note that in the cut mutant many more cells are stained. They are in the regions where es organs normally develop (see Fig. 2). Bottom: Lateral abdominal cluster of a wild-type (left) and cut mutant embryo (right) stained with a ch neuron specific antibody. Note that in the cut mutant the two additional cells are in the location of es neurons. Both mutant staining patterns suggest the transformation of es into ch organs.

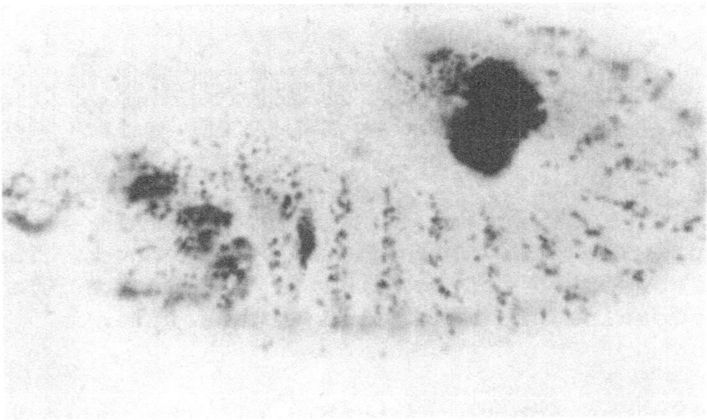

Fig. 9. Staining of es organ precursors with an anti-cut antibody of a 7-71/2 hr. old embryo, which is at a time when many precursors are still in the process of division (Bodmer et al., 1989). Compare with the BrdU staining pattern of a similar stage embryo in Fig. 3. This micrograph was kindly provided by K. Blochlinger.

The molecular cloning of cut revealed that it has homology to a homeobox suggesting that it may act as a transcription factor regulating later differentiation genes for the es organ cells (Blochlinger et al., 1988). This is supported by anti-cut antibody staining where labelling is restricted to the nucleus (Blochlinger et al., 1988). The homology with known homeodomains (see Scott et al., 1989) is rather low, but the nine most conserved amino acids (Gehring, 1987) are also conserved in cut (Fig. 10). It is, therefore, possible that cut may be the first member of a new class of homeobox genes which act in neurogenesis.

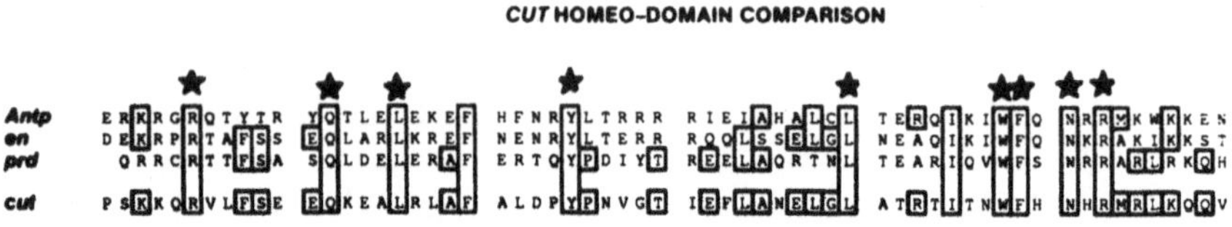

Fig. 10. Comparison of the homeodomain sequence of cut with other homeodomain sequences. The 9 most conserved amino acids in all Drosophila homeodomains are indicated with asterisks.

In the embryo, probably all the various sensory organs that have external sensory structures in the cuticle are affected by cut, because these cuticular structures (e.g. bristles) are absent in mutant embryos. On the other hand, es organ specification in the adult fly may depend on more than only cut. Studies of surviving adult flies that were genetically manipulated to be mosaic for cut and wild-type tissue revealed an altered phenotype for mutant es organs (Bodmer et al., 1987). Frequently, bristle structures could still form despite clear indications of es to ch transformations, which resulted in sensory organs that had a hybrid es/ch character (Fig. 12).

Behavioral studies showed that the correct functioning of es organs is abolished in mutant adult patches. Mosaic flies that had on one side a wild-type and on the other a mutant prothoracic leg were subjected to the test of proboscis extension in response to touching the tarsi with a sucrose solution (Table 1). It was found that only the wild-type tarsus was capable of tasting the sucrose which resulted in the extension of the proboscis. Therefore, cut may not have the identical roles in the adult compared with the embryo, but it is required for functional es organ development.

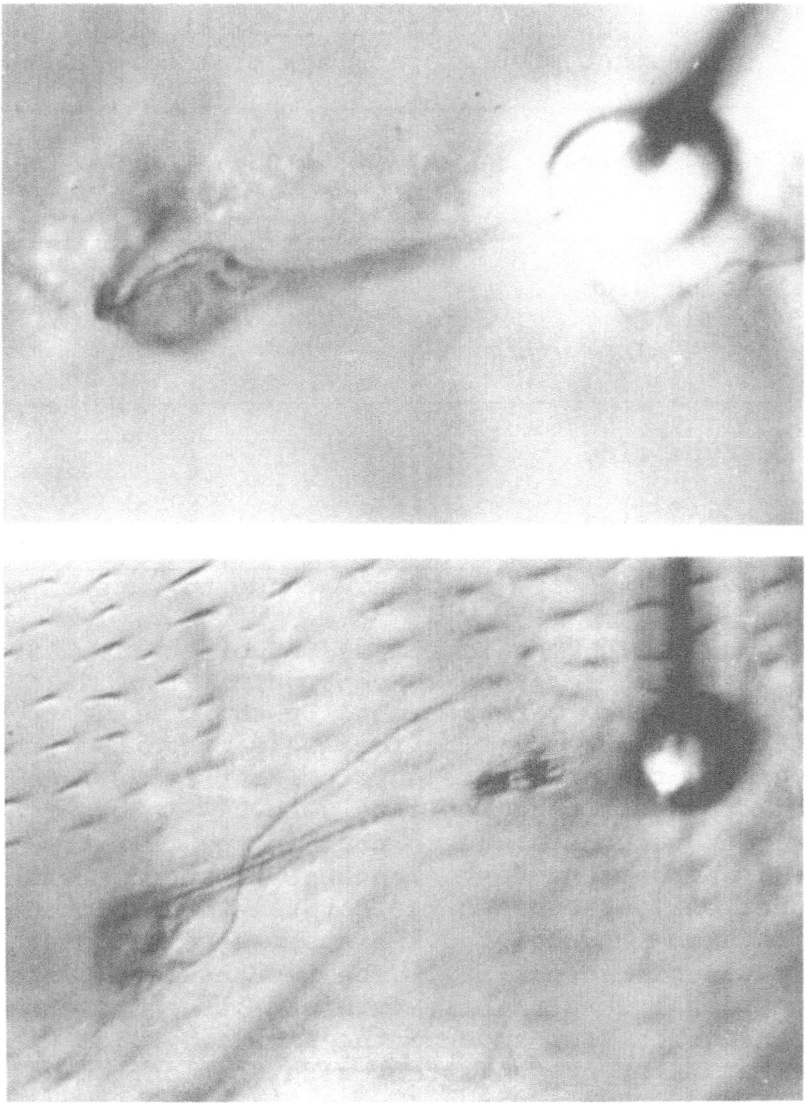

Fig. 11. Wild-type (top) and <u>cut</u> mutant (bottom) bristle-type es organ of an adult fly stained with a neuron specific antibody. Note that in the mutant the dot at the tip of the dendrite is transformed into a ch-like scolopale but without elimination of the es organ characteristic bristle.

Table 1. Proboscis extension in response to 0.1 M sucrose application to the tarsi of wild-type and cut mutant prothoracic legs of mosaic flies.

genotype of leg	# of flies with proboscis responses
wild-type side	8
y,cut^{L-1} side	1*
wild-type side	2
y,cut^{C145} side	0

* hyperactive fly, responsive also to water

Methods : The mosaic flies (Bodmer et al., 1987) were starved for food (except water) for 12-16 hr. Then they were attached by their wings to a plasticin surface. The legs were prevented from touching the head by a physical barrier. Legs and proboscis were satiated with H_2O in order to prevent responses due to touching or thirst. To test whether the taste response to sucrose was functional, the prothoracic tarsi were touched with a small drop of 0.1 M sucrose solution at the tip of a capillary. The response was scored functional when the fly extended its proboscis as soon as the drop touched the tarsus. Flies were tested first on the mutant side then on the wild-type side. Most flies gave the same response when tested repeatedly. 19 control flies that were wild-type tested positive on both prothoracic legs.

Since the ommatidia in the adult eye have completely different structural and developmental features (Ready, 1989), cut mosaic eyes were analyzed for whether or not they behaved as es organs. Eyes mutant for cut showed no obvious morphological aberrations, except for the es type bristles between the ommatidia, which were transformed as expected. Recording of the electroretinogram of mutant eyes showed a significant response to a flash of light (Fig. 12 middle) which was comparable to the wild-type eye of the mosaic fly (Fig. 12 top). The small difference in the light-off response (arrows) could be accounted for by the genetic markers on the same chromosome (y,w) when compared with flies that only had these markers (Fig. 12 bottom). These data indicate that the photoreceptors are not dependant on cut for their function, and by this criterion they are not es sensory organs.

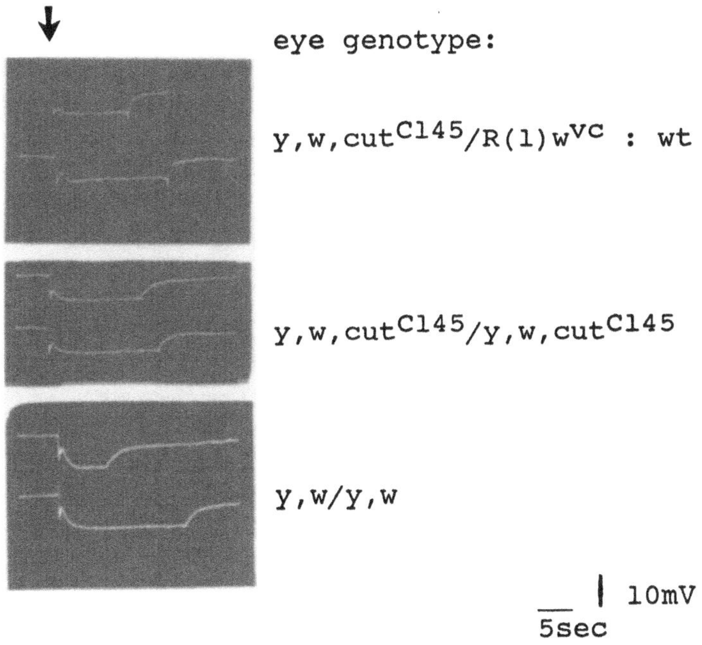

In the embryo as well as in the adult sensory organs that require cut activity also depend on AS-C to develop at all. In the model (Fig. 4) AS-C is thought to act upstream from cut in the hierarchy of sensory organ development. To test this hypothesis I analyzed the phenotype of embryos that where mutant for both AS-C and cut. If cut were upstream, one would expect that the transformation of es organs to ch organs would rescue them from being deleted by AS-C, since AS-C does not affect ch organs (Dambly-Chaudiere and Ghysen, 1987). However, the finding is that the double mutant has the identical phenotype as the AS-C mutant alone and no transformed es organs can be detected (R. Bodmer and Y.N. Jan, unpublished observations). This suggests that cut acts downstream of AS-C in support of the model. In summery, the cut gene has structurally and functionally the quality of a selector gene similar to the early homeotic genes and specifies es organ identity.

In fact, many of the early developmental genes are also are expressed again later during neurogenesis (see Doe and Scott, 1988). These genes show a different pattern and periodicity of expression in the nervous system than during the blastoderm stage when the plan for segmentation is set up. This suggests that they may have different functions in segmentation than in neurogenesis. For instance, engrailed (en) which

Fig. 12 (previous page). Electroretinogram in response to a flash of light (arrow) obtained from a wild-type/cut heterozygous eye (top) and from a y, w, cut^{C145} homozygous eye (middle) of a mosaic fly. The bottom traces are from a y, w control fly. Note that the steady-state potential (downward deflection) as well as the transient on and off responses of the cut eye are similar to the control eyes.

is determining the posterior compartment in every segment and is expressed in that region, is also expressed later in the CNS (DiNardo et al., 1985) and in two neurons in the PNS (Fig. 13). These two PNS cells presumably do not derive from the posterior compartment. In moderate mutations of <u>en</u> (in the promoter region), the <u>en</u> positive PNS cells are more numerous (R. Bodmer, unpublished observations). It is tempting to speculate that <u>en</u> has also a function during neurogenesis.

The homeotic gene <u>Ultrabithorax</u> (<u>Ubx</u>) of the Bithorax Complex (BX-C) is one gene responsible for correct segmental identities and is expressed in parasegment 5-13, but predominantly in the posterior portion of thoracic segment 3 and anterior portion of abdominal segment 1 (parasegment 6). Later, <u>Ubx</u> is also expressed in the corresponding regions of the CNS, but in a more complex pattern (Beachy et al., 1985; White and Wilcox, 1984; Fig. 14 top). <u>Ubx</u> is expressed in most PNS neurons of the abdominal segments, except in those neurons that are derived from the posterior compartment (i.e. the 5 lateral ch neurons, Fig. 14 bottom right). On the other hand, <u>Ubx</u> is expressed in the homologous neurons of the posterior compartment of the second and third thoracic segment (i.e. the 3 dorsal ch neurons, Fig. 14 bottom). In the absence of the whole BX-C, these neurons are transformed into posterior compartment neurons that are characteristic of the first thoracic segment (3 lateral ch neurons pointing anteriorly, R. Bodmer, unpublished). It may be that <u>Ubx</u> expression in the posterior compartment neurons of the second or third thoracic segment is required for them to differentiate with the proper number, location and orientation. <u>Ubx</u> expression in the posterior compartment neurons of the abdominal segments is repressed by another gene from the BX-C, <u>Abd-A</u>, (W. Bender and F. Karch, pers. comm.). <u>Abd-A</u> expression may change their fate again and cause

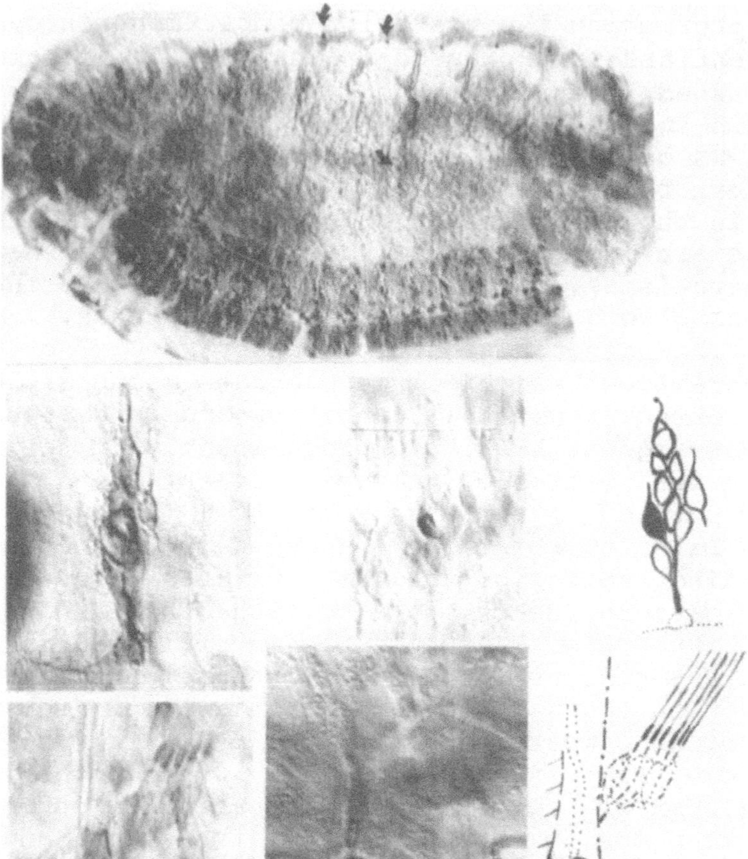

Fig. 13. Staining pattern of an antibody directed against the en protein (DiNardo et al., 1985) in a 12-14 hr. old embryo. A number of cells in the CNS and two neurons in each hemisegment in the PNS are stained (arrows, top micrograph), one in the dorsal (middle) and one in the lateral abdominal cluster (bottom). Note that the 5 ch neurons (dotted outlines, compare with Fig. 2) which derive from the posterior compartment are not stained.

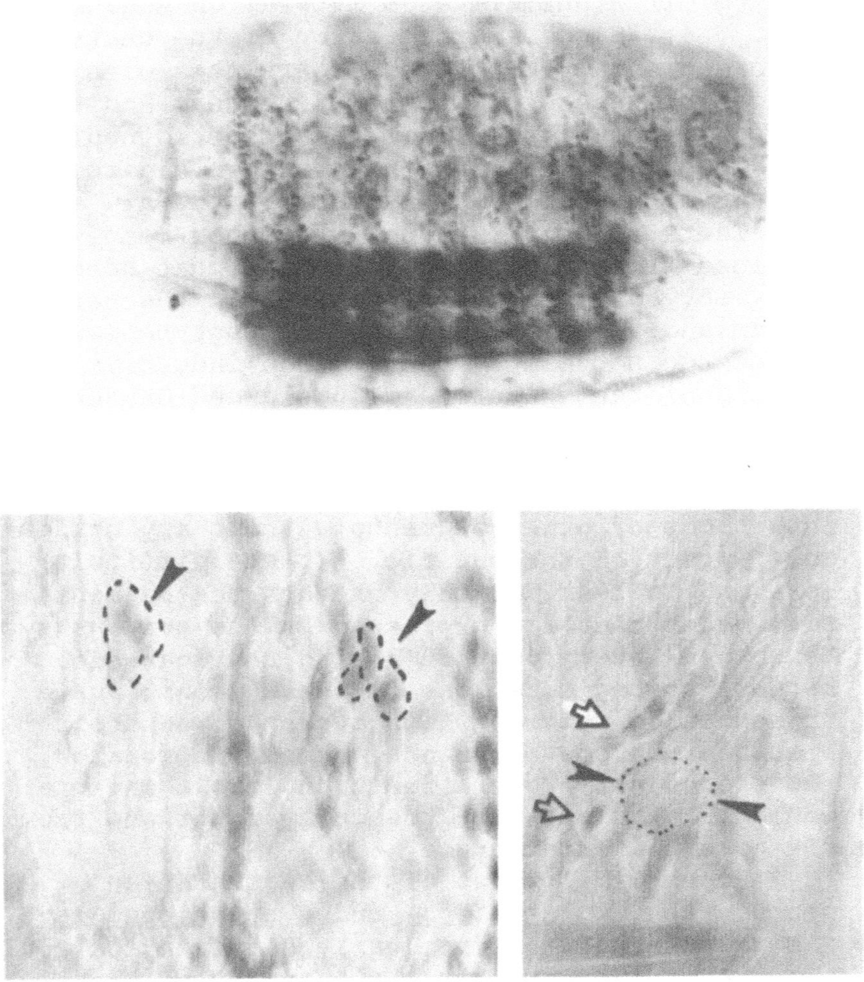

Fig. 14. Staining pattern of 12-14 hr. old embryo in the CNS and the PNS (top) of an antibody raised against the <u>Ubx</u> protein (White and Wilcox, 1984). Stained neurons in a lateral abdominal clusters (bottom right) are indicated with open arrows and the posterior compartment derived ch neurons with arrowheads (bottom, compare with Fig. 2). In the dorsal thoracic segments 2 and 3 (T2, T3) these neurons are stained but not in the abdominal segments (A).

them to differentiate as 5 lateral ch neurons (Fig. 14). Since an alteration of the characteristic segmental PNS pattern correlates with a change in expression of the BX-C genes and this expression persists within the affected neurons it is possible that these genes have a direct influence on specifying neuronal pattern.

There is some evidence which has been obtained from imaginal development that homeotic genes may have separate functions in the nervous system and in the epidermis, and that neural phenotypes do not simply result from inductive effects of the epidermis (Teugels and Ghysen, 1983; Ghysen et al., 1985; Ghysen and Lewis, 1986). Some BX-C mutations caused segmentally characteristic CNS structures to be transformed into those of other segments without any noticeable effect on the cuticle. It was difficult, however, to test directly whether the segmentation genes that are expressed in the nervous system had a later function during neurogenesis, because strong mutations in these genes caused segmentation defects which overshadowed their possible neurodevelopmental role. Expression alone does not prove a functional role and one would like to separate the early functions from the possible later ones.

Recently, for two of the segmentation genes, fushi tarazu (ftz) and even-skipped (eve), a separate function from early segmentation could be demonstrated in the embryonic CNS of the fly (Doe et al., 1988, 1989). Since the early expression of eve is temporally separate from the later CNS expression, it was possible to take advantage of a good temperature-sensitive eve mutation. Shifting eve embryos to the non-permissive temperature just prior to eve CNS expression resulted in normal segmentation but selectively abolished CNS expression. This had the effect that some neurons (especially the RP2 neuron) which normally express eve show an altered axon morphology but were still present. It is possible that they are transformed into other types of neurons which means that eve may also be involved determining the neuronal identity. For ftz a different approach took advantage of the identification of separate promoter elements for expression during segmentation and expression during CNS neurogenesis (Hiromi et al., 1985). An engineered ftz gene which was expressed normally during segmentation but lacked CNS expression was introduced into a ftz mutant background. These embryos show near normal segmentation and a similar phenotype with respect to the RP2 neuron morphology as do eve mutants or double mutants which lack both eve and ftz CNS function. In addition, eve expression is removed in RP2 in embryos that lack ftz CNS function, which suggests that ftz acts through eve in controlling RP2 axon morphology. It is interesting to note that during segmentation it is eve which controls ftz, suggesting that at least some aspect of the genetic control of neurogenesis uses some of the same genes as early during development but they may follow different hierarchical principles.

Whether these early developmental genes described above function as true selector genes in terms of switching on developmental pathways, such that in their absence precursor cells take on a different identity, is not entirely established.

(5) Other Genes Involved in Neurogenesis

There are genes that act at various levels downstream of selector genes. One such gene, numb, has recently been discovered and cloned, and it is thought to act downstream of the precursor level (Uemura et al., 1989). It has been found that numb abolishes most peripheral neurons when stained with neuron-specific antibodies as is shown in Fig. 15. This phenotype is similar to da mutant embryos (Fig. 5, bottom panel). However, when numb mutant embryos are stained with antibodies that mark all cells in the PNS not just the neurons (Uemura et al., 1989), or when labelled with the replication marker BrdU, the number of peripheral cells is not significantly changed (Fig. 16). This result is very different from what has been found in da embryos (Fig. 6). It has been shown that the neurons are transformed into cells that have characteristics of sensory organ support cells. Therefore, the identity of the precursor per se is not altered. Rather, it is the neuronal progeny of the precursors that need numb for their correct differentiation. It is likely in es organs, for example, that the immediate mother cell of the neuron is transformed into its sibling, the mother cell of the bristle forming support cells (Uemura et al., 1989). Therefore, numb is thought to act at the secondary precursor level indicated in the model (Fig. 4) and its function is not restricted to any particular class of precursor cells.

The predicted amino acid sequence of numb suggests that it may be a DNA binding protein similar to zinc finger transcription factors. The numb protein might be needed for turning on the genes that are responsible for some of the precursor progeny in the PNS to differentiate into neurons. So at that level numb could be considered also as a selector gene.

In conclusion, with the discovery of more and more genes that are involved in neurogenesis it is now possible test different models of how the nervous system is formed. These genes can be analyzed in terms of where and when they are act. DNA sequence from cloned genes of this type can yield further information about possible mechanisms of action of these genes by comparing them with known genes. Thus, in the near future many more insights may be gained into the mechanisms and events that lead from a uniform layer of epidermal cells to the specification of neuronal precursors and eventually to the differentiated nervous system with all its intricate diversity and complex morphology.

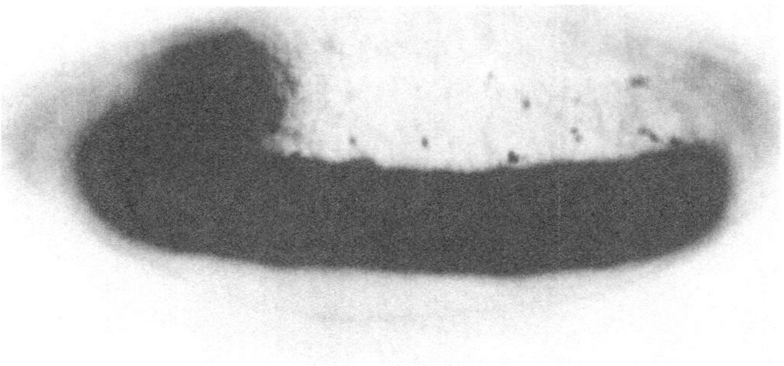

Fig. 15. 14 hr. numb embryo stained with an antibody specific for all neuronal nuclei. Note that most peripheral neurons are absent but that the CNS remains unaffected (compare with wild-type and da pattern in Fig. 5). This micrograph was kindly provided T. Uemura and Y.N. Jan.

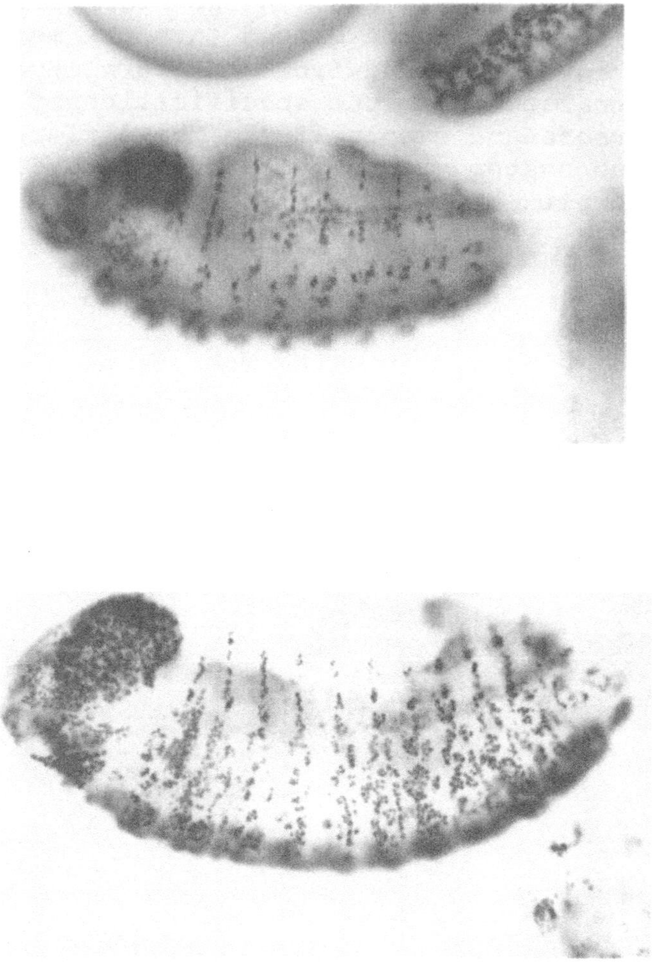

Fig. 16. Wild-type (top) and numb mutant embryos (bottom) that have been labelled with BrdU at 6-7 hr. for 1/2 hr. but instead of being immediately fixed (as in Fig. 3 and 6) these embryos were allowed to develop further until they reached about 12-14 hr. (see Bodmer et al., 1989). The number of cells in the PNS in the numb embryo is comparable to the wild-type (which may even appear less because of the more dorsal view of the top micrograph).

Acknowledgements: I would like to thank Drs. Lily and Yuh Nung Jan for their continuous support and their permission for communication of results that have been obtained in their lab. I also would like to thank them as well as Drs. Karen Blochlinger and Ethan Bier for helpful comments.

REFERENCES

Bate, C.M. (1978). Development of sensory systems in arthropods. In Handbook of sensory physiology, vol. IX, pp. 1-53, M. Jacobson, ed. (Berlin, Heidelberg, New York: Springer-Verlag).

Beachy, P.A., Helfand, S.L., and Hogness, D.S. (1985). Segmental distribution of the bithorax complex proteins during Drosophila development. Nature 313, 545-551.

Bier, E., Ackerman, L., Barbel, S., Jan, L.Y., and Jan, Y.N. (1988). Identification and characterization of a neuron-specific nuclear antigen in Drosophila melanogaster. Science 240, 913-916.

Blair, S.S., and Palka, J. (1985). Axon guidance in the wing of Drosophila. TINS 8, 284-288.

Blochlinger, K., Bodmer, R., Jack, J., Jan, L.Y., and Jan, Y.N. (1988). Primary structure and expression of a product from cut, a locus involved in specifying sensory organ identity. Nature 333, 629-635.

Bodmer, R., and Jan, Y.N. (1987). Morphological differentiation of the embryonic peripheral neurons in Drosophila. Roux's Arch. Dev. Biol. 196, 69-77.

Bodmer, R., Barbel, S., Sheperd, S., Jack, J., Jan, L.Y., and Jan, Y.N. (1987). Transformation of sensory organs by mutations of the cut locus of D. melanogaster. Cell 51, 293-307.

Bodmer, R., Carretto, R., and Jan, Y.N. (1989). Neurogenesis of the peripheral nervous system in Drosophila embryos: DNA replication patterns and cell lineages. Neuron 3, 21-32.

Brand, M., and Campos-Ortega, J.A. (1988). Two groups of interrelated genes regulate early neurogenesis in Drosophila melanogaster. Roux's Arch. Dev. Biol. 197, 457-470.

Campos-Ortega, J., and Hartenstein, V. (1985). The embryonic development of Drosophila melanogaster (New York: Springer-Verlag.

Campos-Ortega, J.A. (1988). Cellular interactions during early neurogenesis of Drosophila melanogaster. TINS 11, 400-405.

Caudy, M., Grell, E.L., Dambly-Chaudire, C., Ghysen, A., Jan, L.Y., and Jan, Y.N. (1988a). The maternal sex determination gene daughterless has zygotic activity necessary for the formation of peripheral neurons in Drosophila. Genes Dev. 2, 843-852.

Caudy, M., Vaessin, H., Brand, M., Tuma, R., Jan, L.Y., and Jan, Y.N. (1988b). daughterless, a gene essential for both neurogenesis and sex determination in Drosophila, has sequence similarities to myc and achaete-scute complex. Cell 55, 1061-1067.

Dambly-Chaudiere, C., and Ghysen, A. (1986). The sense organs in the Drosophila larva and their relation to the embryonic pattern of sensory neurons. Roux's Arch. Dev. Biol. 195, 222-228.

Dambly-Chaudiere, C., and Ghysen, A. (1987). Independent subpatterns of sense organs require independent genes of the achaete-scute complex in Drosophila larvae. Genes Dev. 1, 297-306.

Dambly-Chaudiere, C., Ghysen, A., Jan, L.Y., and Jan, Y.N. (1988). The determination of sense organs in Drosophila: interaction of scute with daughterless. Roux's Arch. Dev. Biol. 197, 419-423.

DiNardo, S., Kuner, J.M., Theis, J., and O'Farrell, P.H. (1985). Development of the embryonic pattern in D. melanogaster as revealed by accumulation of the nuclear engrailed protein. Cell 43, 59-69.

Doe, C.Q., and Goodman, C.S. (1985a). Early events in insect neurogenesis. I. Developmental and segmental differences in the pattern of neuronal precursor cells. Dev. Biol. 111, 193-205.

Doe, C.Q., and Goodman, C.S. (1985b). Early events in insect neurogenesis. II. The role of cell interactions and cell lineage in the determination of neuronal precursor cells. Dev. Biol. 111, 206-219.

Doe, C.Q., and Scott, M.P. (1988). Segmentation and homeotic gene function in the developing nervous system of Drosophila. TINS 11, 101-106.

Doe, C.Q., Hiromi, Y., Gehring, W.J., and Goodman, C.S. (1988). Control of the Drosophila segmentation gene fushi tarazu. Science 239, 170-175.

Doe, C.Q., Smouse, D., and Goodman, C.S. (1989). Control of neuronal fate by the Drosophila segmentation gene even-skipped. Nature 333, 376-378.

Garcia-Bellido, A. (1975). Genetic control of wing disc development in Drosophila. in: Cell patterning, CIBA Foundation Symp. 29, Elsevier, Amsterdam, pp. 161-182.

Garcia-Bellido, A. (1979). Genetic analysis of the achaete-scute system of Drosophila melanogaster. Genetics 91, 491-520.

Gehring, W.J. (1987). Homeoboxes in the study of development. Science 236, 1245-1252.

Ghysen, A., Jan, L.Y., and Jan, Y.N. (1985). Segmental determination in Drosophila central nervous system. Cell 40, 943-948.

Ghysen, A., and Lewis, E.B. (1986). Roux's Arch. Dev. Biol. 195, 203-209.

Ghysen, A., Dambly-Chaudiere, C., Aceves, E., Jan, L.Y., and Jan, Y.N. (1986). Sensory neurons and peripheral pathways in Drosophila embryos. Roux's Arch. Dev. Biol. 195, 281-289.

Ghysen, A., and O'Kane, C. (1989) Neural enhancer-like elements as specific cell markers in Drosophila. Development 105, 35-52.

Ghysen, A., and Dambly-Chaudiere, C. (1989a) Early events in the development of Drosophila nervous system. J. Physiol. (Paris), in press.

Ghysen, A., and Dambly-Chaudiere, C. (1989b). The genesis of Drosophila peripheral nervous system: A sequential process. TIGS, in press.

Goodman, C.S., Raper, J.A., Ho, R.K., and Chang, S. (1982). Pathfinding of neuronal growth cones in grasshopper embryos. In "Development Order: Its origins and regulation" (S. Subtelny and P.B. Green, eds.), pp. 275-316. Alain R. Liss, New York.

Goodman, C.S., Bastiani, M.J., Doe, C.Q., du Lac, S., Helfand, S.L., Kuwada, J.Y., and Thomas, J.B. (1984). Cell recognition during neuronal development. Science (Washington, D.C.) 225, 1271-1279.

Gratzner, H.G. (1982). Monoclonal antibody to 5-Bromo-and 5-Iodo-deoxyuridine: a new reagent for detection of DNA replication. Science 218, 474-475.

Hartenstein, V., and Campos-Ortega, J.A. (1986) The peripheral nervous system of mutants of early neurogenesis in Drosophila melanogaster. Roux's Arch. Dev. Biol. 195, 210-221.

Hartenstein, V. (1988). Development of Drosophila larval sensory organs: spatiotemporal pattern of sensory neurones, peripheral axonal pathways and sensilla differentiation. Development 102, 869-886.

Hertweck, H. (1931). Anatomie und Variabilitaet des Nervensystems und der Sinnesorgane von Drosophila melanogaster (Meigen). Z. Wiss. Zool. 139, 559-663.

Hiromi, Y., Kuroiwa, A., and Gehring, W.J. (1985). Control elements of the Drosophila segmentation fushi tarazu. Cell 43, 603-613.

Ingham, P. (1988). The molecular genetics of embryonic pattern formation in Drosophila. Nature 335, 25-34.

Jack, J.W. (1985). Molecular organization of the cut locus of Drosophila melanogaster. Cell 42, 869-876.

Jan, Y.N., Bodmer, R., Ghysen, A., Dambly-Chaudiere, C., and Jan, L.Y. (1986). Mutations affecting the peripheral nervous system in Drosophila. J. Cell. Biochem. Proceeding of the UCLA symposium for Molecular Entomology, pp. 45-56.

Jan, L.Y., and Jan, Y.N. (1982). Antibodies to horseradish peroxidase as specific neuronal markers in Drosophila and grasshopper embryos. PNAS 72, 2700-2704.

Lawrence, P.A. (1966). Development and determination of hairs and bristles in the milkweed bug, Oncopeltus fasciatus (Lygaeidae, Hemiptera). J. Cell Sci. 1, 475-498.

Murphey, R.K. (1985). Competition and chemoaffinity in insect sensory systems. TINS 8, 120-125.

Nuesslein-Volhard, C., and Wieschaus, E. (1980). Mutations affecting segment number and polarity in Drosophila. Nature 287, 795-801.

Ready, D.F. (1989). A multifaceted approach to neural development. TINS 12, 102-110.

Romani, S., Campuzano, S., Macagno, E.R., and Modolell, J. (1989). Expression of achaete and scute genes in Drosophila imaginal discs and their function in sensory organ development. Genes Dev. 3, 997-1007.

Scott, M.P., and Carroll, S.B. (1987). The segmentation and homeotic network in early Drosophila development. Cell 51, 689-698.

Scott, M.P., Tamkun, J.W., and Hartzell, G.W. (1989). The structure and function of the homeodomain. BBA Reviews on Cancer, in press.

Teugels, E., and Ghysen, A., (1983). Independence of the number of legs and leg ganglia in Drosophila bithorax mutants. Nature 304, 440-442.

Thomas, J.B., Bastiani, M.J., Bate, M., and Goodman, C.S. (1984). From grasshopper to Drosophila: A common plan for neural development. Nature 310, 203-207.

Truman, J., and Bate, C.M. (1988). Spatial and temporal patterns of neurogenesis in the central nervous system of Drosophila melanogaster. Dev. Biol. 125, 145-157.

Uemura, T., Sheperd, S., Ackerman, L., Jan, L.Y., and Jan, Y.N. (1989). numb, a gene required in determination of cell fate during sensory organ formation in Drosophila embryos. Cell 58, 349-360.

White, R.A.H., and Wilcox, M. (1984). Protein products of the bithorax complex in Drosophila. Cell 39, 163-171.

Zacharuk, R.Y. (1985). Antennae and sensilla. In Comprehensive insect physiology, biochemistry and pharmacology, vol. 6, pp. 1-69, G.A. Kerkut and L.I. Gilbert, eds. (New York: Pergamon Press).

From *Insect Neurochemistry and Neurophysiology • 1989* Edited by A. B. Borkovec and E. P. Masler

MODULATION AND REPAIR OF THE INSECT BLOOD-BRAIN BARRIER

J.E. Treherne, E.A. Howes, P.K. Schofield and P.J.S. Smith

AFRC Unit of Invertebrate Chemistry and Physiology, Department of Zoology, University of Cambridge CB2 3EJ, U.K.

INTRODUCTION

Insects are unusual among invertebrate animals in possessing extremely effective blood-brain barrier systems (Abbott & Treherne, 1977; Lane & Treherne, 1981). The necessity for such protection of the neuronal micro-environment is generally supposed to result from the peculiar chemical composition of insect blood (cf. Treherne, 1985). In the more highly-evolved groups of insects, for example, the ionic concentration of the plasma is quite inappropriate for neuronal function, where high levels of potassium can greatly exceed those of sodium ions (cf. Florkin & Jeuniaux, 1974). Even in species such as the cockroach (Periplaneta americana) with 'conventional' - high-sodium - blood, there can be large fluctuations in the ionic composition of the plasma (Lettau et al., 1977). Furthermore, insect blood plasma often contains extremely high concentrations of amino acids (Wyatt, 1961; Florkin & Jeuniaux, 1974) - some of which are neuroactive (Iversen et al., 1975) - as well as toxins, such as nicotine which can be tolerated at high concentrations in the blood of the tobacco hornworm, Manduca sexta (Morris, 1977).

The available evidence indicates that, unlike the vascular barrier of mammalian brain, the insect blood-brain barrier is formed by superficial glial elements (Schofield et al., 1984a,b; Schofield & Treherne, 1984). These limit intercellular movements of small water-soluble ions and molecules between the blood plasma and the neuronal microenvironment. Glia lying at deeper levels are also

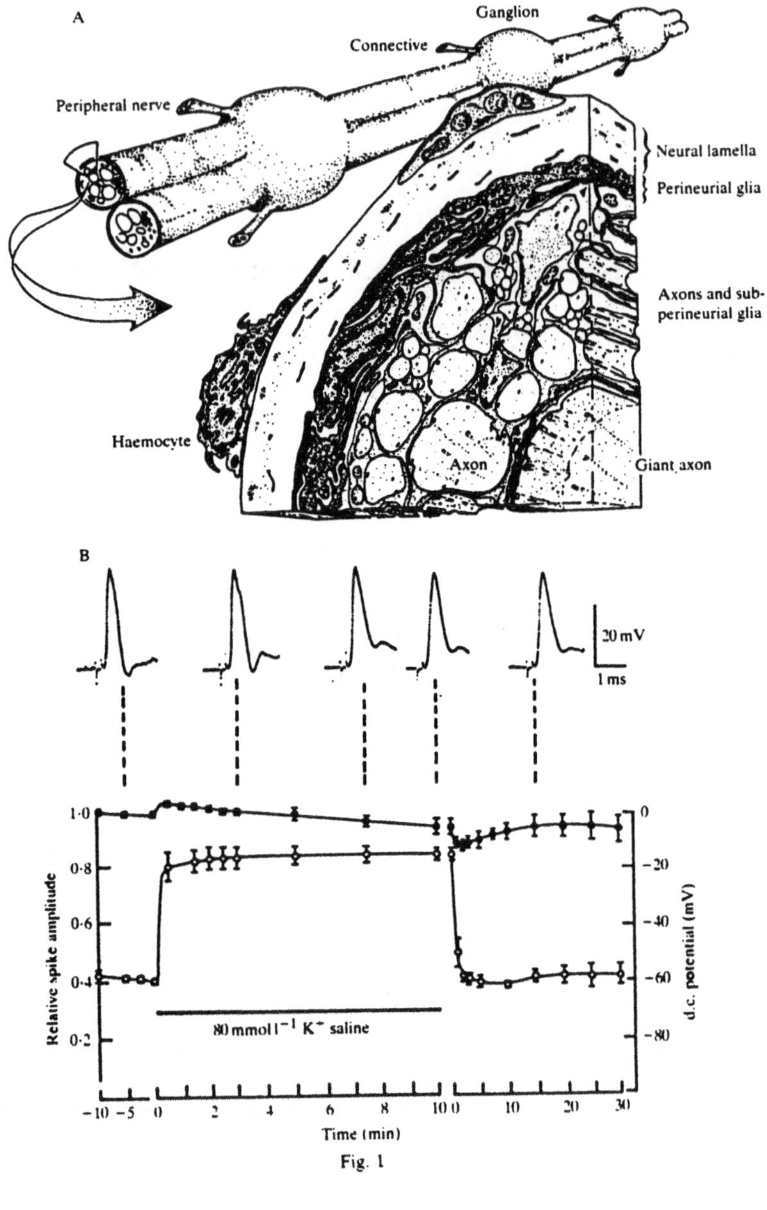
A
Ganglion
Connective
Peripheral nerve
Neural lamella
Perineurial glia
Axons and sub-perineurial glia
Haemocyte
Axon
Giant axon
B
20 mV
1 ms
Relative spike amplitude
d.c. potential (mV)
80 mmol l⁻¹ K⁺ saline
Time (min)

Fig. 1

involved in short-term extracellular cation regulation (Treherne & Schofield, 1981; Treherne, 1985).

In our recent research we have focussed on two dynamic aspects of barrier function: the hormonal modulation of the permeability properties of the blood-brain interface and the cellular mechanisms involved in its repair and maintenance in the central nervous system of the cockroach.

ORGANIZATION OF INSECT GLIA

The periphery of the insect central nervous system is delimited by a relatively leaky connective tissue layer, the neural lamella (cf. Treherne, 1985), which lies above a superficial layer of flattened neuroglia, the perineurium (Fig. 1A). Current evidence indicates that the perineurial glia constitute the blood-brain barrier in cockroach central nervous connectives: due, it is postulated, to occluding junctional complexes at the inner ends of the tortuous intercellular clefts that traverse the perineurium. This is shown by the inability of electron-dense extracellular tracers to penetrate beyond the region of the perineurial clefts (Lane, 1981, 1985) and by electrical responses - recorded from identified

---

Fig. 1. (A) Semi-diagrammatic drawing of a portion of the cockroach abdominal nerve cord, with a transverse section through the connective between the fourth and fifth abdominal ganglia. The surface of the central nervous system is delimited by the acellular neural lamella, which is leaky to small water-soluble cations and molecules. Beneath the neural lamella is a superficial layer of flattened glia, the perineurium, which (due to junctional complexes at the inner margins of the tortuous intercellular clefts) constitutes the blood-brain barrier. The underlying axons are closely wrapped by sub-perineurial glia which delineate a complex three-dimensional network of extracellular channels containing matrix material. (B) Effects of high-potassium saline on relative spike amplitude using sucrose-gap recordings. Due to the blood-brain barrier there is no appreciable effect on the action potential amplitude (●). The change in d.c. potential (o) represents the depolarization of the outwardly directed perineurial membranes, access to the inwardly directed ones being restricted by the perineurial junctional complexes. The vertical lines through the data points indicate the extent of the standard error of the mean (n = 3). (From Smith et al., 1987).

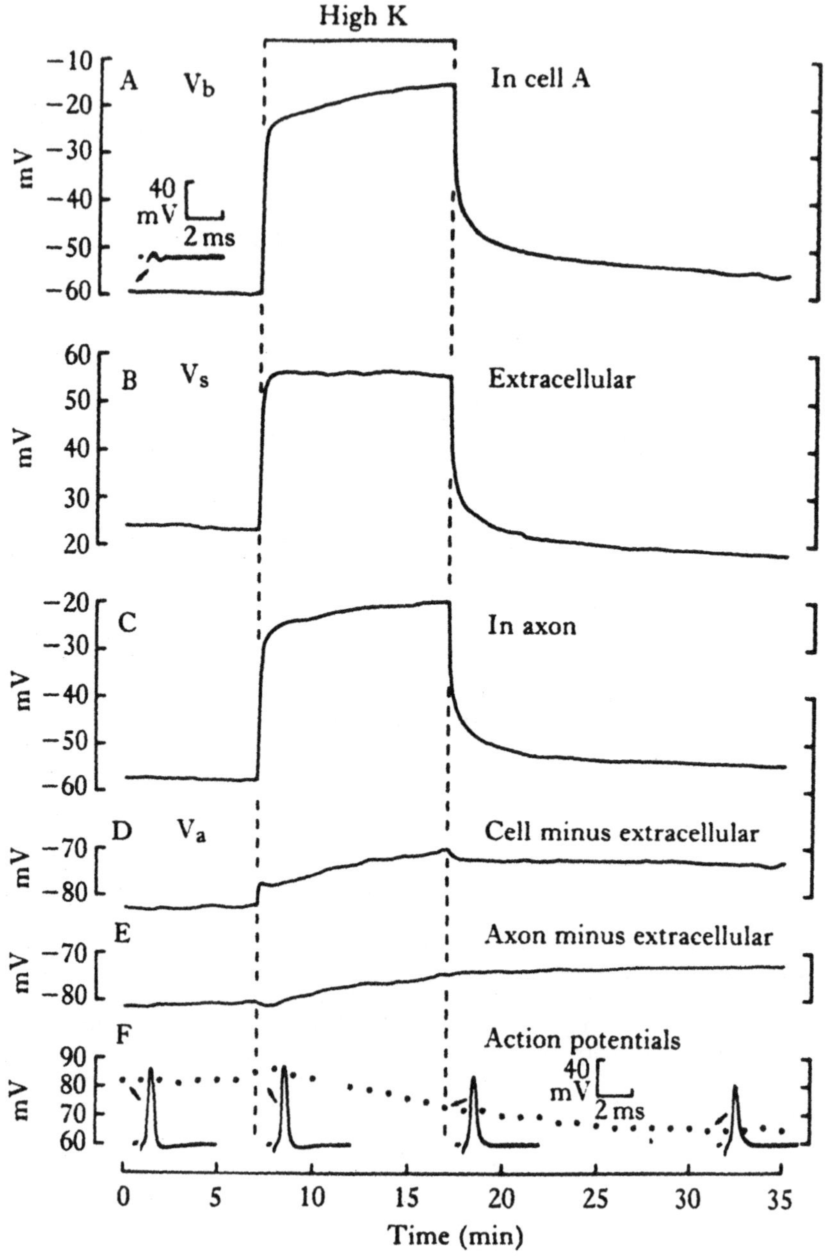
High K
A Vb
In cell A
mV
−10
−20
−30
−40
−50
−60
40 mV
2 ms
B Vs
Extracellular
60
50
40
30
20
C
In axon
−20
−30
−40
−50
−60
D Va
Cell minus extracellular
−70
−80
E
Axon minus extracellular
−70
−80
F
Action potentials
90
80
70
60
0
5
10
15
20
25
30
35
Time (min)

neuroglia, extracellular spaces and axons - following alteration in external cation levels. When the external potassium concentration is raised, for example, there is a rapid depolarization of the outwardly- directed perineurial membranes as the cations diffuse rapidly through the neural lamella and into the perineurial clefts. As a result of the restricted diffusion of potassium through the perineurial junctional complexes there is only a small depolarization of the inwardly directed perineurial membranes and of the underlying axons (Fig. 2). When intercellular access between the perineurial cells is increased by hypertonic urea treatment (Schofield et al., 1984b), or by selective chemical lesioning of the neuroglia (Smith et al., 1984). Depolarization now occurs at the axonal level, producing a rapid decline in action potential amplitude, caused by the penetration of potassium ions into the extracellular system and sodium inactivation (see forward to Fig. 6).

Fig. 2. Simultaneous microelectrode recordings of the effects of exposure of the connective to high K (130 mM) upon the potential differences (p.d.s.) measured within a perineurial cell, subsequently identified by peroxidase injection, and in the extracellular space. (A) p.d. across the outwardly-directed perineurial membrane (V_b), (B) in the interstitial space (V_s), (C) in an axon. (D) Voltage across the inwardly-directed membrane (V_a) was obtained from V_s and V_b. (E) Axon resting potential was obtained from V_s and the p.d. (F) Action potentials were recorded in the axon following extracellular stimulation of the nerve cord. No electrical responses were measured in the perineurial cell during stimulation (inset in A). From Schofield et al., 1984a.

OCTOPAMINE SENSITIVITY OF THE BLOOD-BRAIN BARRIER

With a transperineurial resistance of >900 Ω cm^{-2} (Schofield & Treherne, 1984) the blood-brain barrier of cockroach abdominal connectives is a relatively 'tight' epithelium, approaching values for mammalian brain capillaries (Crone & Oleson, 1982). The permeability of the cockroach perineurium is, however, not a static parameter. Schofield and Treherne (1985) have shown that it can be modulated in response to the ionic conditions of the haemolymph which, in the cockroach, exhibit marked daily fluctuation (Fig. 3).

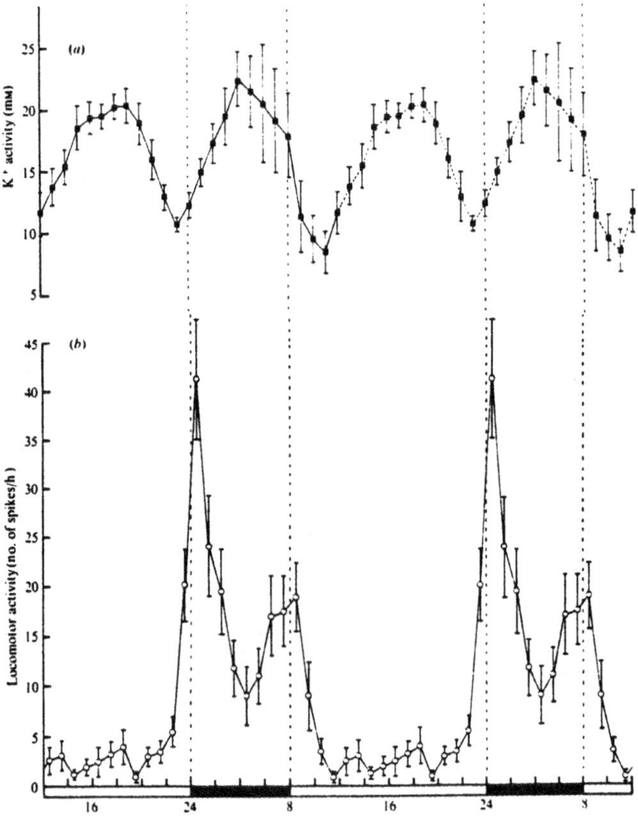

Fig. 3 (a) Composite curve of changes in potassium activity (mM) of the haemolymph through the day from 14 free-walking cockroaches, and (b) their locomotory activity. Bars are ± 1 x S.E. L.D. 16:8 cycle is doubled to show rhythms more clearly. From Lettau <u>et al</u>., 1977.

Modulation appears to be mediated by octopamine: a biogenic amine which is released into the blood of the cockroach (Davenport & Evans, 1984) and is known to exert a number of physiological effects in insects, including modulation of neuromuscular transmission and the neurohormonal control of carbohydrate and lipid levels (cf. Evans, 1985). It can also affect the potassium permeability of the blood-brain interface (Schofield & Treherne, 1985, 1986). This is shown by the influence of octopamine on potassium-induced depolarizations of the outwardly-directed membranes of the perineurium as well as on perineurial resistance (Fig. 4). That these responses are mediated by octopamine receptors is indicated by the reduced perineurial potassium depolarizations in the presence of DL-synephrine, blockage by phentolamine and partial blocking by propranolol (Schofield & Treherne, 1986).

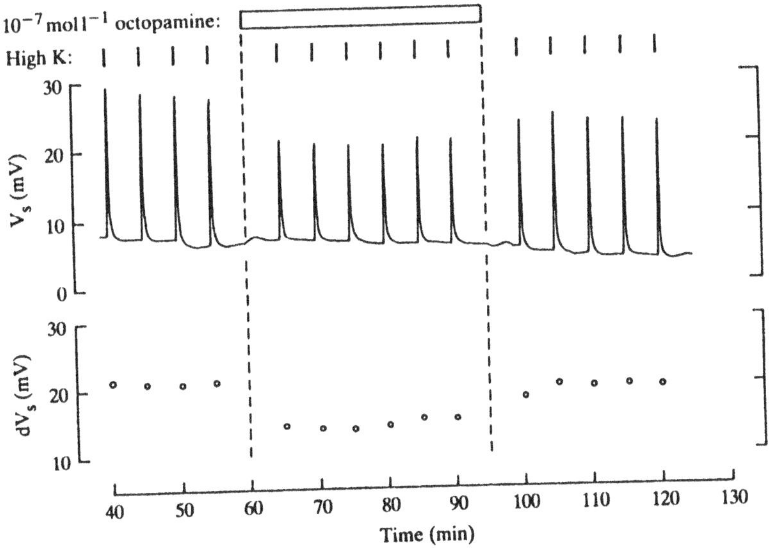

Fig. 4. Effects of 10^{-7}M DL-octopamine upon potassium depolarizations of the outwardly-directed perineurial membranes of cockroach abdominal connectives. The depolarizations in response to 15 s high potassium (66.5 mM) pulses, were measured (V_s) with the tip of the microelectrode in an interstitial position beneath the perineurium. Beneath is shown the changes in potential (dV_s) induced by the potassium pulses. From Schofield and Treherne, 1986.

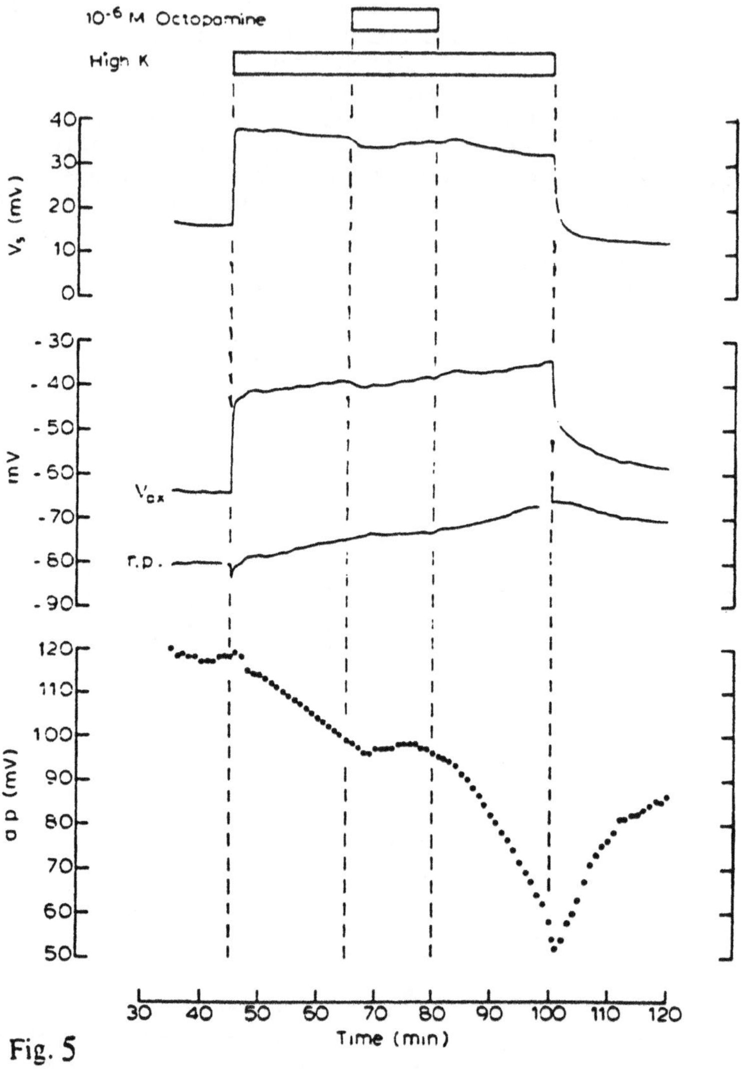
10-6 M Octopamine
High K
Vs (mV)
40
30
20
10
0
mV
-30
-40
-50
-60
-70
-80
-90
Vax
r.p.
a p (mV)
120
110
100
90
80
70
60
50
30
40
50
60
70
80
90
100
110
120
Time (min)

Fig. 5

Octopamine appears to exert its effect by lowering the potassium conductance of the outwardly-directed perineurial membrane (Schofield & Treherne, 1986) and, in so doing, substantially reduces the leakage of this cation to the axon surfaces (Fig. 5). The effect effect on K conductance shows a sharp threshold between 10^{-8} and 10^{-7} mol. l^{-1}: a range which corresponds to the daily changes in octopamine concentration measured in the blood of free-walking cockroach (Davenport & Evans, 1984). The nightly rise in blood octopamine would counter the increased tendency of potassium ions to cross the blood-brain barrier, due to the increase in blood concentration of this cation (Fig. 3), and so reduce the depolarization and consequent inactivation of axonal sodium channels. The reduced locomotory activity during the day time would also accord with this notion for blood octopamine remains at a basal level despite a diurnal increase in blood potassium (Fig. 3). Interference with this natural cycle would be involved in pestistatic and pesticidal actions of formamadines, for these have been found to mimic octopamine in its effects upon potassium permeability of the blood-brain barrier (Schofield, 1987).

Fig. 5. Effect of 10^{-6}M DL-octopamine upon K entry into a connective, as revealed by simultaneous recording from the interstitial system and an axon. By subtracting the interstitial potential (V_s) from the potential recorded in the axon (V_{ax}), the resting potential across the axon membrane (r.p.) is obtained. Propagated action potentials (a.p.) were obtained by extracellular stimulation of adjacent connectives. Upon raising the external K concentration from 3 to 67 mM (high-K), there was a slow depolarization of r.p., and steady decline in a.p. amplitude, as K leaked into the connective. Five minutes after the introduction of 10^{-6}M octopamine into the bath, the indicated K entry was virtually abolished: in the next 10 min, r.p. depolarized by 0.2 mV, and a.p. amplitude decreased by only 1 mV. In contrast, during the 10 min before the amine, r.p. decreased by 2.9 mV and the spike fell by 10 mV. This indicates a slowing of K entry, especially when the non-linear nature of the K effects is considered (Schofield & Treherne, 1985).

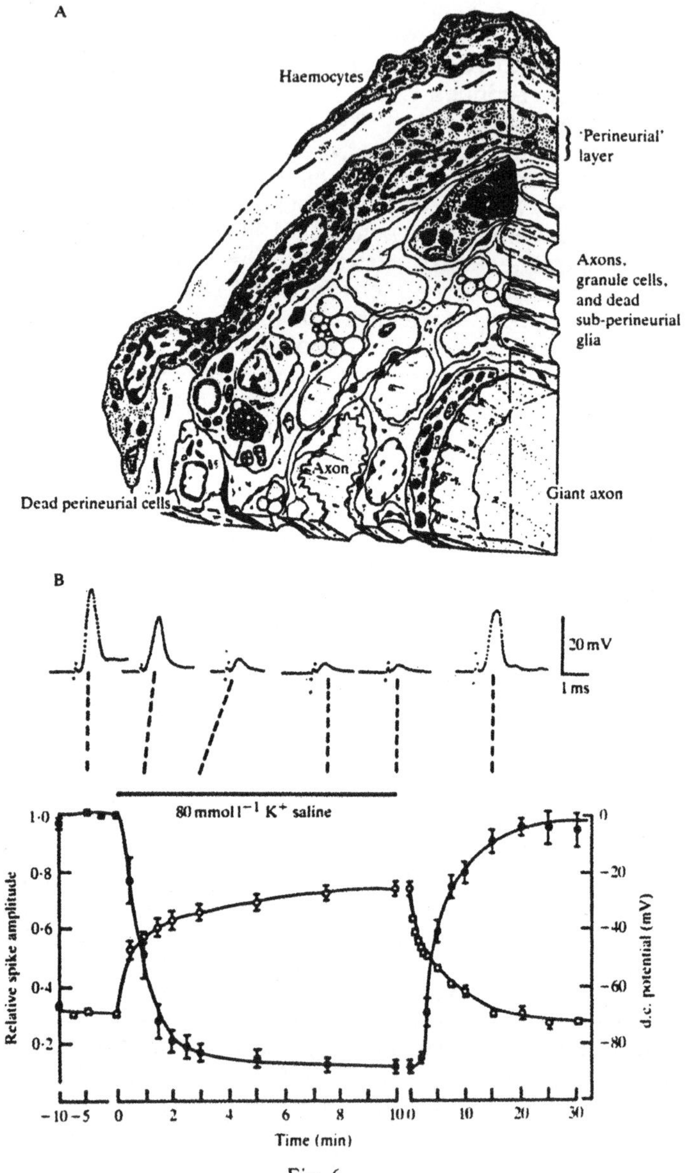
A
Haemocytes
'Perineurial' layer
Axons, granule cells, and dead sub-perineurial glia
Axon
Dead perineurial cells
Giant axon
B
20 mV
1 ms
80 mmol l⁻¹ K⁺ saline
Relative spike amplitude
d.c. potential (mV)
Time (min)

Fig. 6.

SELECTIVE GLIAL DISRUPTION

To follow the cellular reorganization of the repairing blood-brain barrier we have developed a technique of selective glial disruption, using a DNA intercalating agent, ethidium bromide (Nelson & Tinoco, 1984), as a gliotoxin. Unlike surgical lesioning (Treherne et al., 1984) this results in a swift and ordered glial repair (Smith et al., 1984). The toxin is applied locally, in vivo, to penultimate cockroach abdominal connectives. As only the mid-portion of the connectives are exposed to ethidium bromide, the neuronal nuclei in the ganglia are undamaged. The axons in the lesion zone thus retain their normal structure and excitability; extracellular matrices and the neural lamella persist: only neuroglia in the central zone of the connectives degenerate (Smith et al., 1984). Within 1 day the perineurial blood-brain barrier is disrupted - the underlying axonal surfaces being freely accessible to extraneously-applied potassium ions - and the subperineurial and adaxonal glia are severely damaged (Fig. 6).

Fig. 6. (A) Semi-diagrammatic representation of the ultrastructure of a penultimate abdominal connective at 48h after selective glial disruption with ethidium bromide. At this stage haemocytes accumulate in relatively large numbers and then penetrate the neural lamella. The perineurial layer now contains numerous granule-containing cells, derived from the invading haemocytes. (B) Effects of high-potassium saline on the electrical responses of the connectives 48h after exposure to ethidium bromide. Unlike the situation in normal connectives (Fig. 1B) there is now a rapid collapse of the action potentials (●), the change in d.c. potential (o) now corresponding to axonal depolarization due to the relatively rapid penetration of potassium ions. The perineurial blood-brain barrier is now disrupted. The vertical lines through the data points indicate the standard errors of the means (n = 5). (From Smith et al., 1987)

REGENERATION OF THE BLOOD-BRAIN INTERFACE

Repair of the perineurial glia commences with the appearance of cells which are not seen in undamaged preparations. These are characterized by the presence of numerous electron-dense cytoplasmic granules (Smith et al., 1984). Within 2 days of selective disruption the granule-containing cells begin to form a mosaic beneath the neural lamella (Fig. 6). By 4 days cells in this region have developed extensive, interdigitating processes with septate and tight junctional complexes linking adjacent cellular membranes. The granular appearance has been lost. At this stage there is a relatively free access of potassium ions across the barrier (as judged by the decline in the amplitude of axonal action potentials recorded in the lesion zone following exposure to high-potassium saline) although extraneously-applied lanthanum is now excluded from the underlying extracellular system (Smith et al., 1984).

By 11 days, the blood-brain barrier to potassium ions is re-established and the superficial cells regain the typical flattened appearance of the normal perineurium, although potassium depolarizations of the outwardly-directed perineurial membranes are still smaller than those recorded in untreated preparations (Smith et al., 1984). After 28 days, the ultrastructure of repaired connectives is indistinguishable from that of control connectives and potassium induced depolarizations approximate to the normal condition.

EXOGENOUS REACTIVE CELLS

The available evidence indicates that the granule-containing cells which appear in the early stages of repair are derived from circulating blood cells. Within 24h of selective glial lesioning, haemocytes aggregate on the surface of the neural lamella (Treherne et al., 1987) and subsequently penetrate the neural lamella (Smith et al., 1984); monoclonally-labelled haemocytes accumulate in the damaged perineurial tissues - and also at deeper levels - in the early stages of repair (Howes et al., 1987b).

ENDOGENOUS REACTIVE CELLS

The invading haemocytic cells appear to work in concert with endogenous ones. The involvement of endogenous reactive cells is indicated by several lines of evidence. It is show, for example, by the aberrant repair observed in cultured preparations - where there can be no haemocyte involvement (Howes et al., 1987a) - and following injection into the blood of physiologically-inert particles (fluorescent microspheres) which are ingested by haemocytes and thus prevented from entering the lesion (Smith et al., 1986). In both cases glial repair is slower and more disorganized than in vivo following selective glial disruption. A participation of endogenous reactive cells can also be inferred from the substantial increase in cell numbers which occur in undamaged tissues, adjacent to the lesion zone, following selective glial disruption (Treherne et al., 1987): an increase which is unlikely to have resulted from invading haemocytes. Granule-containing cells are not seen outside the lesion zone, either by electron microscopic examination or in monoclonal antibody-binding studies (Howes et al., 1987b).

CELL KINETICS AND PROLIFERATION DURING REPAIR

Recent experiments using a thymidine analogue, 5-Bromo-2-deoxyuridine (BUdR), have enabled us to follow the kinetics of cell division after selective glial lesioning (Smith et al., 1989). The results show that perineurial gliosis is initiated within 2 days, reaches a maximum after 6 and declines within 17 days (Fig. 7). Although the haemocytic invasion and the resultant granule-containing cells are largely confined to the lesion, perineurial gliosis nevertheless spreads extensively into adjacent undamaged regions (Smith et al., 1989). However, as also indicated by other experimental evidence (Smith et al., 1987), cell division is strongly influenced by the arrival of invasive haemocytes in the lesion. BUdR labelling of the nuclei is substantially reduced when the access of these exogenous cells is experimentally restricted (Fig. 8).

As a consequence of vigorous perineurial gliosis, there is a substantial increase in overall cell numbers in repairing connectives (Fig. 9). Despite the swift cellular replacement and restoration of intercellular impermeability (to extracellular tracers, within 4 days, and to potassium ions by 11), the reorganization of the perineurium is, in fact, a protracted business, with a continuing rise in cell numbers up to 28 days (Treherne et al., 1987) and a relatively slow re-establishment of the normal electrophysiological properties (as indicated by the small potassium-induced extraneuronal potentials measured after 11 days) (Smith et al., 1984).

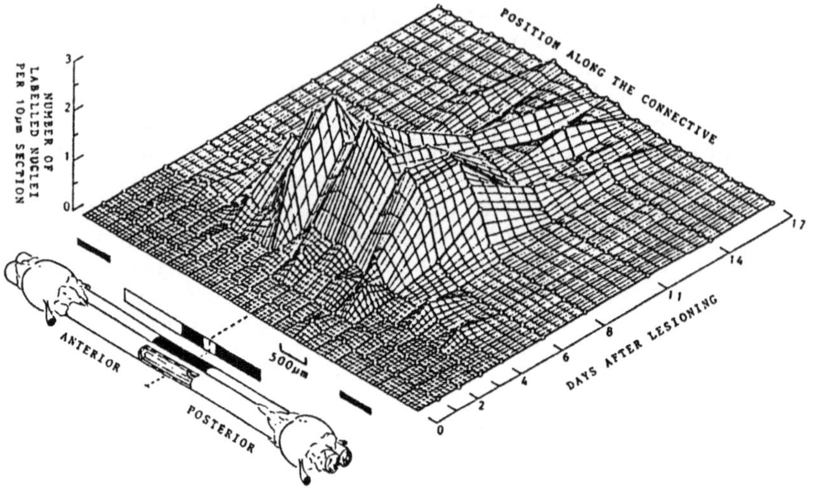

Fig. 7. Perineurial gliosis resulting from selective glial lesioning with ethidium bromide, pulse labelled with BUdR. The area of the lesion is marked by the shaded portion in the diagram below. The opposite side of the connective is also treated but in the diagram is shown cut away to illustrate the distribution of perineurial nuclei around the outer margins of the cord. The number of nuclei per 10µm section, labelled with BUdR (Y-axis), is plotted for every 200µm step from the 4th abdominal ganglion to the fifth (X-axis), and at various intervals post-lesion up to 17 days (Z-axis). The long bar between the diagrammatic cord and the X-axis, extending on either side of the mid-point of the cord (dashed line) indicates the points where, between days 4 and 6, a significant increase in the numbers of labelled nuclei occurs at the 1% level. The unshaded area of the same bar indicates where, between days 6 and 8, there is a significant decrease in labelling (1 - 5% significance level). The data points are mean values from at least 5 preparations. Six preparations were counted at days 3, 8 and 14; 7 at day 17. From Smith et al., 1989.

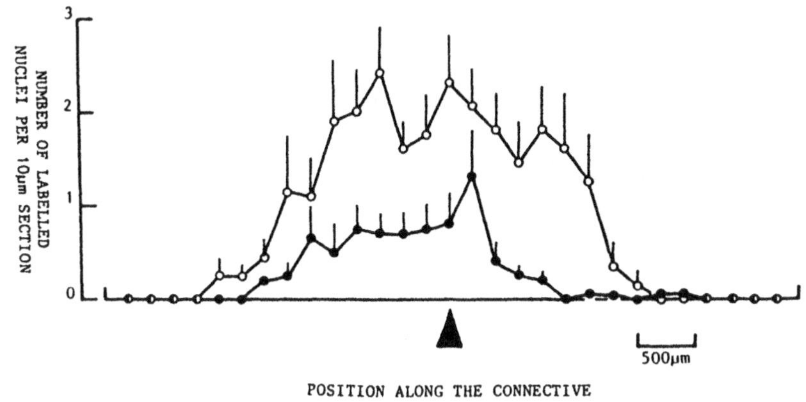

Fig. 8. The number of labelled nuclei per 10μm section from preparations at 6 days post-lesion, pulse labelled with BUdR. The open circles are for normal repairing connectives (n = 5). The closed circles indicate preparations where the involvement of the haemocytes in repair has been blocked by the injection of fluorescent microspheres into the blood (n = 5). The arrow-head indicates the centre of the connective. From Smith et al., 1989. Standard errors are shown on one side of the mean values.

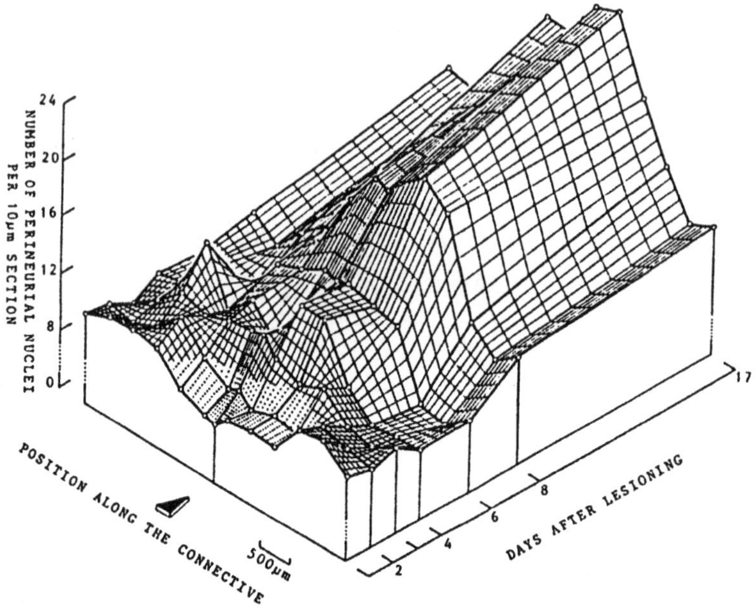

Fig. 9. As in Fig. 7 but for the total number of perineurial nuclei present in the sections from the same preparations, stained in this case with Feuglen's reagent. The stippled zone in the middle of the cords between days 1 and 3 indicates where nuclear counts are less than in controls. The difference, however, is not significant. The centre of the cord is shown by the arrow-head. Nuclear numbers from the ganglia were not counted. From Smith et al., 1989.

Our BUdR studies have revealed the following sequence of repair. The invading exogenous reactive cells - the haemocytes - are transformed within the lesion to granule-containing cells which appear in maximal numbers by 2-3 days (Fig. 10). These are swiftly deployed beneath the neural lamella to commence the structural replacement of damaged glia. Their presence within the lesion appears to trigger the proliferation of perineurial glia. Sub-perineurial gliosis follows the peak of perineurial proliferation. This coincides with the re-establishment of the blood-brain barrier by the repairing perineurial cells: implying that a switch occurs in individual cell behaviour, converting - from proliferation to differentiation - at around 8-9 days post-lesion.

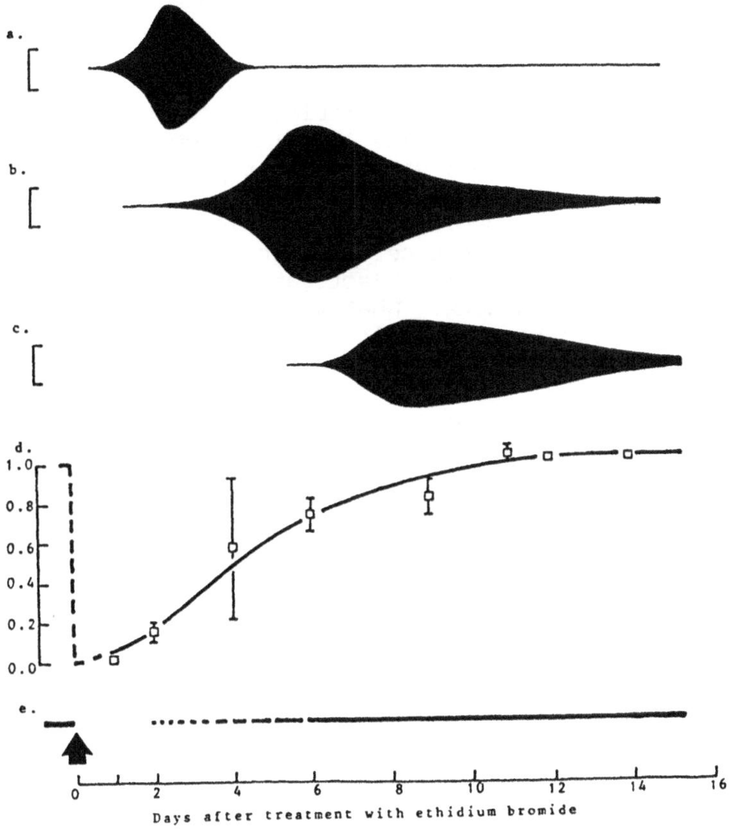

Fig. 10. Summary of the events occurring during the first two weeks of neural repair in the insect CNS after selective glial damage. The shaded areas give the relative magnitude of these events at different times post-lesion for granule-cell numbers (a), perineurial gliosis (b) and subperineurial gliosis (c). The data for granule-cell numbers (a) comes from Smith & Howes (1987); for relative spike amplitudes 5 min after perfusion with 80mM (K^+) in a sucrose gap (d) from Smith et al. (1984; 1986) and lanthanum permeability (e) from Smith et al. (1984). Time zero and the arrow-head signify the time of lesioning. Relative spike amplituded in high potassium saline for 5 min (± S.D.) gives a measure of the blood-brain barrier permeability to small cations. If the barrier is intact the spike amplitude relative to controls should give a value of 1. The large variation at 4 days reflects the onset of repair in some preparations more rapidly than in others. From Smith et al., 1989.

STRATEGIES OF BRAIN REPAIR

A notable feature of the repair and regeneration of the cockroach CNS is an intimate interaction between a mesodermal cell class, circulating blood cells, and neuroectodermal tissue. The invading haemocytes appear to serve two major roles: structural replacement of damaged neuroglia and the triggering - and orchestration - of subsequent cell proliferation. The status of the proliferating cells is uncertain: they could be transformed granule-containing cells or activated endogenous ones from outside the lesion zone. Whatever their origin, the extent of their proliferation implies the existence of powerful and sustained mitogenic signals which are capable of activating cell division in existing G_1-phase cell populations. The available evidence suggests that the signals could be provided by the haemocytically-derived cells. In this respect, this parallels what is known of the equivalent events in vertebrates, where blood-monocytes have been shown to enter mammalian brain within 2 days following surgical lesioning (Adrian & Schelper, 1981) and to transform within 5 to macrophage cells (DuBois et al., 1985): timings which correspond to those in the cockroach CNS (Smith et al., 1984). It has also been shown that mammalian blood cells, principally T-lymphocytes, can release glial growth factors (Fontana et al., 1980; Benveniste et al., 1985) and that there are endogenous glial morphogenetic and mitogenic factors in mammalian brain (Guilian & Baker, 1985; Guilian & Yang, 1986; Guilian et al., 1985, 1986). The notion that such endogenous factors might be released by the arrival of exogenous reactive cells - as we have postulated for the insect CNS - accords with the discovery that mammalian microglia, which are probaby derived from monocytic stem cells (Rio-Hortega, 1932; Perry et al., 1985) can release growth factors that act, selectively, on macroglia (Guilian & Baker, 1985).

In addition to diffusible growth factors, there is abundant evidence for the influence of substrate guidance and cell adhesion molecules in neural structuring in vertebrate systems (e.g. Mirsky et al., 1986; Edgar, 1989). Similarly, in an invertebrate system - the leech CNS - evidence is accumulating for a crucial role of substrate factors in neural regeneration (Chiquet & Acklin, 1986; Chiquet & Nicholls, 1987). In the insect the importance of such guidance is suggested by the disorganized glial repair and delayed restoration of blood-brain barrier function observed following surgical lesioning (Treherne et al., 1984) as compared with the swift and ordered repair obtained after selective glial disruption in which the extracellular matrices are intact (Smith et al., 1984). The vigorous regenerative capacity of the insect CNS should provide ideal preparations for the discovery of cell guidance clues as well as elucidating the mechanisms for integration of exogenous and endogenously derived diffusible signals in glial regeneration.

REFERENCES

Abbott, N.J. & Treherne, J.E. (1977). Homeostasis of the brain microenvironment: a comparative account. In Transport of Ions and Water in Animals (eds. B.L. Gupta, R.B. Moreton, J.L. Oschman & B.J. Wall), pp.481-410. Academid Press: London.

Adrian, E.K., Jr. & Schelper, R.L. (1981). Microglia monocytes and macrophages. In Glial and Neuronal Cell Biology, 11th International Congress of Anatomy, Par A (ed. E.A. Vidrio & A. Federoff), pp.113-124. New York: A.R. Liss Inc.

Benveniste, E.N., Merrill, J.E., Kaufman, S.E., Golde, D.W. & Gasson, J.C. (1985). Purification and characterization of a human T-lymphocyte-derived glial growth factor. Proc. natn. Acad. Sci. U.S.A. 82, 3930-3934.

Chiquet, M. & Acklin, S.E. (1986). Attachment to ConA or extracellular matrix initiates rapid sprouting by cultured leech neurons. Proc. natn. Acad. Sci. U.S.A. 83, 6188-6192.

Chiquet, M. & Nicholls, J.G. (1987). Neurite outgrowth and synapse formation by identified leech neurones in culture. J. exp. Biol. 132, 191-206.

Crone, C. & Oleson, S.P. (1982). Electrical resistance of brain microvascular epithelium. Brain Res. 241, 49-55.

Davenport, A.P. & Evans, P.D. (1984). Stress-induced changes in the octopamine level of insect haemolymph. Insect Biochem. 14, 135-143.

DuBois, M., Bowman, P.D. & Goldstein, G.W. (1985). Cell proliferation after ischemic injury in gerbil brain. An immunocytochemical and autoradiographic study. Cell Tissue Res. 242, 17-23.

Edgar, D. (1989). Neuronal laminin receptors. Trends Neurosci. 12, 248-251.

Evans, P.D. (1985). Octopamine. In Comprehensive Insect Physiology, Biochemistry and Pharmacology 11 (eds. G.A. Kerkut & L.I. Gilbert), pp.499-530. Pergamon: Oxford.

Florkin, M. & Jeauniaux, C. (1974). Haemolymph composition. In The Physiology of Insecta (ed. M. Rockstein), pp.356-307. Academic Press: New York.

Fontana, A., Grieder, A., Arrenbrecht, St. & Grob, P. (1980). In vitro stimulation of glial cells by a lymphocyte-produced factor. J. neurolog. Sci. 45, 55-62.

Giulian, D., Allen, R.L., Baker, T.J. & Tomozawa, Y. (1986). Brain peptides and glial growth. I. Glial promoting factors as regulators of gliogenesis in the developing and injured central nervous system. J. Cell Biol. 102, 803-811.

Giulian, D. & Baker, T.J. (1985). Peptides release by ameboid microglia regulate astroglial proliferation. J. Cell Biol. 101, 2411-2415.

Giulian, D., Tomozawa, Y., Hindman, H. & Allen, R.L. (1985). Peptides from regenerating central nervous system promote specific populations of microglia. Proc. natn. Acad. Sci. U.S.A. 82, 4287-4290.

Giulian, D. & Young, D.G. (1986). Brain peptides and glial growth. II. Identification of cells that secrete glial promoting factors. J. Cell Biol. 102, 812-820.

Howes, E.A., Smith, P.J.S. & Treherne, J.E. (1987a). Glial repair in the cultured central nervous system of an insect. Cell Tissue Res. 247, 111-120.

Howes, E.A., Chain, B.M., Smith, P.J.S. & Treherne, J.E. (1987b). Blood cells contribute to glial repair in an insect. Tissue & Cell 19, 877-880.

Iversen, L.L., Iversen, S.D. & Snyder, S.H. (1975). Amino acid transmitters. Handbook of Psychopharmacology Vol.4. Plenum: New York.

Lane, N.J. (1981). Invertebrate neuroglia: junctional structure and development. J. exp. Biol. 95, 7-33.

Lane, N.J. (1985). Structure of components of the nervous system. In Comprehensive Insect Physiology, Biochemistry and Pharmacology 5, (eds. G.A. Kerkut & L.I. Gilbert), pp.1-42. Pergamon: Oxford.

Lane, N.J. & Treherne, J.E. (1981). Functional organization of arthropod neuroglia. In Insect Biology in the Future (eds. M. Locke & D.S. Smith), pp.765-795. Academic Press: London.

Lettau, J., Foster, W.A., Harker, J.E. & Treherne, J.E. (1977). Diel changes in potassium activity in the haemolymph of the cockroach Leucophaea maderae. J. exp. Biol. 71, 171-186.

Mirsky, R., Jessen, K.R., Schachner, M. & Goridis, C. (1986). Distribution of the adhesion molecules N-CAM and L1 on peripheral neurons and glia in adult rats. J. Neurocytol. 15, 799-815.

Morris, C.A. (1977). Insect blood-brain barrier system: exchange, metabolism and neuropharmacology of noxious organic molecules. Ph.D. theses: University of Cambridge, U.K.

Nelson, J.W. & Tinoco, I., Jr. (1984). Intercalation of ethidium bromide into DNA & RNA oligonucleotides. Biopolymers 23, 213-233.

Perry, V.H., Hume, D.A. & Gordon, S. (1985). Immunohistochemical localization of macrophages and microglia in the adult and developing mouse brain. Neuroscience 15, 313-326.

Rio-Hortega, P. del. (1932). Microglia. In Cytology and Cellular Pathology of the Nervous System, vol.2 (ed. W. Penfield), pp.481-584. New York: Paul B. Hocker.

Schofield, P.K. (1987). Formamidine effects upon the glia that form the insect blood-brain barrier. Pesticide Biochem. & Physiol. 29, 152-156.

Schofield, P.K., Swales, L.S. & Treherne, J.E. (1984a). Potentials associated with the blood-brain barrier of an insect: recordings from identified neuroglia. J. exp. Biol. 109, 307-318.

Schofield, P.K., Swales, L.S. & Treherne J.E. (1984b). Quantitative analysis of cellular and paracellular effects involved in disruption of the blood-brain barrier of an insect by hypertonic urea. J. exp. Biol. 109, 333-340.

Schofield, P.K. & Treherne, J.E. (1984). Localization of the blood-brain barrier of an insect: electrical model and analysis. J. exp. Biol. 109, 319-332.

Schofield, P.K. & Treherne, J.E. (1985). Octopamine reduces potassium permeability of the glia that form the insect blood-brain barrier. Brain Res. 360, 344-348.

Schofield, P.K. & Treherne, J.E. (1986). Octopamine sensitivity of the blood-brain barrier of an insect. J. exp. Biol. 123, 432-439.

Smith, P.J.S., Howes, E.A., Leech, C.A. & Treherne, J.E. (1986). Haemocyte involvement in the repair of the insect central nervous system after selective glial disruption. Cell Tissue Res. 243, 367-374.

Smith, P.J.S. & Howes, E.A. (1987). Neural repair in an insect central nervous system. Cell Tissue Res. 247, 129-135.

Smith, P.J.S., Howes, E.A. & Treherne, J.E. (1987). Mechanisms of glial regeneration in an insect central nervous system. J. exp. Biol. 132, 59-78.

Smith, P.J.S., Howes, E.A. & Treherne, J.E. (1989). Gliogenesis in insect central nervous system repair; 5-Bromo-de-deoxyuridine incorporation in vivo. In preparation.

Smith, P.J.S., Leech, C.A. & Treherne, J.E. (1984). Glial repair in an insect central nervous system: effects of selective glial disruption. J. Neuroscience. 4, 2698-2711.

Treherne, J.E. (1985). Blood-brain barrier. In: Comprehensive Insect Biochemistry, Physiology and Pharmacology, vol.5, Nervous system: structure and motor function (eds. G.A. Kerkut & L.I. Gilbert, pp.115-158.

Treherne, J.E., Harrison, J.B., Treherne, J.M.& Lane, N.J. (1984). Glial repair in an insect central nervous system: effects of surgical lesioning. J. Neurosci. 4, 2689-2697.

Treherne, J.E., Smith, P.J.S. & Edwards, H. (1987). Neural repair in an insect: cell recruitment and deployment following selective glial disruption. Cell Tissue Res. 247, 121-128.

Treherne, J.E. & Schofield, P.K. (1981). Mechanisms of ionic homeostasis in the central nervous system of an insect. J. exp. Biol. 95, 61-73.

Wyatt, G.R. (1961). The biochemistry of insect haemolymph. Ann. Rev. Entomol. 6, 75-102.

From *Insect Neurochemistry and Neurophysiology • 1989* Edited by A. B. Borkovec and E. P. Masler

DRUG DESIGN

Tomas G. Kempe
Department of Chemistry and Biochemistry
University of Maryland
College Park, Maryland 20742, USA

INTRODUCTION

The title of this talk is "Drug Design". I will focus on the design of peptides. It is a very large topic and an attempt to cover it completely is not possible. A very simplified approach to guide in the selection of strategies will be given under this title.

Peptide drug design, in most cases, involves the modification of a natural peptide to a more stable analogue; i.e., in order to have any practical and potential use, the peptide must have a longer half-life *in vivo* giving a longer duration of action. The potency of the analogue may be increased as a result of more resistance to enzymatic metabolism or from higher intrinsic activity, or both. However, in some instances, the naturally occurring peptides are stable enough to be used without modifications. There are several advantages to choosing the naturally occurring ones for drug development, as will be highlighted in this review, since efficient production methods, like recombinant DNA techniques can be used. Even though natural peptides are relatively unstable (a short half-life and duration of action), they can be useful if a new approach is adopted; e.g., managing the expression of the peptide *in vivo*. This approach is equivalent to gene therapy. At this time, such an approach is not entirely possible or ethically acceptable for humans. However, the approach could be applied to insects today through gene transfer using insect viruses. The expression of a peptide in vivo may result in the disruption of insect behavior, metabolism, reproduction, etc. The disruptions may reduce the population and, therefore, we have a potential insect control agent that is more environmentally sound than the ones presently used.

I will be presenting two examples of drug developments. They cover both natural peptide drugs and highly modified peptide drugs and the problems and advantages for each. My personal bias is that the choice of a natural sequence made by genetic engineering offers the best alternative; an example is human insulin. If, however, a highly modified peptide has to be used, it is wise to strictly define the goals of such a project, since that approach rarely results in a drug. An example is the enkephalin research.

With respect to the problem of design, I think the most important factor is the discovery of a peptide structure and its biological function. Once that is established, the method of synthesis should be considered. Today we have two workable ways of making peptides: chemical synthesis and biosynthesis, i.e., recombinant DNA techniques. This choice is important since modified peptides containing non-natural amino acids or derivatives cannot be made by recombinant DNA techniques. For example, the modified pentapeptide sequence of enkephalin containing D-amino acids has greater potency than the natural L-analogues. The technology does not exist to synthesize the modified enkephelin by recombinant DNA. However, the potency increase obtained by non-natural D-amino acids substitution can also be obtained if the enkephalin sequence is extended. The 31 amino acid peptide, β-endorphin, with N-terminal pentapeptide enkephalin sequence and an added 26 amino acids is at least as active an analgesic as the most active enkephalin analogues.

The type of medical treatment should also be considered in the design. Long term treatment with modified analogues could create an immune response. Eventually, the presence of the antibodies may create an autoimmune disorder.

Insulin is the most important peptide drug today and over the last 50 years. The development of insulin to a better drug reached a peak in the early 1980's when the human insulin sequence could be produced by recombinant DNA techniques. With this accomplishment it became possible to supply patients with an insulin product that is identical to the one lacking in their bodies. The natural sequence was preferred by the pharmaceutical industry. This product would minimize the long term side effects associated with the treatment of diabetes. The use of insulin will probably be used until it becomes possible to treat diabetes by gene therapy.

The enkephalins which were discovered by Hughes et al. (1975) and Simantov and Snyder (1976) are the natural endogenous ligands for morphine. The discovery triggered an intensive research in the synthesis of analogues and their pharmacology. There was hope at that time that this natural opioid peptide would not trigger dependence and tolerance. Unfortunately, it has now been confirmed that enkephalins and morphine do not differ in that respect.

The research in designing more potent enkephalins has been very extensive. It is probably the most studied peptide group with hundreds of analogues made. After much extensive work, we have a much better understanding of what is important in the design of peptides. The understanding of enzymatic degradation and its prevention is a crucial contribution from enkephalin research. Although very active analogues (by oral administration) have been made, no drugs have resulted from this research.

The enkephalins and endorphins are particularly interesting since they represent the link between peptides and a peptide mimic, i.e., morphine. Organic molecules which have been identified as having the same or similar modes of action to peptides are called peptide mimics or peptidomimetics. The morphine-endorphin-enkephalins belong to this class. Efforts to identify other classes are continuously being made because organic non-peptidic molecules are easier to formulate into drugs. The molecules can either have the same mode of action (agonist) or inhibit the action of a peptide (antagonist). Because of the stability and resorption characteristics of organic molecules, it would be advantageous to be able to convert peptide structures into non-peptide structures with retained activity. Today, however, there are no methods available for the theoretical transformation of a peptide structure to a classical organic molecule, and such an approach, therefore, does not represent a realistic approach for drug design.

The recent discovery of a complex organic molecule which antagonizes the action of the peptide cholecystokinin is another, but rare, example of a peptide mimic (Chang et al.,(1985); Evans et al., 1986); Bock et al., 1988). Due to the complexity of the organic structure and the high cost of production, such a molecule can only be contemplated for use in human medicine. For the same reason, the cost to produce the peptide-like, angiotensin II converting enzyme inhibitors, captopril (Ondetti, 1977) and enalapril (Patchett, 1980), rules out any consideration to use them in insect pest control. Moreover, the compounds could never be economical because the peptide-like structure limits the absorption of the molecules and high doses of the drug have to be used.

To apply the use of peptides or peptide analogues as potential insect control agents, we may learn from previous work in peptide design. A recent review article by Menn and Borkovec (1989) covers the area of insect neuropeptides, enzyme inhibitors and insect viruses as potential insect control agents. Since cost will be a major factor, the a realistic alternative is production via recombinant DNA technology. The problems associated with the delivery of a peptide in the field would be another factor, and only short peptides could be considered. Therefore, novel delivery systems have to be developed and this may be possible through transfer of genetic information via insect viruses to express highly complex peptides and proteins in vivo (Luckow and Summers, 1987; Maeda, 1989).

GENERAL PROPERTIES OF PEPTIDES AND PROTEINS

Major problems in the development of peptides and their application in therapy are: (1) a short plasma half-life due to clearance and degradation, and (2) a poor oral bioavailability. These factors in turn require larger doses of peptides that are costly to produce. The understanding of the processes which determine the distribution and fate of the peptide is, therefore, important for the design and testing of new drugs and novel delivery systems. Peptides and proteins do not usually cross cell membranes unless specific mechanisms make it possible. As a result, the distribution of these molecules is restricted to those compartments in the body which can be accessed by the transport process. This is important since many peptide degrading enzymes are highly compartmentalized. If a peptide or protein had access to all of these enzymes, degradation would be rapid. The clearance of a drug through kidneys and liver also reduces the amount of active drug available and it is generally difficult to control (Humphrey and Ringrose, 1986). Although considerable increases in potency may result from the design of stable analogues, the non-enzymatic processes of clearance continue to operate and half-lives are usually not extended beyond one or two hours.

The α-carbons of all amino acids except glycine are chiral and mammalian peptides have all the amino acids in the L form. Changing from L-amino acids to D-amino acids may have dramatic effects on biological properties and on stability to enzymatic hydrolysis as will be seen in the enkephalin structure-activity studies.

Peptides vary in molecular weights from dipeptides (MW 250) to polypeptides like insulin (MW 6000). They are often flexible linear molecules, but they can contain loops and complex cross-linkages of disulfide bonds. Proteins are larger molecules with the same basic composition as peptides. A protein usually has a more defined three-dimensional structure than the smaller peptides and are generally more stable towards enzymatic cleavage due to a more rigid conformation. It is of interest to note that more proteins are introduced as drugs today, the reason due to stability and a more defined pharmacological profile compared with the shorter counterparts, the peptides. Recently introduced proteins are: recombinant human growth hormone (hGH), recombinant human tissue plasminogen activator (TPA), streptokinase, urokinase, recombinant human erythropoietin (EPO), recombinant human α-interferon, recombinant human antiheamophilic factor (AHF), soluble T-cell receptors for HIV virus (CD4) and vaccines. Examples of peptides are: recombinant and semi-synthetic human insulin, synthetic oxytocin analogues, synthetic calcitonin and analogues, synthetic luteinizing hormone releasing hormone and antagonists (LHRH), synthetic angiotensin II converting enzyme inhibitors, synthetic glucagon, synthetic parathyroid hormone, and synthetic somatostatin.

ENZYMATIC CLEAVAGE OF PEPTIDE BONDS

Peptide and protein metabolism usually involves the enzymatic cleavage of an amide bond. The main reason for attempting to modify a peptide sequence is to avoid such enzymatic inactivation. The ideal molecule would look like an organic molecule and not be capable of enzymatic inactivation. Such aspects have been highlighted by Farmer (1980).

The amide bond hydrolysis is catalyzed by a wide range of enzymes, each with its own specificity for amino acids or pairs of amino acids. These enzymes have well formed binding pockets and require the peptide and side-chains to be in the correct conformation. Examples of peptidases or classes of peptidases and their targets are given in Table 1.

Common strategies in the design of analogues are the protection of the C-terminal carboxylic group from the action of carboxy-peptidases by amidation. The action of amino-peptidases is prevented by acylation of the N-terminal amino group. Common amino protecting groups are acetyl, formyl or pyroglutamyl. Stabilization toward exopeptidases of both types can be simultaneously accomplished by N-terminal and C-terminal cyclization, via an intramolecular amide bond. The conformational constraints obtained by cyclization often reduce the susceptibility of the peptide to endopeptidases as well.

A systematic process of replacing selected amino acids by derivatives that are not well recognized by proteases is also an important strategy for metabolic stabilization. N-Methyl and α-methyl amino acids as well as D-amino acids are classes of particularly successful substituents (Veber and Freidinger, 1985; Allen et al., 1984).

Proteins with a well defined three-dimensional structure are less accessible to peptidase attack, although certain links may be present in a manner that makes them extremely susceptible to specific enzymes such as activation of proenzymes and prohormones (e.g., prothrombin, plasminogen, trypsinogen, and angiotensinogen) and inactivation of locally acting proteins (interferons and interleukins).

Table 1. Examples of peptidases of different specificity.

```
          angiotensinogen

Asp-Arg-Val-Tyr-Ile-His-Pro-Phe-His-Leu-Leu-Val-Phe-Ser---
                                       ^
                                     renin

          angiotensin I

Asp-Arg-Val-Tyr-Ile-His-Pro-Phe-His-Leu
                               ^
                angiotensin II converting enzyme


          angiotensin II

Asp-Arg-Val-Tyr-Ile-His-Pro-Phe
   ^                       ^
aminopeptidase          carboxypeptidase

   Arg-Val-Tyr-Ile-His-Pro
      ^       ^
  trypsin   chymotrypsin
```

ROUTES OF ADMINISTRATION

Two of the major problems dealing with peptide drugs have been mentioned, degradation and clearance. The third problem is administration. In human medicine there are standard methods of delivery. However, in insect pest control, it will probably be the biggest problem in the field.

A short summary of available routes of administration is given to highlight the problems. Intravenous, intramuscular and sub-cutaneous injection via a syringe or a pump are the most commonly used routes of administration of peptides. For molecules up to the size of insulin, sub-cutaneous injection will normally provide input into the systemic circulation over a period of 30 to 60 minutes. Certain somatostatin analogues have proved to be exceptions to this rule and prolonged input has been observed at the appropriate dose (Baker et al., 1984). Proteins may leave the site of injection very slowly over periods of longer than one day (Schultze and Heremans, 1966). Oral administration is usually not used for peptide drugs since peptides are generally very poorly absorbed from the gastrointestinal tract even with stable analogues. Specific pathways do, however, exist for di- and tri-peptides which may be efficiently transported across the mucosal membrane of the small intestine. This has been observed for antibacterial phosphonic peptide analogues which have a high oral bioavailability in man (Allen et al., 1979). The administration through the nose (nasal route) will probably be used more extensively in the future due to successful results and less discomfort to the patients. This route has recently been the subject of considerable investigation (Chien and Chang, 1985). Peptides below MW 1000 can be delivered efficiently and rapidly without adjuvants and that limit can be extended to peptides of at least MW 6000, e.g., insulin, with adjuvants (McMartin et al., 1987). The administration through rectum has been investigated and will probably be of limited use for peptides. The uptake of peptides by this route appears to require the use of adjuvants (Nishihata et al., 1983).

IMMUNOGENICITY

The immunogenicity of peptides is usually not a problem but has to be monitored. Because the body recognizes only the natural peptide, antibodies to a modified peptide may be generated and an autoimmune disorder may develop. Human insulin, when administered subcutaneously is immunogenic, but much less immunogenic than porcine or bovine insulin. There are speculations that the route of administration may cause the effect. If it were true, this shows another important aspect of the route of administration. It is possible that the high concentration of insulin at the injection site allows the molecule to react with itself, e.g., dissociation-association of the disulfide bonds to form dimers or other rearrangements of the disulfide bonds (insulin contains three disulfide bonds). If that appears to be the problem, a more diluted dosage or a continuous administration of diluted insulin may reduce the immune response.

Although immunogenicity arises from the fact that foreign features have been introduced into the molecule, metabolic processing of the molecule plays an essential role in the initial triggering of an immune response. The immunogen must be taken up to antigen presenting cells. If it becomes partially degraded into fragments, it must be displayed on the surface of the antigen presenting cell in order to trigger a primary immune response (Grey and Chestnut, 1985).

ENKEPHALINS AND ENDORPHINS

A large number of peptides have been isolated from the brain. Such peptides are called neuropeptides and are present in defined neuronal pathways. When they are able to change the membrane potentials of neurons, they may act as neurotransmitters or neuromodulators. Several peptides fulfill those criteria but the most thoroughly studied group is the opioid peptides or endorphins (endogenous morphine). A whole family of such peptides is known, with the common structure, Tyr-Gly-Gly-Phe-X, where X is either Leu, Met, or an extended peptide chain (Hughes et al., 1975; Simantov and Snyder, 1976). The pentapeptide sequence, named enkephalin, is the minimum sequence for opioid activity. An extension in the C-terminal direction can be extensive and may represent β-endorphin of 31 amino acids.

The biological actions of enkephalins, endorphins and related peptides are similar to those of morphine, though important differences between them have been described. The original assay used to isolate enkephalins utilized the ability of these substances, like morphine, to inhibit electrically stimulated contraction of guinea pig ileum (GPI) or mouse vas deferens (MVD). The activity of the peptides in these assays is considerably greater than that of morphine.

Like morphine, enkephalins and β-endorphin induce analgesia, i.e., resistance to pain. They are far more potent in this respect if injected intraventricularly (into the ventricles of the brain) than if injected intravenously, presumably because they cannot easily pass the blood-brain barrier (Pardridge, 1983). Analgesia induced by β-endorphin is of longer duration than that induced by the enkephalins. On a molar basis, β-endorphin is 20-50 times more potent than morphine, depending on the particular assay used (Loh et al., 1976; Tseng et al., 1976).

β-Endorphin has a range of other biological actions. It induces hypothermia and salivation in treated animals. Other behavioral effects in rats and cats include prolonged catatonia. In this condition, the animal is reduced to a rigid state, unresponsive to external stimuli, although completely conscious.

The biosynthesis of these neuropeptides follow a general scheme. A gene is transcribed into a messenger RNA which is translated into a protein, the preprohormone, with a signal sequence for specific transport and storage. The signal sequence is cleaved with proteolytic enzymes to give a prohormone which upon further cleavage gives one or more functional peptides. All neuropeptide prohormones are commonly flanked by pairs of Lys or Arg, or both. Such pairs are attacked by trypsin-like endopeptidases, generating fragments having one of these amino acids at the C-terminal. The amino acid is subsequently trimmed off by a carboxy-peptidase to give the functional neuropeptide. The sequential action of these two types of enzymes can account for the formation of most known opioid peptides.

Three different prohormones for the endorphins have been described. They are of similar size, about 260 to 280 amino acids each, and they are polyfunctional. Proopiomelanocortin is a prohormone for β-endorphin, and also for the non-opioid peptides adrenocorticotropic hormone (ACTH) and melanocyte-stimulating hormone (MSH) (Nakanishi et al., 1979). β-Endorphin (1-31) is known to be susceptible to further proteolytic cleavage with retention of opioid activity. The 1-27 and 1-26 fragments have been isolated from the pituitary and are much less analgesic (Smyth et al., 1980). The 1-17 fragment, γ-endorphin, has unique behavioral properties. While it maintains opioid activity, it may also act as an endogenous neuroleptic (DeWied et al., 1978). This peptide has also been used therapeutically with some success in the treatment of acute schizophrenia (DeWied, 1979).

Proenkephalin A (Gubler et al., 1982) is also polyfunctional. There are 7 enkephalin sequences in the prohormone. Two of the Met-enkephalin sequences are elongated at the C-terminus with Arg-Phe or with Arg-Gly-Leu. An amidated peptide, metorphamid has also been isolated (Weber et al., 1983).

Proenkephalin B (Kakidani et al., 1982) has three Leu-enkephalin sequences. It is the prohormone for α-neoendorphin and dynorphins A and B, with 11, 17 and 13 amino acids, respectively. Larger molecules have also been isolated and are considered proteolytic intermediates.

RECEPTORS

Receptors which bind labeled morphine, morphine agonist or, in some cases, antagonists such as naloxone (Kosterlitz and Watt, 1968) have been known for some time (Chang et al., 1979 and 1980). They are found in the brain and other morphine target tissues. β-Endorphin and the enkephalins will displace labeled ligands from such receptors, and show greater affinity for them than morphine itself. Labeled β-endorphin and enkephalins will also bind specifically to such receptors.

These receptors can be classified into at least three types: μ, κ and δ. The enkephalins possess a preferential affinity for δ receptors; dynorphin and α-neoendorphin bind preferentially to κ receptors; β-endorphin has high affinity for μ and δ receptors; and morphine binds exclusively to the μ receptor. Table 2 summarizes binding affinity of certain peptides in relation to morphine (Paterson et al., 1983). These observations provide a basis for differential actions of the various opioid peptides, but the precise roles of each of these peptides, and their receptors, are not yet clear. The opioid peptides appear to have inhibitory effects on their target cells, inhibiting the action potential discharges in individual nerve cells.

THE PHYSIOLOGICAL ROLE OF ENKEPHALINS AND ENDORPHINS

These peptides have a remarkably wide range of biological activities, but their true physiological roles remain unclear. Many of their actions take place in the central nervous system, as shown by the fact that intraventricular injection is necessary for a full response. In addition, it seems likely that they play an important role there as neurotransmitters and neuromodulators (Snyder, 1980). Enkephalins and β-endorphin have been shown to inhibit the release of neurotransmitters, especially dopamine, from brain tissue and to inhibit the release of acetylcholine from neuromuscular junctions, hence the inhibition of contraction of smooth muscle. A role as inhibitory transmitters agrees well with the observed biological effects of the peptides in reducing responses to pain and other stimuli. It has been proposed that release of these or related opiate-peptides into the circulation or CNS represents the fundamental basis of acupuncture.

ENZYMATIC DEGRADATION

The transient nature of pharmacological actions of the enkephalins was noted early in opioid peptide studies. The transient nature was observed to be correlated with the rapid deactivation by plasma and tissue homogenates (Hughes et al., 1975; Pert et al., 1976). An experiment wherein radiolabeled Tyr was released (Lane et al., 1977) showed that the primary mode of degradation is cleavage of

Table 2. Receptor selectivity of opioid peptides and morphine by binding analysis.

Substance	Relative selectivity for: μ-receptor	κ-receptor	δ-receptor
β-endorphin	1.0	0.02	0.9
Met-enkephalin	0.1	0.01	1.0
Leu-enkephalin	0.06	0.01	1.0
Dynorphin A	0.2	1.0	0.1
Morphine	1.0	0.03	<0.01

the Tyr-Gly bond. The Tyr-Gly bond is susceptible to attack by many nonspecific aminopeptidases. These aminopeptidases are non-functional at 4°C or in the presence of bacitracin or puromycin. The action of enkephalins is enhanced when assay conditions are accordingly modified (Fournie-Zaluski et al., 1979).

Recently, aminopeptidases have been isolated from both human brain (Traficante et al., 1980) and plasma (Coletti-Previero et al., 1981) with marked preference for Met- and Leu-enkephalin. A D-Ala substitution analogue at position 2 was resistant to the enzyme as will be described later. In contrast, a carboxy-peptidase, enkephalinase, isolated from mouse striatum, specifically cleaves Gly-Phe bond in the enkephalin sequence (Guyon et al., 1979). It is a membrane-bound enzyme localized in the vicinity of the opioid receptors, and it may, therefore, play a role in the physiological control of endogenous opioid activity. The enzyme is not inhibited by puromycin but is blocked by thiorphan (D,L-3-mercapto-2-benzylpropanoyl)glycine which is a substrate analogue of enkephalin (Roques et al., 1980). Thiorphan fits into the active site of the enzyme. It protects enkephalins under both in vitro and in vivo conditions. Additionally, it displays significant antinociceptive activity, attributed to the potentiation of endogenous enkephalins released in response to noxious stimuli (Corbett et al., 1982).

STRUCTURE-ACTIVITY RELATIONSHIP

Once the enkephalin structure was determined, an enormous amount of structure-activity studies on the enkephalins and their analogues started. I will describe some of this work; this body of work serves as a model for strategy in modifications of newly identified peptides. At USDA, we have applied these concepts to structure-activity studies of the pheromone biosynthesis activating neuropeptide (PBAN) in insects (Raina et al., 1989).

In assessing the potency of analogues, three in vitro test systems have been widely employed: the opioid binding assay (Pert and Snyder, 1973), inhibition of electrically stimulated guinea pig ileum (GPI) (Kosterlitz et al., 1970), and mouse vas deferens (MVD) (Hughes et al., 1975), with agonist activity expressed by IC or ID values relating to 50% inhibition or displacement. In vivo assays of analgesia are done by the rat tail-flick assay (RTF, radiant heat focused on tip of tail) or by the mouse hot-plate assay (MHP, animal placed on a hot plate maintained at 55-70°C), and the effective dose at 50% response (ED_{50}) recorded.

The behavior of peptide analogues in analgesic tests is less frequently available and in many cases is seen only after central administration by intracerebroventricular and related routes. This is due to the instability of most of the enkephalins and their analogues. With many analogues a lack of correlation between individual in vitro assays and between in vitro and in vivo assessments of potency are observed. This observation can be explained if there were different types of opioid receptors, the collection of which vary from tissue to tissue. Moreover, evaluation of assay data is complicated since endopeptidases seem to play a major role in determining the observed activity (Frederickson et al., 1981).

The effects of amino acid substitutions in the enkephalin pentapeptide sequence are described below.

	Tyr-	Gly-	Gly-	Phe-	X	X= Met-enkephalin, Leu-enkephalin
position	1	2	3	4	5	

Tyrosine at position 1. The Tyr residue at position 1 in the enkephalin sequence is essential for opioid activity (Beddell et al., 1980). Modifications through acylation, deamination, removal of phenolic hydroxyl group in Tyr (Phe substitution) and D-Tyr substitution have pronounced reduction in activity. The N-Methyl-Tyr analogues are as active as Met-enkephalin in the GPI test, whereas the N,N-dimethyl analogues have no activity (Morley, 1980). Therefore, the hydrogen-bonding may be important for activity. It is of interest to note that N-allyl and N,N-diallyl analogues have antagonistic properties (Belton et al., 1983), as does naloxone which is an N-allyl substituted morphine analogue.

Glycine at position 2. One of the most important discoveries in enkephalin structure-activity studies is the enhanced activity exerted by D-amino substituents at position 2 in both Met- and Leu-enkephalin in combination with, or without, C-terminal amide structures (Pert et al., 1976). Potencies in GPI studies relative to Leu-enkephalin methyl ester (activity = 1) are as follows: D-Ser (11.0), D-Met (8.0), D-Ala (5.6), D-Leu (1.3), D-Phe (1.3), and D-Pro (0.002). In comparison, the corresponding L-amino acid analogues have lower activities (Dutta et al., 1977). Other combinations of position 2 and position 5 substitution analogues are even more potent, as described under multiple modifications in enkephalins.

Glycine at position 3. Modifications at position 3 in the pentapeptide sequence of enkephalins result in drastic reduction in activity. Position 3 is, therefore, left unmodified (Morley, 1980). However, a deletion of Gly at position 3 in the tetrapeptide sequence Tyr-D-Ala-Phe-Met(NH_2) results in an activity similar to Met-enkephalin amide. The N-terminal sequence of the tetrapeptide, Tyr-D-Ala-Phe-, is the same as dermorphin, Tyr-D-Ala-Phe-Gly-Tyr-Pro-Ser amide, a potent new type of opioid peptide. Dermorphins have been isolated from the skin of certain South American frogs (Broccardo et al., 1981).

Phenylalanine at position 4. The substitution of Phe at position 4 with other substituents generally results in analogues with lower or no activity (Morley, 1980). There are, however, some remarkable exceptions; substitutions combined with D-amino substituents at position 2 and the N-Methyl-Phe substituent at position 4 give some of the most potent enkephalins known (Roemer et al., 1977).

Methionine and Leucine at position 5. A majority of single substitution analogues at position 5 result in reduced activity (Morley, 1980). The introduction of C-terminal amide structures increases activity several-fold compared with the parent enkephalin structure with a free carboxylic group at position 5 (Chang and Cuatrecasas, 1979; Mathur et al., 1979). Reduction of the carboxylic group to carbinol also increases activity, presumably due to increased stability to carboxypeptidases (Morley, 1980). The great boost in activity is seen when position 2 and position 5 are both substituted with enzyme resistant groups. This suggests that the activity of many enkephalin analogues may actually be a measure of enzymatic stability rather than increased intrinsic activity.

Multiple changes of amino acids in enkephalin. A systematic variation of substituents in position 2 and position 5 in the Met-enkephalin sequence, Tyr-Gly-Gly-

Phe-Met, has resulted in very active analogues. The choice of substituents seems to have been random, and with enough critical mass, some very active analogues were discovered. The observation that combined substitutions at positions 2 and 5 were beneficial on activity as well as some selective substituents at position 4 guided the search for better analogues. This all agrees well with inactivation by enzymatic degradation of enkephalins. In the design of analogues, a thorough investigation on enzymatic degradation could save a lot of time in the design process. Such degradation can be monitored by selective radiolabeling of individual amino acids in a short sequence. Some additive effects which rarely are discovered in structure-activity studies of peptides were found in the analogues of Met-enkephalin with substitutions at the 2, 4 and 5 positions. What resulted was the discovery of one of the most active enkephalin analogues, namely Sandoz FK 33-824, Tyr-D-Ala-Gly-MePhe-Met-(O)-ol (Roemer et al., 1977). Modifications in this sequence appear to be additive, Table 3.

FK33-824 showed high activity at opioid receptors in vitro and demonstrated preference for the μ-receptor compared with the δ-receptor (Kosterlitz et al., 1980). It was 100 to 1000 times more potent than morphine as an analgesic when given by the intraventricular route and was active after systemic administration (Roemer et al., 1977). In a double-blind study against experimental pain, the compound gave dose-dependent analgesia (Stacher et al., 1979; VonGraffenried et al., 1978). Naloxone precipitated a marked withdrawal syndrome in monkeys self-administering this compound (Roemer et al., 1977). This compound is no longer being pursued clinically as an analgesic candidate. This shows the difficulty in drug design when the best of analogues do not make it to market.

The N-methyl-Tyr analog is even more active than the parent peptide (Table 3) in mice, and it is orally equipotent with morphine (Roemer and Pless, 1979). Moritoki et al. (1981) investigated the effects of N-terminal methylation and C-terminal reduction in enkephalin. The N-Methyl-Tyr-carbinol analogue is 66 times more active than morphine and the Tyr-carbinol analogue is 0.96 times as potent as morphine. Thus the dramatic increase is obtained by N-terminal methylation.

Table 3. Analgesic activity of Met-enkephalin and some analogues. Tail-flick test ED50.

	iv 30 min (mg/kg)	sc 30 min (mg/kg)	po 120 min (mg/kg)
Tyr-D-Ala-Gly-Phe-Met-ol	13.0	59.0	>640
-D-Ala- Met(O)-ol	3.5	3.9	319
-D-Ala- -MePhe-Met(O)-ol	0.4	1.4	102

Met = methionine with carboxylic group (COOH)
Met-ol = methionine with carboxylic group reduced to alcohol (-CH_2-OH)
Met(O)-ol = methionine with carboxylic group reduced to alcohol and sulfur oxidized to sulfoxide
MePhe = N-Methyl-Phenylalanine

The introduction of proline or prolinamide at position 5 increases receptor binding and stability to carboxypeptidases (Bajusz et al., 1978). Here, again, a combination of position 2 and position 5 substitution analogues gives an orally active enkephalin analogue, Table 4 (Bajusz et al., 1978).

Tolerance toward the analgesic effect of the proline derivative in mice develops rapidly (Wei, 1981; Miglecz et al., 1979). Naloxone effectively antagonized the analgesic effect of the peptide in RTF and MHP tests so there is no doubt as to its morphine-like nature (Szekely et al., 1978).

Finally, another enkephalin analog which has undergone clinical trials will be presented. The compound is named metkephamid and has D-Ala at position 2 and N-Methyl-Met at position 5. The parent compound can be further modified to even more potent analogues, Table 5 (Gesellchen et al., 1981).

The parent compound produces naloxone-reversible depression in GPI and MVD assays, which is characteristic of an opioid peptide (Frederickson et al., 1981 and 1982). The peptide has a high affinity for both δ- and μ-receptors with analgesic activity of 1/3 to 10 times that of morphine, depending on the test system and route of administration. When given by the intraventricular route, metkephamid was about 100 times more potent than morphine. It was reported that metkephamid in rats produced only slightly more dependence than control, which is interesting, but contrary to other analogues studied. Frederickson speculates that high affinity for the δ-receptor gives this lack of dependence and a different side effect spectrum (Hynes and Frederickson, 1982). The compound was in clinical trials during 1986, but no recent information can confirm if the trial is still ongoing. The clinical studies indicate that 70 to 140 mg of metkephamid are approximately equivalent to 100 mg meperidine which is the generic analgesic for demerol, dolantin and dolosal. The difference in cost between these two compounds is tremendous, the peptide being an estimated 100 to 1000 times more expensive per dose. It must be hard to justify such a huge difference in price, with little or any beneficial therapeutic value.

Hopes that the natural enkephalin peptides or their analogues might provide the long-sought nonaddicting analgesic have not been met. Tolerance and physical

Table 4. The combination of position 2 with D-amino acid substituents and position 5 with prolinamide gives another set of superpotent enkephalin analogues.

1 2 3 4 5	potency* iv	potency* icv
Tyr-Gly-Gly-Phe-Met	0.0	0.02
-D-Ala- -Pro(NH_2)	0.22	3.9
-D-Met- -Pro(NH_2)	5.5	49.8

*The potency is related to morphine = 1 in the rat tail-flick test (RTF).

Table 5. Metkephamid and analogues.

	MVD IC/50 (nMol)	MHP ED/50 (mg/kg, sc)
Tyr-D-Ala-Gly-Phe-MeMet(NH_2)	12.2	0.36
-D-Thr-	1.2	0.3
-p-fluoro-Phe-	0.77	0.022

dependence can be produced by chronic perfusion of rat cerebral ventricles with enkephalin or by repeated injections of β-endorphin (Loh et al., 1976). Furthermore, rats made tolerant to morphine proved cross-tolerant to β-endorphin as well as synthetic Met-enkephalin. Rats, pretreated with endorphins, precipitated withdrawal syndrome when treated with the morphine antagonist naloxone, similar to that of rats treated with normorphine (Blasig and Herz, 1976). Heroin addicts preferred morphine to the super agonist enkephalin analogue FK 33-824, Tyr-D-Ala-MePhe-Met(O)-ol (Roemer et al., 1977), because enkephalin evokes intoxication without euphoria.

DISCOVERY OF INSULIN AND ITS DEVELOPMENT TO A HUMAN DRUG

The discovery of insulin by Banting and Best (1922) and its rapid application to the successful treatment of diabetes mellitus is one of the medical triumphs of this century. Diabetes is a condition in which blood glucose is permanently elevated and large amounts are detected in the urine.

The primary structure of insulin (Ryle et al., 1955) consists of two polypeptide chains, A and B, linked by two disulfide bonds. The biosynthesis of the peptide starts with a preproinsulin (Chan et al., 1976; Patzelt et al., 1978), which is sequentially cleaved by proteolytic enzymes into proinsulin and insulin. Insulin, like other peptide hormones, contain a hydrophobic N-terminal sequence, called the signal or leader sequence, in the preprohormone which serves as a hydrophobic signal sequence for the transport of the nascent protein through the microsomal membranes of the rough endoplasmatic reticulum (Blobel and Dobberstein, 1975).

Insulin, one of the oldest known hormones, has been used as a therapeutic agent for over half a century. Human and pork insulins differ by one amino acid, and human and beef insulins, by three amino acids.

	A-chain:			B-chain:
human insulin	Thr(8)	Ser(9)	Ile(10);	Thr(30)
porcine	+	+	+	Ala
bovine	Ala	+	Val	Ala

Production of human insulin based on pork insulin is called semi-synthesis. The semi-synthesis process for producing human insulin transforms the porcine insulin molecule into an exact copy of human insulin by substituting Thr for Ala at position 30 in the B-chain. The enzyme trypsin is used to remove the C-terminal Ala under conditions which favor peptide bond cleavage. In the presence of a large excess of Thr, the reaction conditions are changed to favor peptide bond formation using the same enzyme, thus giving an insulin B-chain with C-terminal Thr as in the human sequence. Peptide bond cleavage is usually done at acidic pH and peptide bond formation at higher pH. The conditions used are proprietary and have been described without specifics (Barfoed, 1987).

The early semi-synthesis approach for making human insulin from porcine insulin used a combination of enzymatic cleavage and chemical coupling techniques. When trypsin is used to cleave porcine insulin, the C-terminal of the B-chain can be cleaved at positions 29-30 (-Lys-Ala-COOH) or at positions 22-23 (-Arg-Gly-), or both. Upon prolonged cleavage, the entire sequence 23-30 will be trimmed off. In order to reconstitute the entire insulin sequence, a synthetic fragment corresponding to the octapeptide with Thr substituted for Ala at position 30 is chemically coupled to the cleaved porcine insulin to give the exact human sequence. Deprotection and purification of the semi-synthetic human insulin is then done (Obermeier, 1978). A more selective removal of the C-terminal Ala in the B-chain has been used by Morihara and his co-workers (Morihara et al., 1979 and 1980).

The first human insulin made by recombinant DNA techniques was accomplished by separate productions of the insulin A-chain and the insulin B-chain in E. coli. (Goeddel et al., 1979). The genes coding for the 21 amino acid A-chain and the 30 amino acid B-chain were fused to a β-galactosidase gene with a methionine between the protein and peptide, which allowed for cyanogen bromide cleavage of the expressed fusion protein. The two insulin chains were isolated separately and oxidized, giving human insulin with two intermolecular disulfide bonds between the A- and B-chains and one intramolecular disulfide bond in the A-chain. The potential problem of generating random disulfide bonds between the A- and B-chains appears not to be a problem since the overall yield for the correct molecule was about 50%. The side products could be separated by HPLC.

This potential problem of random disulfide bond formation has been circumvented by first producing proinsulin, which preferentially has the correct three-dimensional orientation for correct intramolecular disulfide bond formation (Villa-Komaroff et al., 1974; Wetzel et al., 1981). Proinsulin is thus produced in E. coli, similarly to the separate A- and B-chain approach, and then oxidized. The cross-linked proinsulin can be enzymatically converted to human insulin using trypsin and carboxypeptidase B by removing the spacer peptide (C-chain) between the A- and B-chains (Steiner et al., 1974).

The procedure developed for the manufacture of human insulin by the proinsulin route opens up new possibilities for future investigations in the field of diabetes research. The presence of human proinsulin in human insulin may offer some advantages since the human pancreas secretes proinsulin along with insulin in a ratio of about 1:100. It is now possible to investigate if any combination of insulin, proinsulin and the C-chain of the identical human structure can have synergistic effect in the treatment of diabetes. Since all these components result from the metabolism of proinsulin, they are present in the body and may have additional physiological roles.

The real advantage of human insulin produced by recombinant DNA technology is that it provides an unlimited and relatively inexpensive supply of a complex peptide with very important therapeutic applications. In 1980, FDA received an investigational new drug (IND) application from Eli Lilly for human insulin, the first recombinant DNA product submitted to the FDA. Several factors made this an ideal drug for evaluation as a prototype for recombinant DNA product. First, there was vast clinical experience with insulins; second, human insulin is an endogenous product; and third, the drug is used in a deficiency state. It was the Agency's opinion that extensive animal toxicity studies and prolonged clinical trials on human insulin would be unnecessary provided that the product had correct structure, high purity and low concentration of contaminants.

Human insulins manufactured by Novo Industri A/S and Nordisk Gentofte, two Danish firms, are by the semi-synthesis method.

SUMMARY

The majority of therapeutically useful peptides will be made by recombinant DNA technology. As a result, they will contain natural amino acids. Since stability of short peptides is generally a limiting factor, longer peptides with some stabilization through three-dimensional structures will be developed into drugs. It may be possible to use proteins as prodrugs for peptide release. The preprohormones or prohormones which are precursors for several peptides may be useful if degradation to a functional peptide took place at the site of action, thus minimizing the multitude of pharmacological effects seen in many peptides. Depending on the peptide structure, different host-vector systems can be employed for production. Peptides which contain non-modified amino acids can be made in procaryotes, and modified peptides that are glycosylated, phosporylated, sulfated, acetylated and amidated can be made in eucaryotes.

REFERENCES

Allen J.G., Havas L., Leicht E., Lenox-Smith I. and Nisbet L.J. (1979) Phosphonopeptides as antibacterial agents: Metabolism and pharmacokinetics of alafosfalin in animals and humans. Antimicrob. Agents Chemother. 16, 306-313.

Allen M., McMartin,C., Peters G.E. and Wade R. (1984) The mechanism of degradation Cyclo(Asn-Phe-Phe-D-Trp-Lys-Thr-Phe-GABA) and the relative stabilities of this and other octapeptide Somatostatin analogs in rat. Regul. Pept. 10, 29-35.

Bajusz S., Patthy A., Kenessey A., Graf L., Szekely J.I. and Ronai A.Z. (1978) Is there correlation between analgesic potency and biodegradation of enkephalin analogs. Biochem. Biophys. Res. Commun. 84, 1045-1053.

Baker J.R.J., Kemmenoe B.H., McMartin.C. and Peters G.E. (1984) Pharmacokinetics, distribution and elimination of a synthetic octapeptide analog of somatostatin in rat. Regul. Pept. 9, 213-226.

Banting F.G. and Best C.H. (1922) The internal secretion of the pancreas. J. Lab. Clin. Med. 7, 251-266.

Barfoed H.C. (1987) Insulin production technology. Chem. Engineering Progr. 49-54.

Beddell C.R., Clark R.B., Hardy G.W., Lowe L.A., Ubatuba F.B., Vane J.R., Wilkinson S., Chang K-J., Cuatrecasas P. and Miller R.J. (1977) Structure requirements for opioid activity of analogues of the enkephalins. Proc. R. Soc. (London) Ser. B 198, 248-265.

Beddell C.R., Lowe L.A. and Wilkinson S. (1980) Endogenous opioid peptides -- the enkephalins and endorphins. Prog. Med. Chem. 17, 1-39.

Belton P., Cotton R., Giles M.B., Gormley J.J., Miller L., Shaw J.S., Timms D. and Wilkinson A. (1983) Divergent structure-activity relationships in series of enkephalin agonists and cognate antagonists. Life Sci. 33, (Suppl.I) 443-446.

Blasig J., Papeschi R., Gramsch C. and Herz A. (1976) Central serotonergic mechanisms and development of morphine dependence. Drug Alcohol Depend. 1, 221-232.

Blobel G. and Dobberstein B. (1975) Transfer of proteins across membranes. J. Cell Biol. 67, 835-851.

Bock M.G., DiPardo R.M., Evans B.E., Rittle K.E., Veber D.F., Freidinger R.M., Chang R.S.L. and Lotti V.J. (1988) Cholesystokinin antagonists. Synthesis and biological evaluation of 4-Substituted 4H-[1,2,4]Triazolo[4,3-a][1,4]benzodiazepines. J. Med. Chem. 31, 176-181.

Broccardo M., Erspamer V., Flaconieri Erspamer G., Improta G., Linari G., Melchiorr,P. and Montecucchi P.C. (1981) Pharmacological data on dermorphines, a new class of potent opioid peptides from amphibian skin. Br. J. Pharmacol. 73, 625-631.

Chan S.J., Keim P., Steiner D.F. (1976) Cell-free synthesis of rat preproinsulin: Characterization and partial amino acid sequence determination. Proc. Natl. Acad. Sci.(USA) 73, 1964-1968.

Chance R.E., Kroeff E.P. and Hoffmann J.A. (1981) Chemical, physical, and biological properties of recombinant human insulin. In Insulin, growth hormone, and recombinant DNA technology (Ed. by Gueriguian J.L.). pp.71-80. Raven Press, New York.

Chance R.E., Kroeff E.P., Hoffman J.A., Frank B.H. (1981) Chemical, physical, and biological properties of biosynthetic human insulin. Diabetes Care 4, 147-154.

Chang K-J. and Cuatrecasas P. (1979) Multiple opiate receptors: enkephalins and morphine bind to receptors of different specificity. J. Biol. Chem. 254, 2610-2618.

Chang K.-J., Hazum E. and Cuatrecasas P. (1980) Multiple opiate receptors. Trends Neurosci. 3, 160-162.

Chang R.S.L, Lotti V.J., Monaghan R.L., Birbaum J., Stapley E.O., Goetz M.A., Albers-Schonberg G., Patchett A.A., Liesch J.M., Hensens O.D. and Springer J.P. (1985) A potent nonpeptide cholecystokinin antagonist selective for peripheral tissues isolated from Aspergillus alliaceus. Science 230, 177-179.

Chien Y.W. and Chang S.F. (1985) In Transnasal Medical Systems (ed. Y.W. Chien) Elsevier, Amsterdam.

Chiu Y.Y.H. and Gueriguian (1982) Radioimmunoassays for determination of proinsulin content in purified insulin crystals. In Hormone Drugs (Ed. by Gueriguian et al.). The United States Pharmacopeial Convention, Rockville, Maryland.

Coletti-Previero M.A., Mattras H., Descomps R. and Previero A. (1981) Human plasma amino-enkephalinase: its further characterization and distribution. Neuropeptides (Edinburgh) 2, 31-36.

Corbett A.D., Gillan M.G.C., Kosterlitz H.W., Paterson S.J., Robson L.E. and McKnight A.T. (1984) Selectivities of opioid peptide analogs as agonists and antagonists at the δ-receptor. Br. J. Pharmacol. 83, 271-279.

DeWied D., Kovacs G.L., Bohus B., Van Ree J.M. and Greven H.M. (1978) Neuroleptic activity of the neuropeptide β-LPH(62-77),[des-Tyr(1)-]-gamma-endorphin. Eur. J. Pharmacol. 49, 427-436.

DeWied D. (1979) Schizophrenia as an inborn error in the degradation of β-endorphin -- a hypothesis. Trends Neurosci. 2, 79-82.

Dutta A.S., Gormley J.J., Hayward C.F. Morley J.S., Shaw J.S., Stacey G.J. and Turnbull M.T. (1977) Enkephalin-like peptides: Structure-activity relationships. Acta. Pharmac. Suec. Suppl. 14, 14-15.

Evans B.E., Bock M.G., Rittle K.E., DiPardo R.M., Whitter W.L., Veber D.F., Anderson P.S. and Freidinger R.M. (1986) Design of potent, orally effective, nonpeptidal antagonists of the peptide hormone cholecystokinin. Proc. Natl. Acad. Sci.(USA) 83, 4918-4922.

Farmer P.S. (1980) Bridging thr gap between bioactive peptides and nonpeptides: Some perspectives in design. Drug Design Vol X, 119-143.

Fournie-Zaluski M-C., Pedrisot R., Gacel G., Swerts J-P., Roques B.P. and Schwartz J.C. (1979) Inhibitory potency of various peptides on enkephalinase activity from mouse striatum. Biochem. Biophys. Res. Commun. 91, 130-135.

Frederickson R.C.A., Smithwick E.L., Shuman R. and Bemis K.G. (1981a) Metkephamid, a systemically active analog of methionin enkephalin with potent opioid δ-receptor activity. Science 211, 603-605.

Frederickson R.C.A., Smithwick E.L., Shuman R. and Gesellchen P.D. (1981b) Analgesic testing of opioid peptides. Psycopharmacol. Bull. 17, 106-108.

Frederickson R.C.A., Parli C.J., DeVane G.W. and Hynes M.D. (1982) Preclinical pharmacology of metkephamid (LY127623), a met-enkephalin analogue. In NIDA Research Monograph. Vol.43 (Ed. by Harris L.S.) pp. 150-156.

Gesellchen P.D., Frederickson R.C.A., Tafur S. and Smiley D. (1981) In Peptides (Ed. by Rich D.H. and Gross E.). Pierce Chemical Co. pp. 621-630.

Goeddel D.V., Kleid D.G., Bolivar F., Heyneker H.L., Yansura D.G., Crea R., Hirose T., Kraszewski A., Itakura K. and Riggs A.D. (1979) Expression in E. coli of chemically synthesized genes for human insulin. Proc. Natl. Acad. Sci.(USA) 76, 106-110.

Goldstein A., Tachibana S., Lowney L.I., Hunkapillar M. and Hood L. (1979) Dynorphine-(1-13), an extraordinary potent opioid peptide. Proc. Natl. Acad. Sci.(USA) 76, 6666-6670.

Von Graffenried B., del Poso E., Roubicek J., Krebs E., Poldinger W., Burmeister P. and Kerp L. (1978) Effects of the synthetic enkephalin analogue FK 33-824 in man. Nature 272, 729-730.

Grey H.M. and Chesnut R. (1985) Antigen processing and presentation to T cells. Immuology Today 6, 101-106.

Gubler U., Seeburg P., Hoffman B.J., Gage L.P. and Udenfriend S. (1982) Molecular cloning establishes proenkephalin as precursor of enkephalin-containing peptides. Nature 295, 206-208.

Guyon A., Roques B.P., Guyon F., Foucault A., Perdrisot R., Swerts J.P. and Schwartz J.C. (1979) Enkephalin degradation in mouse brain studied by a new HPLC method: Further evidence for the involvment of carboxypeptidase. Life Sci. 25, 1605-1611.

Hand,A.R. and Oliver,C., eds. (1981) Basic Mechanism of Cellular Secretion. Academic Press, New York.

Hughes J., Smith T., Morgan B. and Fothergill L. (1975) Purification and properties of enkephalin -- The possible endogenous ligand for the morphine receptor. Life Sci. 16, 1753-1758.

Hughes J., Smith T.W., Kosterlitz H.W., Fothergill L.A., Morgan B.A. and Morris H.R. (1975) Identification of two related pentapeptides from the brain with potent opiate agonist activity. Nature 258, 577-579.

Hughes J., Kosterlitz H.W. and Leslie F.M. (1975) Effect of morphine on adrenergic transmission in the mouse vas deferens: Assessment of agonist and antagonist potencies of narcotic analgesics. Br. J. Pharmacol. 55, 541-546.

Huguenin R. and Maurer R. (1980) Resistance of FK 33-824 and other enkephalin analogues to peptidase degradation. Brain Res. Bull. 5, 47-50.

Humphrey M.J. and Ringrose P.S. (1986) Peptides and related drugs: a review of their absorbtion, metabolism and excretion. Drug Met. Rev. 17, 283-310.

Hynes M.D. and Frederickson R.C.A. (1982) Cross-tolerance studies distinguish morphine- and metkephamid-induced analgesia. Life Sci. 31, 1201-1204.

Johnson,I.S. (1982) Autenticity and purity of human insulin (recombinant DNA) Diabetes Care (suppl.2) 5, 4-12.

Kakidani H., Furutani Y., Takahashi H., Noda M., Morimoto Y., Hirose T., Asai M., Inayama S., Nakanishi S. and Numa S. (1982) Cloning and sequence analysis of cDNA for porcine β-neo-endorphin/dynorphin precursor. Nature 298, 245-249.

Kosterlitz H.W. and Watt A.J. (1968) Kinetic parameters of narcotic agonists and antagonists, with particular reference to N-allylnoroxymorhone (naloxone). Br. J. Pharmacol. 33, 266-267.

Kosterlitz H.W., Lord J.A.H., Paterson S.J. and Waterfield A.A. (1980) Effects of changes in the structure of enkephalins and narcotic analgesic drugs on their interaction with μ- and δ-receptors. Br. J. Pharmacol. 68, 333-342.

Lane A.C., Rance M.J. and Walters D.S. (1977) Subcellular localisation of leucin-enkephalin-hydrolysing activity in rat brain. Nature 269, 75-76.

Loh H.H.,Tseng L.F., Wei E. and Li C.H. (1976) B-Endorphin is a potent analesic agent. Proc. Natl. Acad. Sci.(USA) 73, 2895-2898.

Luckow V.A. and Summers M.D. (1987) Trends in the development of Baculovirus expression vectors. Bio/Technology 6, 47-55.

Maeda S. (1989) Expression of foreign genes in insects using Baculovirus vectors. In Ann. Rev. Entomol. (Ed. by Mittler T.E., Radovsky F.J. and Resh V.H.) 34, 351-372.

McMartin C. and Peters G.E. (1986) In Advanced Drug Delivery Systems for Peptides and Proteins (Ed. by David,S.S., Illum,L. and Tomlinson,E.). Plenum Press, New York.

McMartin C., Hutchinson L.E.F., Hyde R. and Peters G.E. (1987) Analysis of structural requirements for the absorption of drugs and macromolecules from the nasal cavity. J. Pharm. Sci. 76, 535-540.

Marki,F., and Albrecht,W. (1977) Biological activity of synthetic human insulin. Diabetologica 13, 293-95.

Markussen J., Damgaard U., Johansen N.L., Jorgensen K.H., Sorensen E. and Thim L. (1981) Human insulin. Paper presented at the 16th Annual Meeting of the Scandinavian Society for the Study of Diabetes, Linkoping, Sweden, May 21-23.

Mathur K.B., Dhotre B.J., Raghubir R., Patnaik G.K. and Dhawan B.N. (1979) Morphine-like activity of some new Met-enkephalin analogs. Life Sci. 25, 2023-2028.

Menn J.J. and Borkovec A.B. (1989) Insect neuropeptides: Potential new insect control agents. J. Agricult. Food Chem. 37, 271-278.

Miglecz E., Szekely J.I. and Dunai-Kovacs Z. (1979) Comparison of tolerance development and dependence capacities of morphine, β-endorphin and [D-Met^2, Pro^5]-enkephalinamide. Psychopharmacology (Berlin) 62, 29-34.

Morihara K., Oka T. and Tsuzuki H. (1979) Semi-synthesis of human insulin by trypsin-catalysed replacement of Ala-B30 by Thr in porcine insulin. Nature 280, 412-413.

Morihara K., Oka T., Tzuzuki H., Tochino Y. and Kanaya T. (1980) Achromobacter protease I-catalyzed conversion of porcine insulin into human insulin. Biochem. Biophys. Res. Commun. 92, 396-402.

Moritoki H., Kiso Y., Kageyama T. and Matsumoto K. (1981) Morphine-like activities of synthetic enkephalin analogues. J. Pharm. Pharmacol. 33, 54-55.

Morley J.S. (1980) Structure-activity relationships of enkephalin-like peptides. Ann. Rev. Pharmacol. Toxicol. 20, 81-110.

Nakanishi S., Inoue A., Kita T., Nakamura M., Chang ACY, Cohen N. and Numa S. (1979) Nucleotide sequence of cloned cDNA for bovine corticotropin-β-lipotropin precursor. Nature 278, 423-427.

Nishihata T., Kim S., Morishita S., Kamada A., Yata N. and Higuchi T. (1983) Adjuvant effects of glycerylesters of acetoacetic acid on rectal absorption of insulin and inulin in rats. J. Pharm. Sci. 72, 280-285.

Obermeier,R. (1978) In Semisynthetic peptides and proteins (Ed. by Offord,R.E. and DiBello,C.) Academic Press, London, pp. 206-211.

Ondetti M.A., Rubin B. and Cushman D.W. (1977) Design of specific inhibitors of angiotensin-convering enzyme: New class of orally active antihypertensive agents. Science 196, 441-443.

Pardridge W.H. (1983) Neuropeptides and the blood-brain barrier. Ann. Rev. Physiol. 45, 73-82.

Patchett A.A., Harris E., Tristram E.W., Wyvratt M.J., Wu M.T., Taub D., Peterson E.R., Ikeler T.J., Broeke J., Payne L.G., Ondeyka D.L., Thorsett E.D., Greenlee W.J., Lohr N.S., Hoffsommer R.D., Joshua H., Ruyle W.V., Rothrock J.W., Aster S.D., Maycock A.L., Robinson F.M., Hirschmann R., Sweet C.S., Ulm E.H., Gross D.M., Vassil T.C. and Stone C.A. (1980) A new class of angiotensin-converting enzyme inhibitors. Nature 288, 280-283.

Paterson S.J., Robson L.E. and Kosterlitz H.W. (1983) Classification of opioid receptors. Br. Med. Bull. 39, 31-36.

Patzelt C., Labrecque A., Duguid J., Carroll R., Keim P., Heinrikson R. and Steiner D.F. (1978) Detection and kinetic behavior of preproinsulin in pancreatic islets. Proc. Natl. Acad. Sci (USA) 75, 1260-1264.

Pert C.B. and Snyder S.H. (1973) Opiate receptors: Demonstration in nervous tissue. Science 179, 1011-1014.

Pert C.B., Pert A., Chang J-K. and Fong B.T.W. (1976) [D-Ala(2)]-Met-enkephalinamide: A potent long-lasting synthetic pentapeptide analesic. Science 194, 330-332.

Pert C., Bowie D.K., Fong B.T.W. and Chang J.-K. (1976) In Opiates and Endogenous Opioid Peptides (Ed. by Kosterlitz H.W.). North-Holland, Amsterdam.

Raina A.K., Jaffe H., Kempe T.G., Keim P., Blacher R.W., Fales H.M., Rily C.T., Klun J.A., Ridgway R.L. and Hayes D.K. (1989) Identification of a neuropeptide hormone that regulates sex pheromone production in female moths. Science 244, 796-798.

Roemer D., Buescher H.H., Hill R.C., Pless J., Bauer W., Cardinaux F., Closse A., Hauser D. and Huguenin R. (1977) A synthetic enkephalin analogue with prolonged parenteral and oral analgesic activity. Nature 268, 547-549.

Roemer D. and Pless J. (1979) Structure-activity relationship on orally active enkephalin analogues as analgesics. Life Sci. 24, 621-624.

Roques B.P., Fournie-Zaluski M.C., Soroca E., Lecomte J.M., Malfroy B., Llorens C. and Schwartz J.C. (1980) The enkephalinase inhibitor thiorphan shows antinociceptive activity in mice. Nature 288, 286-288.

Ryle A.P., Sanger F., Smith L.F., Kitai R. (1955) The disulfide bonds of insulin. Biochem. J. 60, 541-556.

Schultze H.F. and Heremans J.F. (1966) Molecular Biology of Human Proteins,with special reference to plasma proteins.Vol 1. Nature and metabolism of extracellar proteins. Elsevier, Amsterdam.

Simantov R. and Snyder S.H. (1976) Morphine-like peptides in mammalian brain: Isolation, structure elucidation, and interaction with opiate receptors. Proc. Natl. Acad. Sci.(USA) 73, 2515-2519.

Smyth D.G. and Zakarian S. (1980) Selective processing of β-endorphin in regions of porcine pituitary. Nature 288, 613-615.

Snyder S.H. (1980) Brain peptides as neurotransmitters. Science 209, 976-983.

Stacher G., Bauer P., Steinringer H., Schreiber E. and Schmierer G. (1979) Effects of the synthetic enkephalin analogue FK 33-824 on pain treshold and pain-tolerance in man. Pain 7, 159-172.

Steiner D.F., Kemmler W., Tager H.S. and Peterson J.D. (1974) Proteolytic processing in the biosynthesis of insulin and other proteins. Fed. Proc. 33, 2105-2115.

Szekely J.I., Ronai A.Z., Dunai-Kovacs Z., Miglecz E., Berzetei I., Bajusz S. and Graf C. (1977) [D-Met^2, Pro^5]-enkaphalinamide: A potent morphine-like analgesic. Eur. J. Pharmacol. 43, 293-294.

Szekely J.I., Dunai-Kovacs Z., Miglecz E., Ronai A.Z. and Bajusz S. (1978) In vivo antagonism by naloxone of morphine, β-endorphin and a synthetic enkephalin analog. J. Pharm. Exp. Ther. 207, 878-883.

Traficante L.J., Rotrosen J., Siekierski J., Tracer H. and Gershon S. (1980) Enkephalin inactivation by N-terminal Tyrosine cleavage: Purification and partial characterization of a highly specific enzyme from human brain. Life Sci. 26, 1697-1706.

Tseng L-F., Loh H.H. and Li C.H. (1976) B-Endorphin as a potent analgesic by intravenous injection. Nature 263, 239-240.

Veber D.F. and Freidinger R.M. (1985) The design of metabolically-stable peptide analogs. Trends Neurosci. 8, 392-396.

Villa-Komaroff L., Efstratiadis A., Broome S., Lomedico T., Tizard R., Naber S.P., Chick W.L. and Gilbert W. (1978) A bacterial clone synthesizing proinsulin. Proc. Natl. Acad. Sci.(USA) 75, 3727-3731.

Wallis,M., Howell,S.C, Taylor,K.M. (1985) Biochemistry of the polypeptide hormones. Wily-Interscience, London.

Weber E., Esch F.S., Bohlen P., Paterson S., Corbett A.D., McKnight A.T., Kosterlitz H.W., Barchas J.D. and Evans C.J. (1983) Metorphamide: Isolation, structure, and biologic activity of an amidated opioid octapeptide from bovine brain. Proc. Natl. Acad. Sci.(USA) 80, 7362-7366.

Wei E.T. (1981) Enkephalin analogs and physical dependence. J. Pharmacol. Exp. Ther. 216, 12-18.

Wetzel R., Kleid D.G., Crea R., Heyneker H.L., Yansura D.G., Hirose T., Kraszewski A., Riggs A.D., Itakura K., Goeddel D.V. (1981) Expression in Escherichia coli of a chemically synthesized gene for a "mini-C" analog of human proinsulin. Gene 16, 63-71.

From *Insect Neurochemistry and Neurophysiology • 1989* Edited by A. B. Borkovec and E. P. Masler

INSECT NEUROENDOCRINOLOGY: ITS PAST; ITS PRESENT; FUTURE OPPORTUNITIES

Larry L. Keeley, T.K. Hayes and J.Y. Bradfield
Laboratories for Invertebrate Neuroendocrine Research
Department of Entomology
Texas A&M University
College Station, TX 77843

INSECT NEUROENDOCRINOLOGY: ITS PAST

Insects played a major historical role as experimental animals to provide some of the first evidence concerning endocrine functions by the nervous system. Kopec (1917; 1922) extirpated brains from larvae of the gypsy moth, *Lymantria dispar* and demonstrated that debrained larvae were unable to pupate. This finding provided early evidence that the brain of animals secreted factors that promoted physiological events and, hence, acted in an endocrine manner. Despite these early findings, little further evidence was provided for neurosecretion until the definitive pioneering work of Ernst and Berta Scharrer. With the animal kingdom divided between them (Ernst studying the vertebrates; Berta studying the invertebrates), Ernst first reported that nerve cells had secretory activity in the fish *Phoxinus laevis* (Scharrer, 1928). The first description of neuroendocrine activity in invertebrates was reported by Berta for neural cells in the X-organ of *Aplysia* (Scharrer, 1935). This was followed in 1937, by a report identifying secretory cells in the nervous systems of a variety of invertebrates including insects (Scharrer, 1937). Finally, in 1941, a detailed description of insect neurosecretory cells was presented for the corpora cardiaca of the cockroach, *Leucophaea maderae* (Scharrer, 1941). Eventually this husband-wife team described the structural and functional similarities between the brain-corpora cardiaca-corpora allata complex of insects and the hypothalamic-neurohypophysial complex of vertebrates (Scharrer and Scharrer, 1944). Hence, insects played a key role in the birth of neuroendocrinology by serving as models in which to identify and define the phenomenon of neurosecretion, and insects remain useful as model animals for this purpose today.

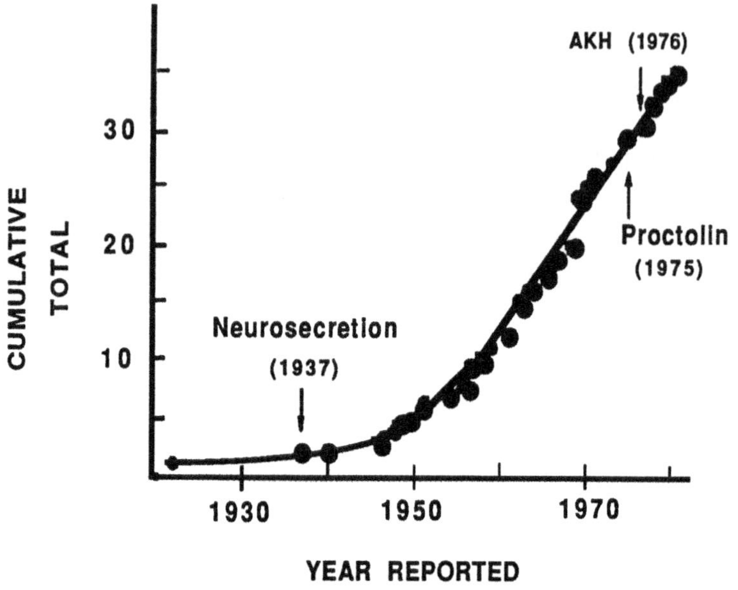

Figure 1. Year of discovery of neurohormone-dependent physiological processes in insects

Insect neuroendocrinology progressed slowly through the 1940s and 1950s. Most of the early research consisted of detailed descriptions of neurosecretory cells with occasional identification of neuroendocrine-dependent physiological processes. Early studies on neuroendocrine regulated physiological processes identified the role of the brain in the termination of diapause (Williams, 1946; 1947) and the initiation of molting (Williams, 1947; 1952) and showed nervous regulation of the corpora allata suggestive of possible neurosecretory relationships (Engelmann, 1957; 1959). Also, neuroendocrine involvement was reported for heartbeat rate (Gersch, 1958), diapause (Hasegawa, 1957) and color change (DuPont-Raabe, 1951).

The discovery of neurohormone-dependent physiological processes began to proliferate in the 1960s and 1970s (Fig.1). Today we recognize over two dozen physiological processes that are affected by neurohormones. Neuroendocrine regulated processes of insects fall into three broad classes: growth (molting and metamorphosis), reproduction and physiological homeostasis (Table 1).

Table 1. Neuroendocrine-dependent physiological processes of insects.

Hormone	Process
Growth and Development	
Prothoracicotropic Hormone	Ecdysteroid synthesis
Allatotropin	Juvenile hormone synthesis
Allatostatin	Inhibition of juvenile hormone synthesis.
Bursicon	Cuticular tanning
Eclosion Hormone	Onset of eclosion behavior
Diapause Hormone	Onset of dormancy
Pupariation Hormones	anterior-retraction; puparium formation; tanning
Reproduction	
Egg Development Neurohormone	Ovarian ecdysteroid synthesis
Allatotropin	Juvenile hormone synthesis
Allatostatin	Inhibition of juvenile hormone
Ovulation Hormone	Oviposition
Antigonadotropin	Suppress ovarian maturation
Pheromone synthesis stimulating hormone	Stimulate pheromone synthesis
Homeostasis	
Adipokinetic Hormone	Mobilize fat body lipid stores
Hyperglycemic Hormone	Mobilize fat body carbohydrate stores
Hypoglycemic Hormone	Enhance carbohydrate uptake
Diuretic Hormone	Stimulate water excretion
Antidiuretic Hormone	Stimulate water reabsorption
Protein Synthesis Stimulating Hormone	Simulate protein synthesis
Proctolin	Increase hindgut contraction
Cuticle Plasticity Factor	Increase cuticle plasticity
Chloride-transport Factor	Stimulates rectal Cl^- transport
Cardioacceleratory Hormone	Stimulates heartbeat rate

By the mid-1970s, most of the presently recognized neuroendocrine-regulated processes of insects were identified. Nevertheless, insect neuropeptide hormones were not isolated or structurally defined before 1975, despite the fact that peptide neurohormones from vertebrate animals had been isolated and their sequences determined nearly two decades earlier (Pierce and du Vigneaud, 1950; du Vigneaud *et al.*, 1953; Popenoe and du Vigneaud, 1954; Sanger, 1959). In 1975, the primary structure of proctolin was published (Starratt and Brown, 1975), and the following year, the structure of the adipokinetic hormone (AKH) for *Locusta migratoria* and *Schistocerca gregaria* was reported (Stone *et al.*, 1976). No further sequences for insect peptide hormones were determined until 1984 when three laboratories published nearly simultaneously the sequence of a cardioacceleratory hormone (CAH-I) from *Periplaneta americana* (Baumann and Penzlin, 1984; Witten *et al.*, 1984; Scarborough *et al.*, 1984). A second cardioacceleratory hormone (CAH-II) was also reported for *Periplaneta* at the same time (Scarborough *et al.*, 1984; Witten *et al.*, 1984). Both CAH-I and -II stimulated hyperglycemia in *Periplaneta* (Scarborough *et al.*, 1984), and both hormones were naturally occurring analogs (bioanalogs) of the locust AKH.

INSECT NEUROENDOCRINOLOGY: ITS PRESENT

Identified Neurohormones

Since the work in the early 1980s, there have been a proliferation of neurohormone structures especially within the AKH family. Two octapeptide bioanalogs of locust AKH-I were found in *Schistocerca nitans*, *S. gregaria* and *L. migratoria* (Gade *et al.* 1986). The two AKH-IIs differed between the two genera by one amino acid. Another AKH bioanalog was isolated and characterized that acted as a hypertrehalosemic hormone (HrTH) in the cockroaches, *Blaberus discoidalis* (Hayes *et al.*, 1986) and *Nauphoeta cinerea* (Gade and Rinehart, 1986), and an AKH from *Manduca sexta* (Ziegler *et al.*, 1985) was structurally identical with an AKH from *Heliothis zea* (Jaffe *et al.*, 1986). A second AKH was found also in *H. zea* (Jaffe *et al.*, 1988a). In addition to these structures, there are also AKH bioanalogs from *Romalea microptera* (Gade *et al.*, 1988), *Gryllus bimaculatus* (Gade and Rinehart, 1987) and tabanid species (Jaffe *et al.*, 1988b). Presently, there are twelve AKH bioanalogs from 10 species of insects and, in addition, the red pigment-concentrating hormone of crustaceans (Fernlund and Josefsson, 1972) (Table 2).

There are frequently more than one AKH bioanalog per species and some species share the same peptide. What physiological role two forms of the same hormone have in the same species is unclear, but they may serve independent functions (Schaffer, 1986; Loughton and Orchard, 1981). For example, AKH-I was more effective than AKH-II at stimulating metabolic events related to lipid mobilization in locusts; whereas AKH-II was more effective than AKH-I at increasing cAMP levels in the fat body (Goldsworthy *et al.*, 1986).

Table 2. Adipokinetic/hyperglycemic family of neuropeptides.

AKH 1	*Locusta* *Schistocerca*	pGlu-Leu-Asn-Phe-Thr-Pro-Asn-Trp-Gly-ThrNH$_2$
AKH II	*Locusta*	pGlu-Leu-Asn-Phe-Ser-Ala-Gly-TrpNH$_2$
AKH II	*Schistocerca*	pGlu-Leu-Asn-Phe-Ser-Thr-Gly-TrpNH$_2$
AKH I	*Manduca* *Heliothis*	pGlu-Leu-Thr-Phe-Thr-Ser-Ser-Trp-GlyNH$_2$
AKH II	*Heliothis*	pGlu-Leu-Thr-Phe-Ser-Ser-Gly-Trp-Gly-AsnNH$_2$
AKH	*Acheta*	pGlu-Val-Asn-Phe-Ser-Thr-Gly-TrpNH$_2$
CAH I	*Periplaneta*	pGlu-Val-Asn-Phe-Ser-Pro-Asn-TrpNH$_2$
CAH II	*Periplaneta*	pGlu-Leu-Thr-Phe-Thr-Pro-Asn-TrpNH$_2$
DCC I	Tabanidae	pGlu-Leu-Thr-Phe-Thr-Pro-Gly-TrpNH$_2$
DCC II	Tabanidae	pGlu-Leu-Thr-Phe-Thr-Pro-Gly-Trp-Gly-TyrNH$_2$
HrTH	*Blaberus* *Nauphoeta*	pGlu-Val-Asn-Phe-Ser-Pro-Gly-Trp-Gly-ThrNH$_2$
HrTH	*Carausius*	pGlu-Leu-Thr-Phe-Thr-Pro-Asn-Trp-Gly-ThrNH$_2$
RPCH	*Pandalus*	pGlu-Leu-Asn-Phe-Ser-Pro-Gly-TrpNH$_2$

During the past five years another major group of insect neurohormones has been defined. A series of myokinins are described for *L. maderae* cockroaches (see Review: Holman *et al.*, 1988). There are now eight octapeptide leucokinins, two sulfur-containing leucosulfakinins, and two additional, related peptides (leucomyosuppressin and leucopyrokinin) (Table 3). All of these peptides are stimulatory to the cockroach hindgut except for leucomyosuppressin which is inhibitory (Holman *et al.*, 1986).

In addition to the *L. maderae* myokinins and proctolin, there is evidence that the molluscan cardioacceleratory peptide FMRF-amide and FMRF-amide related peptides exist in insects (Nambu et al., 1988; Schneider and Taghert, 1988). Finally, a FLRF-amide related peptide with cardioinhibitory activity and high sequence identity with leucomyosuppressin is isolated from *S. gregaria* (Robb et al., 1989)

Presently, we know the most about two groups of insect neuropeptides – the AKH/HrGH family which is widely recognized among many insect species, and the myokinin family which is only described for one species, so far. The physiological actions of the AKH/HrGH family are well documented, but the myokinin family is not well understood except for their ability to affect hindgut contractility.

Table 3. Myokinin family of neuropeptides from *Leucophaea maderae*.

LK1	Asp-Pro-Ala-Phe-Asn-Ser-Trp-GlyNH_2
LK2	Asp-Pro-Gly-Phe-Ser-Ser-Trp-GlyNH_2
LK3	Asp-Gln-Gly-Phe-Asn-Ser-Trp-GlyNH_2
LK4	Asp-Ala-Ser-Phe-His-Ser-Trp-GlyNH_2
LK5	Gly-Ser-Gly-Phe-Ser-Ser-Trp-GlyNH_2
LK6	pGlu-Ser-Ser-Phe-His-Ser-Trp-GlyNH_2
LK7	Asp-Pro-Ala-Phe-Ser-Ser-Trp-GlyNH_2
LK8	Gly-Ala-Asp-Phe-Tyr-Ser-Trp-GlyNH_2
LSK1	Glu-Gln-Phe-Glu-Asp-Tyr(SO_3H)-Gly-His-Met-Arg-PheNH_2
LSK2	pGlu-Ser-Asp-Asp-Tyr(SO_3H)-Gly-His-Met-Arg-PheNH_2
LPK	pGlu-Thr-Ser-Phe-Thr-Pro-Arg-LeuNH_2
LMS	PGlu-Asp-Val-Asp-His-Val-Phe-Leu-Arg-PheNH_2

LK = leucokinin; LSK = leucosulfakinin; LPK = leucopyrokinin; LMS = leucomyosuppressin.

Both families of hormones most likely have broader physiological actions than the single effect used as the basis for the bioassay. For example, the AKHs mobilize fat body lipid stores in locusts to elevate hemolymph lipids (Mayer and Candy, 1969); whereas the HrGHs mobilize fat body glycogen in cockroaches to elevate hemolymph trehalose (Steele, 1963; Bowers and Friedman, 1963). In Colorado potato beetles, proline is a major source of flight energy, and locust AKH stimulates fat body proline synthesis and elevates hemolymph glucose (Weeda, 1981); hemolymph proline is not elevated. In all cases these hormones appear to affect the synthesis of whatever metabolite serves as the primary energy source for the flight muscles or other peripheral tissues.

In addition to stimulating the synthesis of energy substrates, the AKH/HrGH peptides also affect other aspects of energy-related metabolism. For example, *Blaberus* HrTH stimulates the synthesis of hemes *aa3* and *b* that we recognized some years ago as a synthetic event during fat body mitochondriogenesis in young adult male *B. discoidalis* (Keeley, 1978). The mitochondriogenesis occurs during a period of fat body cellular reorganization that occurs during the first week of adult life in *B. discoidalis* and prepares the adipocytes for adult biosynthetic processes (Keeley, 1981). Active heme synthesis precedes and accompanies increases in mitochondrial cytochromes *aa3*, electron transport and respiratory activities (Keeley, 1972; 1977; 1978;). It now appears that HrTH may be the putative cytochromogenic hormone (Hayes and Keeley, 1981).

In addition to stimulating the respiratory capacity of the fat body through effects on heme synthesis, we also find that *Blaberus* HrTH stimulates juvenile hormone (JH)-dependent protein synthesis (Keeley *et al.*, 1988). The effect is not specific on the synthesis of any particular polypeptide in the export protein fraction but affects a general increase in the biosynthetic capacity of the fat body after stimulation of specific protein synthesis by JH-3. It is speculated that the HrTH-dependent cytochrome synthesis may elevate the overall bioenergetic potential of the fat body through increases in respiratory capacity and ATP synthesis.

Finally, we find that HrTH also stimulates heartbeat rate in *B. discoidalis*. This agrees with the cardioacceleratory effects of *Periplaneta* CAH-I and -II, which was the basis for the bioassay used to isolate these two peptides (Scarborough *et al.*, 1984).

The conclusions from this multiplicity of actions by the AKH/HrGH peptides is that they have broad regulatory effects on energy metabolism in insects. They stimulate the synthesis of the metabolites used by peripheral tissues for energy production. In addition, they affect other processes of fat body energy metabolism such as heme synthesis for mitochondrial maturation which has the effect of determining the basal metabolic rate of the fat body and setting the metabolic "tone" for the tissue. Changes in fat body basal metabolism could have wide-ranging effects on the physiological integrity of the entire insect since the fat body is the source for circulating lipids, carbohydrates and proteins, including the storage and vitellogenic proteins. Factors that influence fat body exergonic metabolism influence all other biosynthetic processes of the tissue and subsequently affect the total physiological balance of the insect. Therefore, the AKH/HrGH family of hormones is highly significant to the metabolic integrity of insects which may explain its apparent ubiquity among the insect species.

It is uncertain what effects the multiplicity of leucokinins have in *L. maderae*. One can only speculate that they may exert subtle differences in their actions on gut contractility, if, indeed, that is their primary physiological action. It can be speculated that some of these peptides may promote contractions of particular regions of the gut, while others may affect specific digestive enzyme secretion, and finally, some may control diuresis/antidiuresis. The latter effect has been confirmed recently for the leucokinins in a heterologous system using Malpighian tubules from *Aedes aegypti* mosquitoes (Hayes *et al.*, 1989). Leucokinins 2 through 8 depolarized the Malpighian tubules membranes, and LK-8 inhibited fluid secretion at low concentrations but stimulated secretion at high concentrations.

Other insect neurohormones are also isolated that are well known from their physiological actions. Two types of prothoracicotropic hormones are found in *Bombyx mori* silkworms. The larger PTTH (22K) promotes molting in debrained *B. mori* pupae; the smaller PTTH (4K) has no effect on *B. mori* pupae but promotes molting by debrained *Samia cynthia ricini* pupae (Suzuki *et al.*, 1986). The 4K PTTH has been isolated and its structure determined (Nagasawa *et al.*, 1986). The 4 K-PTTH consists of two subunits linked together by disulfide crosslinks reminiscent of vertebrate insulin. The A subunit has a structure that is 50% identical with human insulin A-chain; the B subunit is 38% identical with human insulin B-chain. The cysteine residues are distributed identically with insulin supporting the argument that the 4K PTTH is an insect representative of the insulin family of peptides. Furthermore, the PTTH B-chain exists in four different sequences (Nagasawa *et al.*, 1986) so that, like the AKH/HrGH and myokinin families, there appear to be multiple bioanalogs of the same peptide within the same species. The functional significance for multiple bioanalogs of PTTH in *B. mori* is uncertain at this time.

Manduca PTTH also exists in two forms (Bollenbacher *et al.*, 1984). The big (29 kDa) PTTH appears to be released in a single burst and stimulates a maximal production of ecdysteroids related to the onset of molting; whereas, the small (7 kDa) PTTH results in a subtle increase in hemolymph ecdysteroids that elicits the change in behavior from feeding to wandering at the time of epidermal cell recommitment and onset of pupation (Bollenbacher *et al.*, 1984). Whether the 22K and 4K PTTHs have comparable actions in *B. mori* is uncertain. Although the 4K PTTH does not promote development in debrained *B. mori* pupae, this does not necessarily eliminate the possibility of its having the more subtle behavioral effect in last instar larvae, yet, at the same time, resemble the active form of PTTH for the initiation of adult development in *S. ricini* .

Melanization and reddish coloration hormone (MRCH) of *B. mori* also exists as three bioanalogs (Matsumoto et al., 1985). MRCH bioanalogs show four regions of residue identity at the C-terminus with human insulin-like growth factor II.

Finally, based upon their physiological actions, there are other classes of insect neurohormones with representatives that are structurally defined. These are the eclosion hormones, diuretic hormones, allatotropins and pheromone synthesis-regulating neurohormones. Eclosion hormones have been isolated from *B. mori* (Kono *et al.*, 1987) and *M. sexta* (Kataoka *et al.*, 1987; Marti *et al.*, 1987). The *Manduca* and *Bombyx* hormones differ in their sequences, and four bioanalogs are described for *Bombyx* eclosion hormone. In addition, there are now two sequences for diuretic hormones: one from *M. sexta* (Kataoka *et al.*, 1989a) and one from *L. migratoria* (Proux *et al.*, 1987). The *Locusta* diuretic hormone has immunological cross-reactivity with antibodies to vertebrate Arg-vasopressin. A structural sequence is described for an allatotropin from adult *M. sexta* that has no sequence similarity with any other known peptides (Kataoka *et al.*, 1989b). The allatotropin is a small peptide (13 residues) and is blocked at the C-terminus. Finally, a pheromone-biosynthesis activation hormone has been isolated from *H. zea* (Raina *et al.*, 1989). This hormone is active in six other moth species as well as *H. zea*, and has similarity with the N-terminal sequence of the MRCH of *B. mori*.

Neurohormone Genes and Prohormones

It is important to elucidate neurohormone gene structures and hormonal precursors encoded by these genes to understand how hormone synthesis proceeds and is regulated in insects. Understanding of neurohormone genes and their resulting prohormones is at its inception in insects when compared to vertebrate neuroendocrinology.

Several insect neurohormone genes have been isolated and described during the past year. A gene was isolated from a *Drosophila* genomic library using a cDNA to *Aplysia* FMRFamide and synthetic oligonucleotides and the resulting gene used to screen a *Drosophila* head cDNA library to obtain a cDNA for 13 peptides related to FMRFamide (Schneider and Taghert, 1988). In a comparable study, a major FMRFamide-related peptide was isolated from whole adult *Drosophila* and identified as DPKQDFMRFamide (Nambu *et al.*, 1988). Both groups isolated a sequence that mapped to 46C on the right arm of chromosome 2. The resulting DNA sequence (1342 bp, Nambu *et al.*, 1988; 1352 bp, Schneider and Taghert, 1988) was used to deduce a prohormone precursor. Within the precursor, fifteen potential amidated peptides were identified of which ten putative peptides contained the FMRFamide sequence and five included authentic DPKQDFMRFamide. A sequence with partial similarity to corticotropin-releasing factor was also noted (Schneider and Taghert, 1988). The physiological role for FMRFamide-related peptides is uncertain in *Drosophila* and for insects in general.

A gene for drosulfakinin, a CCK-like peptide of *Drosophila*, related to leucosulfakinin of *L. maderae*, was isolated using a mixture of oligonucleotide probes based on conserved amino acid sequences of the

CCK-gastrin family (Nichols, et al, 1988). Three putative peptides were identified within a deduced prohormone sequence that were flanked by prohormone processing sites and contained C-terminal glycyl residues as potential amidation sites. The leucosulfakinins and drosulfakinin-I and -II share 7 identical C-terminal amino acids. In addition, both the leucosulfakinins and two of the drosulfakinins share five of these seven residues with human gastrin II and rat CCK, including sulfated tyrosine. A drosulfakinin-0 was unique. Hybridization studies showed the drosulfakinin gene localized near 81 F of the third chromosome.

Recently, we undertook to isolate the gene for the *Manduca* AKH. This peptide was selected since it represents a member of one of the best known families of insect neuropeptides. It is also the AKH of *M. sexta*, a major experimental species for insect endocrinology studies, and *M. sexta*, itself, represents the Lepidoptera, a large and economically important group of insects. Lastly, this identical peptide is present in both *M. sexta* and *Heliothis zea*, and we have research interests in the endocrinology of *Heliothis*.

We used a genomic library for isolating the AKH gene rather than using cDNA libraries of the head or neurohemal tissues. Although *Manduca* AKH is found in the corpora cardiaca (Ziegler and Gade, 1984), we were unsure of the exact timing of its synthesis and whether ample amounts of AKH-mRNA would be present in the glands at all times. Furthermore, we were not certain that the AKH-mRNA would even be present in the corpora cardiaca since the hormone could be synthesized by brain neurosecretory cells and transported to the glands for storage and secretion. Despite the fact that the locust AKH was synthesized by intrinsic neurosecretory cells of the corpora cardiaca (Hekimi and O'Shea, 1989), we were uncertain of the situation in *M. sexta* at the time we initiated these studies. Since then, the synthesis of AKH has been indicated for intrinsic neurosecretory cells of the corpora cardiaca in *M. sexta* (Ziegler et al., 1988). Nevertheless, for these reasons, we approached the problem from the standpoint of probing a genomic library where the specific tissue of origin and timing of expression were irrelevant.

There are 55,296 possible AKH-precursor mRNA sequences for *Manduca* AKH, and an oligonucleotide pool incorporating all possible codon degeneracies would be useless as a hybridization probe. Since hybridization with pools of oligonucleotides containing more than a few hundred sequences usually generates too many false positive signals, perfect complementarity to the sought-after gene sequence must be compromised in the design of useful probes. For these reasons,we decided to use a new, but possibly risky, approach to restrict the number of probes.

For AKH probe design, *pyro*-Glu is derived from Gln (Smyth *et al.*, 1963) and the amide at the C-terminus is derived from an adjacent Gly (Bradbury *et al.*, 1982). To compensate for four-fold degeneracies, we inserted deoxyinosine (Ohtsuka *et al.*, 1985). When two-fold degeneracy occurred, we inserted both possible bases, and we assumed UC were the first- and second codon positions for serine based on eukaryotic usage frequency (Aota *et al.*, 1988) (Fig. 2).

AKH: <Glu–Leu–Thr–Phe–Thr–Ser–Ser–Trp–Gly–NH_2

Inferred Structure: N–Gln–Leu–Thr–Phe–Thr–Ser–Ser–Trp–Gly–Gly–C

Met– Tyr–Lys–Leu–Thr–Val–Phe–Leu–Met–Phe–Ile
Ala
Phe– Thr–Leu–Gln–Ala–Glu–Ala–Ile–Ile–Val– Phe
Thr
Ser–Ser–Trp–Gly–Gly–Lys–Arg–Ala–Met–Thr–Asn
Ser
Ala–Ala–Ile–Ala–Glu–Asp–Asn–Arg–Cys–Ser–Ile
Ile
Tyr–Lys–Ala–Ile–Gln–Asn–Glu–Ala–Glu–Arg–Phe
Ile
Asn–Lys–Gln–Cys–Met

Figure. 2. Proposed prepro-adipokinetic hormone sequence from *Manduca sexta*.

Using these AKH oligonucleotide probes, we isolated the gene for the *Manduca* AKH (Bradfield and Keeley, 1989). Four signals were identified from 250,000 recombinant plaques of the *M. sexta* genome and three of these were eliminated as false positives. From the remaining plaque, we determined a base sequence that contained the coding sequence for the *Manduca* AKH (Fig. 2), and further sequencing suggested that prepro-AKH was encoded by a short, intronless gene.

We deduced the preproAKH structure from the nucleotide sequence (Fig. 2). An open reading frame for 65 amino acids occurred beginning with the first ATG after a TATA homology. The first series of 19 amino acids represented a signal sequence and consisted of a core of twelve straight hydrophobic amino acids (nos. 6-17) bordered by Ala- at 17 and 19. Residues 20-29 (Gln to Gly) represent the AKH peptide followed by a classic -Lys-Arg- cleavage site. The putative cleavage site is followed by a C-terminal peptide of 34 amino acids. This C-terminal tail is unrelated to over 7000 sequences listed with the National

Biomedical Research Foundation protein sequence database. About 80 nucleotides downstream from the open reading frame is a polyadenylation signal, and twenty nucleotides further is a GT-rich region related to correct polyadenylation. *Manduca* preproAKH appears to contain only one copy of the AKH and no other identifiable hormone sequences.

Few studies have examined prohormone processing in insects. Prohormones are reported for the locust AKH-I and -II. These prohormones were identified by incorporation of [^{3}H]tryptophan into corpora cardiaca *in vitro* followed by isolation of radiolabeled proteins (Hekimi and O'Shea, 1987). Two 8 kDa proteins (P1, P2) were identified as prohormones based upon their incorporation of radiolabeled tryptophan before its incorporation into AKH-I or AKH-II, and the finding that label was transferred from P1 and P2 to AKH-I. P1 and P2 are converted to AKH-I and to two other 6.5 kDa peptides (AKH-Precursor Related Peptides, APRP-1 and APRP-2) of unknown function (Hekimi and O'Shea, 1989). APRP-1 and -2 are co-secreted along with the AKHs. The significance of the APRPs is not known, but the authors speculate that they may be neurohormones of unknown function (Hekimi and O'Shea, 1989). cDNA cloning studies with *S. nitans* have shown that the two preproAKHs of this locust (Schaffer, personal communication) are similar in structure and organization to the *Manduca* preproAKH that we have described here. Therefore, it is likely that the APRPs represent the C-terminal "tail" from the downstream reading frame following the AKH sequence. As suggested by Hekimi and O'Shea (1989) this C-terminal peptide may contain unknown hormones, or it may simply represent a peptide residue cleaved from the hormone during processing and released during secretion.

Hormone Processing and Degradative Enzymes

Discussion of hormone processing and degradative enzymes was combined because these two events represent extremes of controlled proteolytic activity for the regulation of neurohormone levels, and in both cases, little information exists concerning the nature and action of the enzymes. Considering that it is only during the past year that there has been any information concerning prohormone sequences for insect neuropeptides, it is no wonder that there is little knowledge about processing enzymes for insect neurohormones.

One can predict some of the processing enzymes that might be anticipated. Surely, amidation enzymes will be involved because of the frequent blockage of insect neuropeptides at the C-terminus by a glycine-derived amide group. Trypsin-like cleavage enzymes are predicted from the presence of classic cleavage sites consisting of paired basic amino acids (lysine and/or arginine) at the N- and C-termini of hormone sequences within the prohormones. However, other amino acids and amino acid combinations can also act as cleavage sites (Steiner *et al.*, 1983; Fisher and Scheller, 1988) and some of these cleavage sites may not yet be identified. For example, additional regulatory peptides may reside within the "tails" of the proAKHs that are presently unrecognized because of the absence of known cleavage sites.

Some initial investigations are reported concerning the degradation of insect neurohormones, but in-depth investigations of this topic suffer from the difficulty and cost of preparing neuropeptides as substrates that are labeled in the correct positions to demonstrate specific degradative steps. An aminopeptidase is proposed as the most likely degradation pathway for proctolin (Starratt and Steele, 1984; Steele and Starratt, 1985; Isaac, 1987). However, measurement of [tyrosyl-3,5-^{3}H]proctolin degradation in *P. americana* demonstrated that for all tissues examined Tyr-Leu cleavage exceeds aminopeptidase activity (Quistad *et al.*, 1984). Furthermore, the ratio of Tyr-Leu cleavage activity to aminopeptidase activity for a given tissue is proportional to the tissue's natural level of proctolin. Proctolin-rich tissues such as the proctodeum and terminal ganglion have ratios of 13 and 4, respectively; whereas, tissues such as brain and midgut, which have a low proctolin content, show ratios of approximately 2. *Periplaneta* CAH-I is reportedly inactivated by an endogenous metalloendopeptidase of Malpighian tubules (Baumann and Penzlin, 1987). This metalloendopeptidase pathway is comparable to the vertebrate pathway for neuropeptide degradation. *Periplaneta* CAH-II had a half life of 1 hr in vivo and in homogenates of fat body or Malpighian tubules but persisted much longer in hemolymph (Skinner *et al.*, 1987). In *S. gregaria*, AKH was degraded by nerve membrane preparations to pGlu-Leu-Asn and Phe-Thr-Pro-Asn-Trp-Gly-Thr-NH_2, again indicative of endopeptidase activity (Isaac, 1988). It is to be expected that endopeptidases are responsible for inactivation of the AKH/HrGH family members since these peptides are uniformly blocked at both the N- and C-termini. However, both pGlu-aminopeptidase and/or a deamidase could also inactivate the AKH/HrGHs, since we find that with *Blaberus* HrTH, pGlu- removal results in no activity and deamidation results in partial inactivation (Ford et al., 1988). Deamidase and pGlu aminopeptidase both inactivate TRH in vertebrate animals (Redding and Schally, 1969; Taylor and Dixon, 1978).

Structure-Activity Studies

The information contained in the structure of a peptide is translated into biological action through reception at the surface membrane of the target cell. There are two steps involved in the interaction of a neurohormone at the target cell receptor: 1) binding (recognition), and 2) an induced conformational change by the receptor that stimulates the cellular response (transduction). Information about the structural elements of the neurohormone responsible for these two events can be inferred from structure-activity studies where full dose-response curves have been tested. The potency (ED_{50}) of an analog relative to the natural hormone provides information about structural changes in that portion of the hormone responsible for initial receptor binding (the "address"). Changes in maximum activity by an analog in a dose-response curve provides information about structural changes in that portion of the neurohormone that induces transduction (the "message"). Address and message can exist on separate or overlapping portions of the hormone sequence.

Structure-activity studies of insect neuropeptides have focused on the AKH and myotropin peptide families. Sequences for members of both families have been available for several years and all family members are small enough for synthesis of the many analogs that are necessary for comprehensive structure-activity studies. These investigations have focused on determining what sites in a peptide sequence are required for biological activity and where radioactive labels can be incorporated into the structure to maintain biological activity.

Initial studies on the AKH family did not publish full dose-response information. However, some information was obtained on the importance of portions of *Locusta* AKH-I for lipid mobilization. The N-terminal octapeptide contains the critical core residues for activity and the C-terminal dipeptide strengthens the biological response (Stone et. al., 1978). Incorporation of dehydroproline at position six sharply reduces the activity of the hormone (Hardy and Sheppard, 1983). This observation suggests that a ß-turn is important for AKH activity.

Full dose-response studies of the relative activity of bioanalogs of the AKH family in different bioassays have illustrated species specificity. For example, we find that *Manduca* AKH is completely inactive in the *Blaberus* hyperglycemic bioassay (Table 4). The *Manduca* AKH does not block the action of *Blaberus* HrTH in *B. discoidalis*, even at 100 times the concentration of HrTH. Therefore, *Manduca* AKH does not interact with the *Blaberus* fat body. The closer the structural identity of any exogenous bioanalog to the naturally occurring hormone of the species (*Blaberus* HrTH) the more potent the analog. *Periplaneta* and *Locusta* exhibit a narrow range of potencies for the different AKH bioanalogs tested (Gade 1986; Goldsworthy et. al., 1986) . A reduction in the maximum activity of *Locusta* AKH-I suggests that position 5 is important for "message" in *Blaberus*. Position 5 is also important for reading a "message" in *Periplaneta* (Gade 1986) but not in *Locusta*

(Goldsworthy et. al., 1986). The message in *Locusta* appears to depend on position 6.

***Blaberus* HrTH**

pGlu-Val-Asn-Phe-Ser-Pro-Gly-Trp-Gly-Thr-NH_2
1 2 3 4 5 6 7 8 9 10 amide

Table 4. Sequence and bioassay parameters for several bioanalogs of *B. discoidalis* hypertrehalosemic hormone.

NAME	SEQUENCE	ED_{50} (pmole)	Max_{act} (µg/µl)
Blaberus HrTH	pQVNFSPGWGT-NH_2	0.6	39
Pandalus RPCH	pQLNFSPGW-NH_2	20	38
Periplaneta CAH I	pQVNFSPNW-NH_2	20	38
Schistocerca AKH II	pQLNFSTGW-NH_2	20	39
Locusta AKH II	pQLNFSAGW-NH_2	30	37
Locusta AKH	pQLNFTPNWGT-NH_2	6	29[a]
Carausius HrTH II	pQLTFTPNWGT-NH_2	200	32
Periplaneta CAH II	pQLTFTPNW-NH_2	350	ND[b]
Manduca AKH	pQLTFTSSWG-NH_2	NA	NA[c]

[a]Max_{act} of *Blaberus* HrTH was not statistically reached by this analog but activities stopped rising with increasing analog concentration.

[b](ND) An apparent saturation by this analog was not detected. Activity levels rise dramatically only with high doses (>350 pmole) to reach the maximum activity of HTH only at the highest concentration.

[c](NA) No activity detected.

We have explored how biological information is carried in *Blaberus* HrTH by constructing analogs that have a single amino acid replacement from the natural hormone (Ford et. al., 1988). The amino acids used for replacements were generally Ala and Gly which have minimal, neutral side chains. The substitution of either Ala or Gly in an analog results in the removal of a single side chain normally found in *Blaberus* HrTH while the peptide backbone is left intact. We have determined that residues 1,4 and 8 are critical for the "address". Residue

5 was confirmed as important for transduction of the "message". Introduction of a negative charge at the C-terminus by removal of the amide also interferes with the "message". Residues 2, 7 and 10 are most tolerant of replacement by other amino acids. We find that aromaticity is required at positions 4 and 8 (See Hayes et. al., this volume). Position 8 also requires a bicyclic ring structure. Finally, we have evidence that the peptide interacts with its receptor through a folded conformation (Ford et. al., 1988), like a ß-turn.

The HrTH structural analogs used for the initial amino acid replacement studies in *B. discoidalis* detect several differences in the hyperglycemic response of *P. americana* (Gade and Hayes, this volume). Positions 4 and 8 are still critical for the "address"; however, unlike *B. discoidalis*, position 3 is critical and position 1 is not critical for the "address" in *P. americana*. In *B. discoidalis* position 5 seems important for "message", but, in *P. americana*, position 6 also appears important for this function. The bulky Trp at position 8 is bounded on both sides in *Blaberus* HrTH by Gly with no side chain except a hydrogen. Both species tolerate bulk at position 7 of HrTH analogs in either the D or L orientation. Bulk in any orientation interferes with the potency of a HrTH position 9 analog in *B. discoidalis;* whereas, bulk is tolerated in *P. americana* in the L configuration.

Comparison of structure-activity probes in target insects has practical benefits. We have demonstrated that species specificity can be designed into analogs of AKH family members. For example, [N-acetyl-Ala^1]-HrTH stimulates hyperglycemia in *P. americana* but not in *B. discoidalis*. Also, [Ala^3]-HTH stimulates hyperglycemia in *B. discoidalis* but is only minimally active in *P. americana.*

Initial structure-activity studies have also been published for representatives of the myotropic peptides. Proctolin requires all five of its amino acids and charged termini for activity (Starratt and Brown, 1979). However, N-terminal extension with select amino acids or exchange of aromatic amino acids does not destroy activity (Sullivan and Newcomb, 1982). The leucosulfakinins contain an active core within the C-terminal octapeptide (Nachman *et al.*, 1988). Within the active core of leucosulfakinin I, the most critical residues for activity are 8, 10 and 11. The sulfated Tyr is also critical for activity, but its exact position around residue 6 is not as important as its presence in the sequence (Nachman *et al*, 1989). The leucosulfakinins are related to gastrin and cholecystokinin from vertebrates (Nachman *et al.*, 1988), but the amino acid in the position next to the C-terminus must be changed from an acidic Asp residue to a basic Arg for gastrin or cholecystokinin fragments to stimulate contractions of the cockroach hindgut. Finally, in leucopyrokinin, a C-terminal pentapeptide constitutes the active core (Nachman *et al.*, 1986).

Mode of Action

Most studies examining the mode of action of neurohormones have implicated intracellular second messengers as the mediators between the primary hormone-receptor interaction and the metabolic response by the cell. Generally, cyclic AMP (cAMP) has been demonstrated as the principal active second messenger in insects. For example, synthetic AKH stimulates fat body mobilization of lipid reserves via cAMP both *in vivo* and *in vitro* (Spencer and Candy, 1976; Asher *et al.*, 1984). Crude CC extracts with diuretic hormone activity increased both the cAMP content and the excretion rate of locust Malpighian tubules (Rafaeli *et al.*, 1984). Conversely, crude extracts of CC glandular lobes had antidiuretic activity when applied to *in vitro* preparations of locust rectum and increased adenylate cyclase activity and cAMP levels within the tissue (Herault and Proux, 1987). Forskolin mimicked the effect of CC extracts, and the phosphodiesterase inhibitor, *iso*butyl-methylxanthine (IBMX) potentiated the forskolin effect. *Periplaneta* CAH-I was reported to stimulate fat body cAMP levels *in vivo* in *P. americana* (Gade, 1985). Conversely, studies with *P. americana* fat body exposed to *Periplaneta* CAH-I and -II *in vitro* failed to show increases in cAMP levels or adenylate cyclase activity in adipocyte membrane preparations (Orr *et al.*, 1985).

Cyclic GMP may serve as a second messenger for some neuroendocrine-dependent processes. For example, cyclic GMP appears to serve as a second messenger in the degeneration of intersegmental muscles that occurs in response to eclosion hormone after the adult molt in *Antheraea polyphemus* (Schwartz and Truman, 1984). Muscles become sensitive to eclosion hormone after ecdysteroid titers decline, and the eclosion hormone stimulates a dose-dependent increase in cyclic GMP within the muscle.

Regulations of Neurohormone Secretion

Like so many other aspects of insect neuroendocrinology, little is presently known about the regulation of neurohormone secretion. In vertebrate animals, biogenic amines either directly stimulate neurosecretion of adenohypophyseal hormones (Weiner and Ganong, 1978), or they stimulate the release of hypothalamic releasing peptides that subsequently promote adenohypophyseal neurosecretory activity

(McCann, 1981). To date, there is no evidence for the presence of comparable releasing factors in the insect nervous system.

Biogenic amines also appear to affect insect neurosecretion. Treatment of locusts with pesticidal chemicals that act via the cholinergic nervous system stimulates AKH and HrGH secretion (Samarayanaka, 1974). Later studies indicate that the aminergic nervous system serves as the mediator between neurosecretion and the cholinergic actions of the insecticides (Samarayanaka, 1976, 1977).

Octopamine is a biogenic amine that promotes neurosecretion in insects. Octopamine promotes secretion of a hyperlipemic factor, presumably AKH, by locust CC *in vitro* (Orchard, 1982; Orchard and Loughton, 1981); whereas, CC from cockroaches secrete a hypertrehalosemic factor after *in vitro* exposure to octopamine (Downer *et al.*, 1984). Octopamine elevates cAMP levels in the cockroach CC, and addition of dbcAMP stimulates HrTH secretion (Downer *et al.*, 1984). However, other biogenic amines such as dopamine and 5-hydroxytryptamine also elevate cAMP levels in the glands *in vitro*, but no release of hypertrehalosemic factors is observed. Finally, formamidine insecticides such as chlordimeform act as octopamine agonists (Hollingworth and Murdock, 1980), and *in vitro* exposure of locust CC to chlordimeform or desmethylchlordimeform stimulates the secretion of hyperlipemic factors (Singh *et al.*, 1981). These studies suggest that octopamine promotes neurosecretion of AKH in locusts and HrTH in cockroaches. Whether the secretion of other neurohormones is also stimulated by octopamine has not been examined nor has the question of whether other biogenic amines affect different patterns of neurohormone secretion.

Comparative Insect-Vertebrate Neuropeptides

A major area of investigation has developed around the question of whether insects and vertebrates share common peptide hormones. The comparison has been largely in the direction of vertebrate hormones in insects rather than insect hormones in vertebrates; however, several studies have addressed the latter question. For example, proctolin reportedly stimulates contractions by rodent ilieum muscle but does not affect blood pressure in cats or the contraction of rat uterine muscle (Penzlin *et al.*, 1981). Proctolin strongly inhibits aminoenkephalinase activity in rat synaptic preparations, but it is not active on the enzyme in brain slices (Hui *et al.*, 1985). No evidence is presented to indicate the presence of proctolin in the vertebrate nervous system . The question of the presence of insect peptides in vertebrates will undoubtedly receive increasing attention as more insect peptides are identified and suitable antibodies become available for their detection.

The studies on insects have used antibodies prepared against vertebrate neuropeptides to identify immunoreactivity in insect tissues. Although numerous reports of vertebrate-like peptides exist for insects, few of the reactive peptides have been isolated and structurally characterized. Three of the best defined situations are described below.

In only one case has the actual insect peptide been isolated that responds to the antibody for a vertebrate peptide. Antibodies to vertebrate Arg-vasopressin were used as the basis for isolating a diuretic factor from locusts. The resulting insect peptide contains a sequence of five amino acids that correspond identically with Arg-vasopressin (Proux *et al.*, 1987). However, the locust peptide has six amino acids that are identical with vasotocin, the member of the oxytocin-vasopressin family that is most widely distributed among the vertebrate classes, and the member of the family that is considered most primitive (Turner and Bagnara, 1976). Interestingly, this insect member of the oxytocin/vasopressin family has a diuretic action in the insect; whereas, vasopressins are antidiuretic in vertebrates. Furthermore, the locust diuretic hormone is most active as an antiparallel dimer rather than through an intrachain disulfide bridge as occurs in the vertebrate peptides (Proux *et al.*, 1987).

Peptides that respond to antibodies for insulin and glucagon were some of the first vertebrate-like peptides reported in insects (Tager *et al.*, 1976). As described earlier in this review, PTTH has a 40-50% similarity to insulin (Nagasawa *et al.*, 1986). Presumably, the similarities between PTTH and insulin are such that antibodies to insulin may be crossreacting with regions of similarity in *Manduca* PTTH. Furthermore, *Periplaneta* CAH-II has a series of several amino acids that are similar to the N-terminal region of glucagon (Scarborough *et al.*, 1984). Examination of *Manduca* AKH shows even closer similarity with glucagon, and it is probable that this region of similarity is sufficient to bind with antibodies to vertebrate glucagon.

Finally, there are several reports concerning responses in insects to antibodies to gastrin/cholecystokinin (Kramer *et al.*, 1977; Duve and Thorpe, 1981; 1984). Leucosulfakinin has a 55% identity with gastrin II (Nachman *et al.*, 1986), and it is probable that the antibodies to gastrin/cholecystokinin are detecting the sequence identity of leucosulfakinin bioanalogs present in the different insect species examined.

It is difficult to draw firm conclusions as to the meaning of the similarities between the vertebrate and insect neuropeptides. Certainly, it can be concluded that there were particular biologically active sequences of peptides that proved especially successful for intercellular communication by early organisms. These sequences have been conserved and presumably elaborated upon during evolution and passed between distantly related organisms for regulation of similar physiological processes (gastrin/CCK:leucosulfakinin, digestion; glucagon:AKH/HrGHs, metabolite mobilization) and, in other cases, dissimilar processes (insulin:PTTH, metabolism:steroid hormone synthesis). Much work remains to be done to understand evolutionary relationships between the peptide hormones of insects, other invertebrate groups and vertebrates. Certainly some sequences of insects and vertebrates have evolved separately; whereas, others, such as those cited, may have partial homologies.

Four years ago when we wrote our article on the practical potential of insect neuropeptides (Keeley and Hayes, 1987), there was little evidence to indicate the significance of crossreactivity between insect peptides and antibodies to vertebrate hormones. Today, the relationships are much clearer. Obviously, the peptides are different but contain related sequences. Sometimes their physiological effects are similar; other times they bear no apparent relationships. However, in no case has the vertebrate hormone demonstrated identical biological activity in the insect assay. We must conclude that although some structural sequence similarities exist between some of the insect and vertebrate neurohormones, the insect neuropeptides still differ in their total structure. Furthermore, related neurohormones may appear to affect identical processes (e.g. elevation of blood carbohydrate), but the insect process is accomplished in an insect-unique fashion (trehalose synthesis).

INSECT NEUROENDOCRINOLOGY: FUTURE OPPORTUNITIES

Future Research Topics

The previous discussion on the present state of insect neuroendocrine research identifies many of the areas in need of further basic investigations. For example, little is known at present about the structure and expression of neurohormone genes and the structures of preprohormones. There is no information about the processing enzymes that convert prohormones to active hormones. Little is known about the regulation of neurosecretion or neurohormone degradation. At the target cell level, we have no information on receptors and except for some examples of increases in cyclic nucleotides, especially cyclic AMP, there is little evidence to indicate the mode of action of the neurohormones with their target cells. Although we have elaborated numerous insect neurohormones during the three years since the second ICINN, the majority of these peptides are related and belong to only two physiologically defined groups: the metabolite-mobilizing peptides (AKH/HrGHs) and the myokinins. And except for the AKH/HrGH family, most other identified peptides are restricted to only one or two insect species. We can infer from the experience with the AKH/HrGH family, the PTTHs and the myokinins that there are likely to be numerous bioanalogs among the different insect species for any particular peptide, and there are likely to be multiple bioanalogs for a peptide within a single species. One can conclude from this that there are undoubtedly hundreds, and possibly even thousands, of neuropeptides yet to be discovered among just the most common insect species, let alone among the multitude of insect species. Although insect neuroendocrine research must consist of more than the isolation and characterization of new neuropeptides from different species, it is obvious from the previous discussion that elucidation of the varieties of hormone structures contributes to our understanding of comparative endocrinology and hormone evolution.

Practical Applications to Pest Insect Control

Basic understanding of insect neuroendocrine processes allows for speculations concerning the potential for practical applications of neuroendocrine information through advances in biotechnology. Just as juvenile hormone and the ecdysteroids have been examined for their pragmatic potential for pest insect control, so too, it is reasonable to evaluate neurohormones for their practical potential. The remainder of this discussion will address this topic.

Constraints on the Application of Neuroendocrine Biotechnology

There are several significant constraints on the use of neurohormones for practical insect pest control. The foremost constraint is the peptide/protein nature of the hormones *per se*. Proteins and peptides are unstable in the environment. Most are sensitive to: temperature, light, pH, and bacterial or other biological actions. Furthermore, topically applied proteins and peptides are not translocated through the cuticle of insects, and they are degraded to their constituent amino acids by intestinal proteases if consumed orally. The latter two points were confirmed for proctolin (Quistad *et al.*, 1984). Therefore, entry into the target insect by routes used by usual pesticidal chemicals are not available to peptide hormones.

The next two constraints were discussed in the previous section and will be mentioned here only briefly. First, the question of how universal insect neuropeptides might be among the different insect species and among other animals such as vertebrates or other invertebrates must be considered. It is apparent that some neurohormones have a certain degree of structural overlap with vertebrate hormones. However, none of the vertebrate peptide hormones have proven to be identical with their insect counterpart. Furthermore, differences exist for the same hormone between different insect species and responsiveness differs for heterologous combinations of peptides and species. For example, in our laboratory, we find that the AKH of *M. sexta* has no hypertrehalosemic activity in *B. discoidalis* cockroaches (unpublished observation). It could be argued that similarities in structure between insect neurohormones and peptides of other animals would mean that development of chemicals that disrupt insect neurohormones would also be toxic to other animals. However, the relative degrees of sensitivity between species for related peptides suggest that a chemical active on neuroendocrine receptors in one species of insect may not affect comparable receptors in other insect species, let alone in vertebrates or other invertebrates. In fact the amount of species specificity that we are beginning to observe for the HrGHs in different cockroach species (Gade and Hayes, this volume) suggests that it might be possible to develop neuropeptide analogs that show a reasonable degree of species selectivity.

The last major constraint in our opinion concerns the general lack of knowledge about insect neurohormones and neuroendocrinology. The above discussion on the present state of knowledge points out the unknown areas in the subject, and it is only through concerted research efforts that these knowledge gaps will be eliminated. It is important that we, as insect neuroendocrinologists, not only do the research, but also strive to further the field by illustrating its interesting aspects. We must demonstrate to other non-insect scientists, especially those that control funds that support our research, how elucidation of insect neurobiology-neuroendocrinology contributes to the general understanding of neural and endocrine functions in higher animal systems. Furthermore, we must consider and communicate how insect research may eventually contribute to solving insect pest problems through the development of agents and strategies that will act in an environmentally safe manner with insect, or even species, selectivity.

It must be stated emphatically that we do not propose the direct application of peptide hormones for insect control. Any agent used to disrupt the neuroendocrine balance must be environmentally stable and must gain entry to the insect readily and without being damaged. For these reasons, as indicated above, neuropeptides are not suitable for direct application. Rather, neuroendocrine disruption must occur from the application of stable, nonpeptide analogs; by infection of insects with genetically engineered biological expression vectors that contain and express neurohormone genes, or by treating insects with chemicals that affect neurosecretion.

Advantages of Neurohormones for Pest Insect Control

Neurohormones have several features that favor their use for practical insect control. First, there are numerous (20 to 30) physiological processes that are regulated by neurohormones. Second, neurohormones are peptides or proteins and are amenable to advances in protein and genetic engineering biotechnology. Third, neurohormones are diverse and complex which provides a vast potential for modifications by synthetic chemistry to produce agonistic/antagonistic analogs. Lastly, several neuroendocrine events are susceptible to manipulation.

We have already explored the various physiological processes affected by neurohormones. Some insect processes are unique to insects such as molting and metamorphosis and have their own complement of unique neurohormones such as PTTH, eclosion hormone, tanning hormone, allatostatin and allatotropin. Other insect processes such as reproduction and water balance are not unique to insects, but insects still use unique peptide hormones to regulate these processes (egg development neurohormone, ovulation hormone, allatotropin, allatostatin, oostatic factors, pheromone synthesis-regulating hormones and diuretic hormones). Some insect processes such as mobilization of carbohydrate and lipid stores, protein synthesis, muscle contraction and antidiuresis are not unique, and the insect uses regulatory peptides that differ by varying degrees from their counterparts in other animals. Finally, some insects have physiological processes that are unique to themselves such as the egg diapause of *Bombyx mori* or pupariation by the cyclorrhaphous flies, and these insects possess unique neurohormones to regulate the processes (Hasegawa, 1957; Fraenkel *et al.*, 1972).

The degree of species specificity dictates how useful the hormone might be in an insect control program. Insect specific processes would be the ideal situation, and species specificity is most ideal from the environmental standpoint. However, too great a degree of specificity makes the approach unattractive from the industrial standpoint, since economics dictates that a commercially useful approach should control several species of the insect pest complex that damages a particular crop. Therefore, it is desirable to attain fairly broad, insect-specific selectivity.

The objection is often raised that removal of the corpora cardiaca does not result in death of treated insects. Likewise, there are no examples where treatment of an insect with a neurohormone results in malformation or death, as is the case with JH treatments (Spielman and

Williams, 1966). However, it should be pointed out that abnormal titers of neurohormones most likely need to be sustained for extended periods to exert disruptive actions, and it is difficult to sustain high levels of peptide hormones in insects using current technology. Reciprocally, hormone deficiency or surplus may be deleterious if the insect is placed into a competitive situation in a natural environment. For example, CC extirpated insects are maintained under controlled laboratory conditions by themselves, but if they were placed into a natural competitive environment, they might be quite susceptible to stress and mortality factors.

Cloning-Expression Vectors

Neurohormones are peptides and are amenable to genetic engineering and recombinant DNA biotechnology. It was proposed that neurohormone genes may be used to transform crop plants or bacteria so that the hormone will be produced and ingested by the pest insect as it feeds (Schneiderman, 1984). The resulting hormone imbalance would be deleterious to the affected insect. However, in order for this approach to be useful, the hormone will either have to be synthesized as the active hormone or the transformed cells must have the capacity to process the hormone correctly. It is unlikely that small peptides, such as the insect neurohormones, can be synthesized directly as the active hormone. It is also unlikely that plant or bacterial cells will have the capacity to process foreign prohormones into active hormones without further bioengineering with processing enzymes. Finally, it is questionable whether ingested peptides would survive the gut of target pest insects and be absorbed as intact, active peptides.

We suggest an alternative approach that uses insect baculovirus cloning-expression vectors (Smith *et al*., 1983) as a vehicle for carrying neurohormone genes into pest insects. Baculoviruses for lepidopteran larvae could be genetically engineered with a neurohormone gene to express the hormone in superabundant amounts during the replication cycle of the virus in infected cells (Fig. 3). It can be anticipated that the insect cells would recognize the preprohormone and use their intrinsic processing enzymes to convert the prohormone into an active hormone. Baculoviruses infect neurons during the secondary, generalized stage of the infection, and it is likely that infected neural tissues have the capacity to process the prohormone correctly. In this way, the virus would carry the neurohormone genes into the host and, thereby, circumvent the exposure of the hormone to the degradative milieu of the intestine. Super-expression of the neurohormone gene by the viral promoter may result in uncontrolled production of extremely high titers of the hormone and produce a neurohormone imbalance that can not be overcome by the usual levels of degradative proteases for neurohormones.

The advantage of including a neurohormone gene with a baculovirus is that the hormone will improve the effectiveness of the virus as a biocontrol agent. Baculoviruses are often slow in their action and can take as long as 3-10 days to kill infected larvae. During this time, infected larvae continue to feed and cause damage. By coupling the neurohormone gene with an early promoter in the viral replication cycle, it should be possible to produce the hormone early in the infection process and either disrupt the physiology of infected larvae to the point that feeding stops or kill the larvae directly from the disruptive actions of the excess hormone. This combines a virus that is either genus- or species-specific with a specific neurohormone gene. The gene itself can have no deleterious effects and its hormonal product is a natural agent that is unstable in the environment outside the insect. The genetically engineered virus is environmentally ideal as an insecticidal agent.

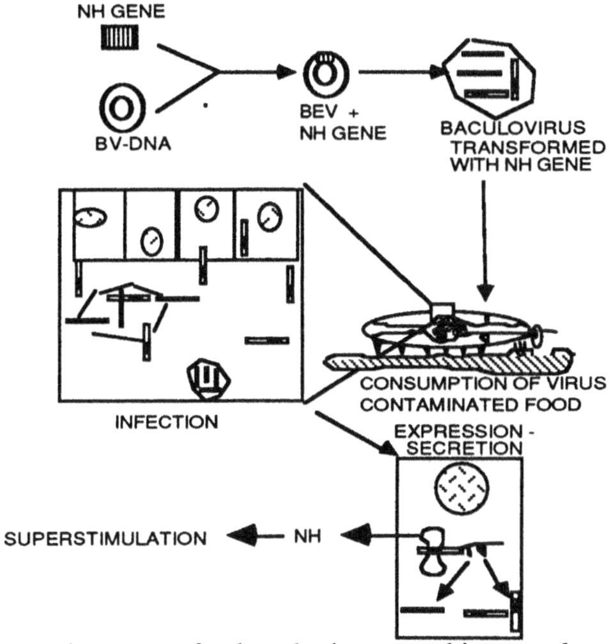

Figure 3. Proposed strategy for baculovirus as a biocontrol agent for lepidopteran larvae by use as a cloning-expression vector for neurohormone (NH) genes.

We propose that diuretic and/or antidiuretic neurohormone genes would be useful in combination with baculovirus vectors (Fig. 3). Baculoviruses are disease agents of many species of larval lepidopterans, including many of the major crop pests of the family Noctuidae. Lepidoptera larvae are soft-bodied and highly susceptible to extreme changes in water balance. Soft-bodied larvae depend on internal fluid (hemolymph) volume to produce the turgidity of the body wall needed for movement. Desiccation by excessive diuresis would result in flaccidity and immobility followed by further dehydration and eventually death. Conversely, antidiuresis could result in edema and the inability to continue feeding on a succulent plant diet.

We have initiated research on the regulation of water balance in lepidopteran larvae, and we find that fifth instar larvae of the noctuid moth *Heliothis virescens* undergo a rapid body weight gain from 100 mg to 400 mg within 24 hr. Without doubt, the rapid weight gain must be accompanied by voracious feeding activity. Malpighian tubules from larvae also show peak rates of fluid excretion at the time of the rapid weight gain (Fig. 4). We have subsequently demonstrated that this rapid fluid excretion is the result of a diuretic hormone secreted during the period of rapid feeding. Other researchers have noted that Malpighian tubules were highly active in lepidopteran larvae, but they concluded that it was an intrinsic property of the tubules and not endocrine dependent (Ryerse, 1978). By comparing combinations of extracts and tubules from feeding and starved *H. virescens* larvae, we have demonstrated that a hormone is responsible for the peak of diuretic activity that occurs in the last larval instar and that this peak relates to intense feeding activity (Chung and Keeley, 1989).

A strategy for control of lepidopteran larvae using neuroendocrine manipulation can be proposed based upon the relationship between endocrine-dependent diuresis and food consumption. We propose that small larvae that are consuming small amounts of food would be susceptible to diuretic hormones and desiccation. Last instar larvae that are feeding voraciously to attain a maximum body weight in preparation for pupation would be affected by antidiuretic agents that would inhibit water excretion and result in fluid retention and edema thus preventing further food consumption. Baculovirus expression vectors transformed with diuretic hormone genes would be useful for control of young lepidopteran larvae; whereas, more mature larval populations would be susceptible to virus containing antidiuretic hormone genes.

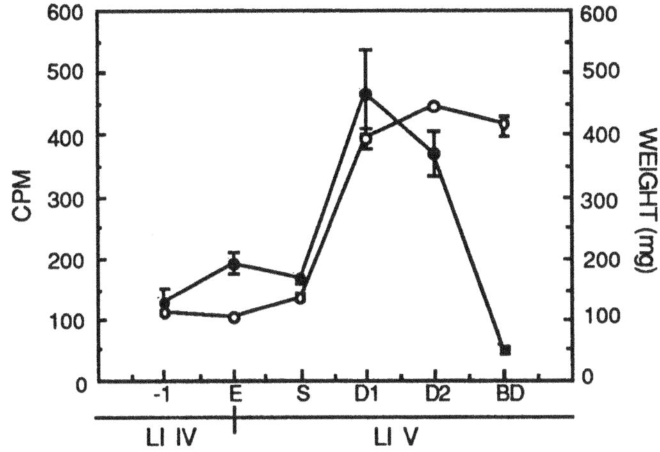

Figure 4. Age-related changes in body weight and fluid secretion rate by *in vitro* Malpighian tubules from fourth and fifth instar larval *Heliothis virescens*. LI = larval instar; E= ecdysis; S = slender stage; D1 = early digging (feeding) stage; D2 = late digging stage; BD = burrowing digging. (●) = diuretic activity; (o) = live weight. Values are means ± SEM for 4-6 replicate animals except for the body weight at ecdysis which represents 2 larvae.

Other aspects of genetic engineering could also prove useful in a viral expression vector for disruption of neurohormone balance. Expression of monoclonal antibody genes prepared against specific neurohormones could result in production of an antibody that would bind to secreted hormone and render the infected larva hormone deficient. Such genes would prove useful in cases such as the eclosion hormone or tanning hormones so that the larva would be incapable of ecdysis or of tanning the new cuticle after ecdysis. Antibodies to PTTH would suppress the onset of molting. Antibodies might also prove useful against diuretic and antidiuretic hormones. Alternatively, the production of antipeptides might prove useful. For example, antihormones might be produced (Bost *et al.*, 1985) or peptides that have been modified by site-directed mutagenesis that can bind the hormone receptor and act antagonistically to the natural hormone.

Analog Development

Insect neurohormones are diverse and complex peptides. As discussed previously, peptides with the same physiological effects may occur in multiple forms among different species, and there may be more than one bioanalog performing the same function within a single species. We find that there are differential susceptibilities between species for the different members of the same peptide family. Generally, the peptide natural to a given species is the most active form of the peptide for that species. [The notable exception to this in insects is the 4K PTTH of *B. mori* that promotes molting in *S. ricini* but not in *B. mori* (Suzuki *et al.*, 1986)]. This suggests that coevolution occurs between the peptide and its receptor to provide optimal "fit". Comparative dose-response studies of multiple bioanalogs can be used to indicate what constitutes an active conformation for a peptide at the active site of the receptor; what differences in conformation can exist in the peptide and what different peptide conformations mean in different species. All of these variables provide information about the active structure for a peptide at its target-cell membrane receptor.

Knowledge about the structure that interacts with the receptor active site can be used in a rational analog design program to develop antagonists or superagonists for a given peptide. As discussed previously, there are address and message portions of a peptide hormone. Design of agonists uses information about both the address and the message to produce an analog that fits into the active site and activates the receptor to perform its transducing effect within the cell. By contrast, antagonists have an active address structure but lack the message so that they fit into the receptor without activity and block interactions between the receptor and its natural hormone.

Although the approach of genetically engineering insect viruses with neurohormone genes is attractive, our long history of pest control with chemicals suggests that analogs may have the most practical promise, once we have identified the types of chemicals that have appropriate activities. At present the factors that determine peptide structures and their activity are not well understood, but the field of protein and peptide bioengineering is moving rapidly and advances that are occurring in the pharmaceutical area, especially, will benefit the discovery and development of insecticidal peptide analogs.

It is difficult to predict what approaches will lead to the design of useful neurohormone analogs. Each peptide has its own unique structure and associated problems that must be dealt with on an individual basis. However, a general scheme can be described based on rational drug design. The use of conformational restrictions is one possibility (Hruby,

1982; Hruby and Mosberg, 1982). Small peptides are flexible in solution but assume an active conformation as they orient to an acceptor molecule. The active conformation can be analyzed by a series of analogs with increasingly rigid structures. Restricted flexibility permits the peptide to fit more readily into its receptors and less readily into binding sites with similar topographies. The more rigid an active structure, the greater the assurance that the structure resembles the active hormonal conformation when bound to the receptor. Flexibility of the basic peptide is limited by intramolecular linkages or by replacing or introducing specific structural moieties (N-or C-alpha-methylation, olefinic replacement of peptide bonds to limit specific rotations, alkylation or dehydration of ß-carbon side chains). The effects of progressively changing the molecular structure of the peptide are monitored by comparing biological activity relative to a predicted 3-D structure determined by modeling with computer graphics to assess conformational rearrangements. As our knowledge expands concerning the correlations between peptide structures and biological activity, it should be increasingly feasible to develop nonpeptide mimics or inhibitors for hormone-dependent processes using the approach of rational peptide design and bioengineering.

Several neuroendocrine events are susceptible to disruption (Fig. 5). Transcription or translation of neurohormones are not good candidates for disruption since they undoubtedly do not involve insect-specific biosynthetic processes. Likewise, interference with processing enzymes is unlikely for the same reasons. However, interference with neurosecretion has already been identified as a possible means of altering neurohormone levels. Formamidine insecticides promote the release of AKH in locusts (Singh *et al.*, 1981) and illustrate that chemicals that affect biogenic amines can be used to modulate insect neurosecretory processes.

Most of the practical approaches for the disruption of neuroendocrine-dependent processes involve post-secretory events since these lend themselves more to insect-specific applications. Disruption of hormone interactions with target cell receptors by neurohormone mimics or inhibitors has already been discussed and has, in our opinion, the highest potential for practical application. Entomologists have a long history of experience using chemicals to control pests and there is a vast reservoir of information on formulation and application of chemical pesticides. Without doubt, development of effective neurohormonal analogs would see their rapid use based upon existing application technology.

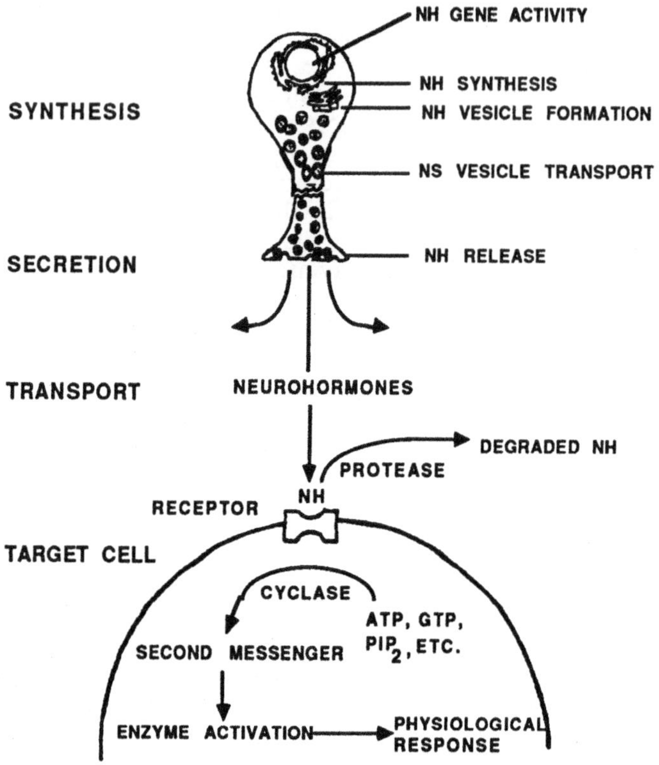

Fig. 5. Sequence of neuroendocrine events.

As with gene expression and processing, it is unlikely that disruption of the target cell, intracellular action of the hormone will have useful application. Modes of action of peptide hormones appear to involve common second messenger systems among diverse organisms. Therefore, interference with cellular information transduction beyond the level of the hormone-receptor interaction is risky and unlikely.

Finally, degradation and cessation of a hormone signal is as essential as the synthesis and presence of the signal in the first place. As discussed previously, hyperstimulation by hormones can be as unbalancing to the homeostasis of an organism as the absence of the endocrine signal. Therefore, degradative enzymes are essential to destroy the hormone and remove the stimulation. Just as the inhibition of acetylcholinesterase by pesticides results in accumulation of acetylcholine at the synapse and hyperstimulation of the nervous system, so too, inhibition of degradative proteases for the neurohormonal peptides could have deleterious actions from the accumulation of neurohormone. Degradative proteases could be as much a target for future antagonistic analogs as the cell receptors. As with acetylcholine, the accumulation of neuropeptide could then be the imbalancing feature that would have deleterious actions. Neuropeptide degradative proteases may be common between insects and other organisms, but the insect protease may have a higher affinity for their peptide which would lend a degree of selectivity

Disruption of degradative proteases would be useful with any of the neurohormones that act rapidly and manifest a direct physiological response. The myokinins and the water and salt regulating hormones would be especially amenable to manipulation by this procedure since their actions are manifested rapidly.

In closing, it is exciting to be part of a dynamic, maturing field such as neuroendocrinology. Founders and pioneers of the field are available for discussions and continue active research. The insect area still has vast unexplored topics for present and future researchers. Partly because of our ties with the Texas Agricultural Experiment Station, we have considered the pragmatic potential of insect neuroendocrinology and are proponents of exploring this possibility. But, even if neuroendocrine biotechnology never has practical application, it will nonetheless, provide new insights into the role of neuropeptides and proteins in higher animals and will broaden our understanding of the varieties and commonalities of how physiological processes are regulated in animals.

ACKNOWLEDGEMENTS

Personal research cited was supported by NSF Grants: PCM 8103277 and DCB 8511058, by NIH grants TMP AI 15190 and NS 20137, by USDA grant 86-CRCR-1-2008 and by the Robert J. Kleberg, Jr. and Helen C. Kleberg Foundation.

LITERATURE CITED

Aota S., Gojobori T., Ishibashi F., Maruyama T. and Ikemura T. (1988) Codon usage tabulated from the Gen Bank genetic sequence data. *Nucleic Acids Res.* **16,** r315-r402.

Asher C., Moshitzky P., Ramachandran J. and Applebaum S.W. (1984) The effects of synthetic locust adipokinetic hormone on dispersed locust fat body cell preparations: cAMP induction, lipid mobilization, and inhibition of protein synthesis. *Gen. comp. Endocr.* **55,** 167-173.

Baumann E. and Penzlin H. (1984) Sequence analysis of neurohormone D, a neuropeptide of an insect *Periplaneta americana*. *Biomed. biochim. Acta* **43,** K13-K16.

Baumann E. and Penzlin H. (1987) Inactivation of neurohormone D by Malpighian tubules in an insect, *Periplaneta americana*. *J. comp. Physiol.* **157B,** 511-517.

Bollenbacher W.E., Katahira E., O'Brien M., Gilbert L.I., Agui N. and Bamhover A. (1984) Insect prothoracicotropic hormone: Evidence for two molecular forms. *Science* **224,** 1243-4245.

Bost K.L., Smith E.M. and Blalock J.E. (1985) Similarity between the corticotropin (ACTH) receptor and a peptide encoded by an RNA that is complementary to ACTH mRNA. *Proc. natn. Acad. Sci. U.S.A.* **82,** 1372-1375.

Bowers W.S. and Friedman S. (1963) Mobilization of fat body glycogen by an extract of corpus cardiacum. *Nature* **198,** 685.

Bradbury A.F., Finnie M.D.A. and Smyth D.G. (1982) Mechanism of C-terminal amide formation by pituitary enzymes. *Nature* **298,** 686-688.

Bradfield J.Y. and Keeley L.L. (1989) *J. biol. Chem.* (In Press).

Chung J.S. and Keeley L.L. (1989) Evidence and bioassay for diuretic factors in the nervous system of larval *Heliothis virescens*. *J. comp. Physiol.* (In Press).

Downer R.G.H., Orr G.L., Gole J.W.D. and Orchard I. (1984) The role of octopamine and cyclic AMP in regulating hormone release from corpora cardiaca of the American cockroach. *J. Insect Physiol.* **30,** 457-462.

DuPont-Raabe M. (1951) Etude experimentale de l'adaptation chromatique chez le phasme *Carausius morosus* Bt. *C.R. Acad. Sci., Paris*, **232**, 886-888.

Duve H. and Thorpe A. (1981) Gastrin/cholecystokinin (CCK)-like immunoreactive neurones in the brain of the blowfly, *Calliphora erythrocephala* (Diptera). *Gen. comp. Endocr.* **43,** 381-391.

Duve H. and Thorpe A. (1984) Immunocytochemical mapping of gastrin/CCK-like peptides in the neuroendocrine system of the blowfly *Calliphora vomitoria* (Diptera). *Cell Tissue Res.* **237,** 309-320.

du Vigneaud V., Ressler C., Swan J.M., Roberts C.W., Katsoyannis P.G. and Gordon S. (1953) Synthesis of an octapeptide amide with the hormonal activity of oxytocin. *J. Amer. chem. Soc.* **75,** 4879-4880.

Engelmann F. (1957) Die Steuerung der Ovarfunktion bei der ovoviviparen Schabe *Leucophaea maderae*. *J. Insect Physiol.* **1,** 257-278.

Engelmann F. (1959) The control of reproduction in *Diploptera punctata* (Blattaria). *Biol. Bull.* **116,** 406-419.

Fernlund P. and Josefsson L. (1972) Crustacean color-change hormone: amino acid sequence and chemical synthesis. *Science* **177,** 173-175.

Fisher J.M. and Scheller R.H. (1988) Prohormone processing and the secretory pathway. *J. biol. Chem.* **263,** 16515-16518.

Ford M. M., Hayes T. K. and Keeley L. L. (1988) Structure-activity relationships for insect hypertrehalosemic hormone: the importance of side chains and termini. In *Peptides: Chemistry and Biology* (Edited by Marshall, G. R.), pp. 653-655, ESCOM, Leiden.

Fraenkel G., Zdarek J. and Sivasubramanian P. (1972) Hormonal factors in the CNS and hemolymph of pupariating fly larvae which accelerate puparium formation and tanning. *Biol. Bull.* **143,** 127-139.

Gade G. (1985) Mode of action of the hypertrehalosaemic peptides from the American cockroach. *Z. Naturforsch.* **40c,** 670-676.

Gade G. (1986) Relative hypertrehalosaemic activities of naturally occurring neuropeptides from the AKH/RPCH family. *Z. Naturforsch.* **41c,** 315-320.

Gade G. and Rinehart K.L. (1986) Amino acid sequence of a hypertrehalosemic neuropeptide from the corpus cardiacum of the cockroach, *Nauphoeta cinerea. Biochem. biophys. Res. Comm.* **141,** 774-781.

Gade G. and Rinehart K.L. (1987) Primary sequence analysis by fast atom bombardment mass spectrometry of a peptide with adipokinetic activity from the corpora cardiaca of the cricket *Gryllus bimaculatus. Biochem. biophys. Res. Comm.* **149,** 908-914.

Gade G., Goldsworthy G.J., Schaffer M.H., Cook J.C. and Rinehart K.L.,. (1986) Sequence analyses of adipokinetic hormones II from corpora cardiaca of *Schistocerca nitans, Schistocerca gregaria*, and *Locusta migratoria* by fast atom bombardment mass spectrometry. *Biochem. biophys. Res. Comm.* **134,** 723-730.

Gade G., Hilbich C., Beyreuther K. and Rinehart K.L. (1988) Sequence analyses of two neuropeptides of the AKH/RPCH-family from the lubber grasshopper, *Romalea microptera. Peptides* **9,** 681-688.

Gersch M. (1958) Neurohormonale Beeinflussung der Herztatigkeit bei der Larve von *Corethra. J. Insect Physiol.* **2,** 281-297.

Goldsworthy G.J., Mallison K. and Wheeler C.H. (1986) The relative potencies of two known locust adipokinetic hormones. *J. Insect Physiol.* **32,** 95-101.

Hasegawa K. (1957) The diapause hormone of the silkworm, *Bombyx mori. Nature* **179,** 1300-1301.

Hardy P.M. and Sheppard P.W. (1983) Synthesis of [4,5-dehydroleu^{2}]- and [3,4-dehydropro^{6}]-locust adipokinetic hormone. *J. chem. Soc. Perkin Trans.* **I,** 723-734.

Hayes T.K. and Keeley L.L. (1981) Cytochromogenic factor: A newly-discovered neuroendocrine agent stimulating mitochondrial cytochrome synthesis in the insect fat body. *Gen. comp. Endocr.* **45,** 115-124.

Hayes T.K., Keeley L.L. and Knight D.W. (1986) Insect hypertrehalosemic hormone: isolation and primary structure from *Blaberus discoidalis* cockroaches. *Biochem. biophys. Res. Comm.* **140,** 674-678.

Hayes T.K., Pannabecker T.L., Hinckley D.J., Holman G.M., Nachman R.J., Petzel D.H. and Beyenbach K.W. (1989) Leucokinins, a new family of ion transport stimulators and inhibitors in insect Malpighian tubules. *Life Sci.* **44**, 1259-1266.

Hayes T.K. and Gade G. (1989) Insect hypertrehalosemic hormone: the importance of side chains and termini for biological activity in *Periplaneta americana.* (This volume).

Hayes T.K., Ford M. and Keeley, L.L. (1989) Importance of aromatic and other amino acid residues for the biological activity of insect hypertrehalosemic hormone (This volume).

Hekimi S. and O'Shea M. (1987) Identification and purification of two precursors of the insect neuropeptide adipokinetic hormone. *J. Neurosci.* **7,** 273-2784.

Hekimi S. and O'Shea M. (1989) Biosynthesis of adipokinetic hormones (AKHs): Further characterization of precursors and identification of novel products of processing. *J. Neurosci.* **9,** 996-1003.

Herault J.P.E. and Proux J.P. (1987) Cyclic AMP: the second messenger of an antidiuretic hormone from the glandular lobes of the migratory locust corpora cardiaca. *J. Insect Physiol.* **33,** 487-492.

Hollingworth R.M. and Murdock L.L. (1980) Formamidine pesticides: octopamine-like actions in a firefly. *Science* **208,** 74-76.

Holman G.M., Cook B.J. and Nachman R.J. (1986) Primary structure and synthesis of a blocked myotropic neuropeptide isolated from the cockroach, *Leucophaea maderae. Comp. Biochem. Physiol.* **85C,** 219-224.

Holman G.M., Wright M.S. and Nachman R.J. (1988) Insect neuropeptides: Coming of age. *ISI Atlas of Science: Plants and Animals* **1,** 129-136.

Hruby V.J. (1982) Conformational restrictions of biologically active peptides via amino acid side chain groups. *Life Sci.* **31,** 189-199.

Hruby V.J. and Mosberg H.I.(1982) Conformational and dynamic considerations in peptide structure-function studies. *Peptides* **3**, 329-336.

Hui K., Hui M.P., Ling N. and Lajtha A. (1985) Proctolin: a potent inhibitor of aminoenkephalinase. *Life Sci.* **36,** 2309-2315.

Isaac R.E. (1987) Proctolin degradation by membrane peptidases from nervous tissues of the desert locust (*Schistocerca gregaria*). *Biochem. J.* **245,** 365-360.

Isaac R.E. (1988) Neuropeptide-degrading endopeptidase activity of locust (*Schistocerca gregaria*) synaptic membranes. *Biochem. J.* **255,** 843-847.

Jaffe H., Raina A.K., Riley C.T., Fraser B.A., Holman G.M., Wagner R.M., Ridgway R.L. and Hayes D.K. (1986) Isolation and primary structure of a peptide from the corpora cardiaca of *Heliothis zea* with adipokinetic activity. *Biochem. biophys. Res. Comm.* **135,** 622-628.

Jaffe H., Raina A.K., Riley C.T., Fraser B.A., Bird T.G., Tseng C.-M., Zhang Y.-S. and Hayes D.K. (1988a) Isolation and primary structure of a neuropeptide hormone from *Heliothis* zea with hypertrehalosemic and adipokinetic activities. *Biochem. biophys. Res. Comm.* **155,** 344-350.

Jaffe H., Raina A.K., Fraser B.A., Keim P., Rao K.R., Zhang Y.-S., Lancaster J.L. and Hayes D.K. (1988b) Isolation of two neuropeptides in the AKH/RPCH-family from horseflies (Diptera). *Biochem. biophys. Res. Comm.* **151,** 656-663.

Kataoka H., Troetschler R.G., Kramer S.J., Cesarin B.J. and Schooley D.A. (1987) Isolation and primary structure of the eclosion hormone of the tobacco hornworm, *Manduca sexta. Biochem. biophys. Res. Comm.* **146,** 746-750.

Kataoka H., Troetschler R.G., Li J.P., Kramer S.J., Carney R.L. and Schooley D.A. (1989a) Isolation and identification of a diuretic hormone from the tobacco hornworm, *Manduca sexta. Proc. natn Acad. Sci. U.S.A.* **86,** 2976-2980.

Kataoka H., Toschi A., Li J.P., Carney R.L., Schooley D.A. and Kramer S.J. (1989b) Identification of an allatotropin from adult *Manduca sexta. Science* **243,** 1481-1483.

Keeley L.L. (1972) Biogenesis of mitochondria: neuroendocrine effects on the development of respiratory functions in fat body mitochondria of the cockroach *Blaberus discoidalis. Arch. Biochem. Biophys.* **153,** 8-15.

Keeley L.L. (1977) Developmental and endocrine regulation of cytochrome levels in fat body mitochondria of the cockroach, *Blaberus discoidalis. Insect Biochem.* **7,** 297-301.

Keeley L.L. (1978) Development and endocrine regulation of mitochondrial cytochrome biosynthesis in the insect fat body.

Delta-[^{14}C]aminolevulinic acid incorporation. *Arch. Biochem. Biophys.* **187,** 87-95.

Keeley L. L. (1981) Neuroendocrine regulation of fat body development and function. In *Energy Metabolism in Insects* (Edited by Downer, R. G. H.), pp. 207-237, Plenum Press, NY.

Keeley L.L. and Hayes T.K. (1987) Speculations on biotechnology applications for insect neuroendocrine research. *Insect Biochem.* **17,** 639-651.

Keeley L.L., Sowa S.M., Hayes T.K. and Bradfield J.Y. (1988) Neuroendocrine and juvenile hormone effects on fat body protein synthesis during the reproductive cycle in female *Blaberus discoidalis* cockroaches. *Gen. comp. Endocr.* **72,** 364-373.

Kono T., Nagasawa H. and Isogai A. (1987) Amino acid sequence of eclosion hormone of the silkworm, *Bombyx mori. Agric. biol. Chem.* **51,** 2307-2308.

Kopec S. (1917) Experiments on metamorphosis of insects. *Bull. Int. Acad. Cracovie B* 57-60.

Kopec S. (1922) Studies on the necessity of the brain for the inception of insect metamorphosis. *Biol. Bull.* **42,** 323-342.

Kramer K.J., Spiers R.D. and Childs C.N. (1977) Immunochemical evidence for a gastrin-like peptide in insect neuroendocrine system. *Gen. comp. Endocr.* **32,** 423-426.

Loughton B.G. and Orchard I. (1981) The nature of the hyperglycaemic factor from the glandular lobe of the corpus cardiacum of *Locusta migratoria. J. Insect Physiol.* **27,** 383-385.

Marti T., Takio K., Walsh K.A., Terzi G. and Truman J.W. (1987) Microanalysis of the amino acid sequence of the eclosion hormone from the tobacco hornworm *Manduca sexta. FEBS Lett.* **219,** 415-418.

Matsumoto S., Isogai A. and Suzuki A. (1985) N-terminal amino acid sequence of an insect neurohormone, melanization and reddish coloration hormone (MRCH): heterogeneity and sequence homology with human insulin-like growth factor II. *FEBS Lett.* **189,** 115-118.

Mayer R.J. and Candy D.J. (1969) Control of haemolymph lipid concentration during locust flight: an adipokinetic hormone from the corpora cardiaca. *J. Insect Physiol.* **15,** 611-620.

McCann S.M. (1981) Monoaminergic and peptidergic control of anterior pituitary secretion. In *Neurosecretion: Molecules, Cells, Systems.* (Edited by Farner D.S. and Lederis K.), pp.139-151. Plenum Press, New York.

Nachman R.J., Holman G.M., Haddon W.F. and Ling N. (1986) Leucosulfakinin, a sulfated insect neuropeptide with homology to gastrin and cholecystokinin. *Science* **234,** 71-73.

Nachman R.J., Holman G.M. and Cook B.J. (1986) Active fragments and analogs of the insect neuropeptide leucopyrokinin: structure-function studies. *Biochem. biophys. Res. Comm.* **137,** 936-942.

Nachman R.J., Holman G.M. and Haddon, W.F. (1988) Structural aspects of gastrin/CCK-like insect leucosulfakinins and FMRF-amide. *Peptides* **9,** 137-143.

Nachman R.J., Holman G.M. and Haddon, W.F. (1989) Effect of sulfate position on myotropic activity of the gastrin/CCK-like insect leucosulfakinins. *Int. J. Peptide Prot. Res.* **37**, 223-229.

Nagasawa H., Kataoka H., Isogai A., Tamura S., Suzuki A., Mizoguchi A. and Fujiwara Y.A. (1986) Amino-acid sequence of a prothoracotropic hormone of the silkworm, *Bombyx mori. Proc. natn. Acad. Sci. U.S.A.* **83,** 5840-5843.

Nambu J.R., Murphy-Erdosh C., Andrews P.C., Feistner G.J. and Scheller R.H. (1988) Isolation and characterization of a *Drosophila* neuropeptide gene. *Neuron* **1,** 55-61.

Nichols R., Schneuwly S.A. and Dixon J.E. (1988) Identification and characterization of a *Drosophila* homologue to the vertebrate neuropeptide cholecystokinin. *J. biol. Chem.* **263,** 12167-12170.

Ohtsuka E., Matsuki S., Ikehara M., Takahashi Y. and Matsubura K. (1985) An alternative approach to deoxyoligonucleotides as hybridization probes by insertion of deoxyinosine at ambiguous codon positions. *J. biol. Chem.* **260,** 2605-2608.

Orchard I. (1982) Octopamine in insects: neurotransmitter, neurohormone, and neuromodulator. *Can. J. Zool.* **60,** 659-669.

Orchard I. and Loughton B.G. (1981) Is octopamine a transmitter mediating hormone release in insects? *J. Neurobiol.* **12,** 143-153.

Orr G.L., Gole J.W.D., Jahagirdar A.P., Downer R.G.H. and Steele J.E. (1985) Cyclic AMP does not mediate the action of synthetic hypertrehalosemic peptides from the corpus cardiacum of *Periplaneta americana. Insect Biochem.* **15,** 703-709.

Penzlin H., Agricola H., Eckert M. and Kusch T. (1981) Distribution of proctolin in the sixth abdominal ganglion of *Periplaneta americana* L. and the effect of proctolin on the ileum of mammals. *Adv. Physiol. Sci.* **22,** 525-535.

Pierce J.G. and du Vigneaud V. (1950) Preliminary studies on amino acid content of a high potency preparation of the oxytocic hormone of posterior lobe of the pituitary gland. *J. biol. Chem.* **182,** 359-366.

Popenoe N.A. and du Vigneaud V. (1954) A partial sequence of amino acids in performic acid-oxidized vasopressin. *J. biol. Chem.* **206,** 353-360.

Proux J.P., Miller C.A., Li J.P., Carney R.L., Girardie A., Delaage M. and Schooley D.A. (1987) Identification of an arginine vasopressin-like diuretic hormone from *Locusta migratoria*. *Biochem. biophys. Res. Comm.* **149,** 180-186.

Quistad G.B., Adams M.E., Scarborough R.M., Carney R.L. and Schooley D.A. (1984) Metabolism of proctolin, a pentapeptide neurotransmitter in insects. *Life Sci.* **34,** 569-576.

Rafaeli A., Pines M., Stern P.S. and Applebaum S.W. (1984) Locust diuretic hormone-stimulated synthesis and excretion of cyclic-AMP: A novel Malpighian tubule bioassay. *Gen. comp. Endocr.* **54,** 35-42.

Raina A.K., Jaffe H., Kempe T.G., Keim P., Blacher R.W., Fales H.M., Riley C.T., Klun J.A., Ridgway R.L. and Hayes D.K. (1989) Identification of a neuropeptide hormone that regulates sex pheromone production in female moths. *Science* **244,** 796-798.

Redding T.W. and Schally A.V. (1969) Studies on inactivation of TRH. *Proc. Soc. exp. Biol. Med.* **131,** 415-418.

Robb S., Packman, L.C. and Evans P.D. (1989) Isolation, primary structure and bioactivity of SchistoFLRF-amide, a FMRF-amide-like neuropeptide from the locust, *Schistocerca gregaria*. *Biochem. biophys. Res. Comm.* **160,** 850-856.

Ryerse J.S. (1978) Developmental changes in Malpighian tubule fluid transport. *J. Insect Physiol.* **24,** 315-319.

Samaranayaka M. (1974) Insecticide-induced release of hyperglycaemic and adipokinetic hormones of *Schistocerca gregaria*. *Gen. comp. Endocr.* **24,** 424-436.

Samaranayaka M. (1976) Possible involvement of monoamines in the release of adipokinetic hormone in the locust *Schistocerca gregaria*. *J. exp. Biol.* **65,** 415-425.

Samaranayaka M. (1977) Role of acetylcholine in organophosphate-induced release of adipokinetic hormone in the locust *Schistocerca gregaria*. *Pestic. Biochem. Physiol.* **7,** 283-288.

Sanger F. (1959) Chemistry of insulin. *Science* **129,** 1340-1345.

Scarborough R.M., Jamieson G.C., Kalish F., Kramer S.J., McEnroe G.A., Miller C.A. and Schooley D.A. (1984) Isolation and primary structure of two peptides with cardioacceleratory and hyperglycemic activity from the corpora cardiaca of *Periplaneta americana*. *Proc. natn. Acad. Sci. U.S.A.* **81,** 5575-5579.

Schaffer M.H. (1986) Functional and evolutionary relationships among the RPCH-AKH family of peptides. *Amer. Zool.* **26,** 997-1005.

Scharrer B. (1935) Über das Hanstromsche Organ X bei Opisthobranchiern. *Pubbl. Staz. Zool. Napoli* **15,** 132-142.

Scharrer B. (1937) Uber sekretorisch tatige Nervenzellen bei wirbellosen Tieren. *Naturwissenschaften* **9,** 131-138.

Scharrer B. (1941) Neurosecretion. II. Neurosecretory cells in the central nervous system of cockroaches. *J. comp. Neurol.* **74,** 93-108.

Scharrer B. and Scharrer E. (1944) Neurosecretion. VI. A comparison between the intercerebralis-cardiacum-allatum system of the insects and the hypothalamo-hypophyseal system of vertebrates. *Biol. Bull.* **87,** 243-251.

Scharrer E. (1928) Die Lichtempfindlichkeit blinder Elritzen. I. Untersuchungen über das Zwischenhirn der Fische. *Z. vergl. Physiol.* **7,** 1-38.

Schneider L.E. and Taghert P.H. (1988) Isolation and characterization of a *Drosophila* gene that encodes multiple neuropeptides related to Phe-Met-Arg-Phe-NH2 (FMRFamide). *Proc. natn. Acad. Sci. U.S.A.* **85,** 1993-1997.

Schneiderman H.A. (1984) What entomology has in store for biotechnology. *Bull. ent. Soc. Am.* **30,** 55-61.

Schwartz L.M. and Truman J.W. (1984) Cyclic GMP may serve as a second messenger in peptide-induced muscle degeneration in an insect. *Proc. natn. Acad. Sci. U.S.A.*. **81,** 6718-6722.

Singh G.J.P., Orchard I. and Loughton B.G. (1981) Octopamine-like action of formamidines on hormone release in the locust, *Locusta migratoria. Pestic. Biochem. Physiol.* **16,** 249-255.

Skinner W.S., Quistad G.B., Adams M.E. and Schooley D.A. (1987) Metabolic degradation of the peptide periplanetin CC-2 in the cockroach, *Periplaneta americana. Insect Biochem.* **17,** 433-437.

Smith G.E., Summers M.D. and Fraser M.J. (1983) Production of human beta interferon in insect cells infected with a baculovirus expression vector. *Molec. cell. Biol.* **3,** 2156-2165.

Smyth D.G., Stein W.H. and Moore S. (1963) The sequence of amino acid residues in bovine pancreatic ribonuclease: revisions and confirmations. *J. biol. Chem.* 227-234.

Spencer I.M. and Candy D.J. (1976) Hormonal control of diacyl glycerol mobilization from fat body of the desert locust, *Schistocerca gregaria. Insect Biochem.* **6,** 289-296.

Spielman A. and Williams C.M. (1966) Lethal effects of synthetic juvenile hormone on larvae of the yellow fever mosquito, *Aedes aegypti. Science* **154,** 1043-1044.

Starratt A.N. and Brown B.E. (1975) Structure of the pentapeptide proctolin, a proposed neurotransmitter in insects. *Life Sci.* **17,** 1253-1256.

Starratt A.N. and Steele R.W. (1984) *In vivo* inactivation of the insect neuropeptide proctolin in *Periplaneta americana. Insect Biochem.* **14,** 97-102.

Steele J.E. (1963) The site of action of insect hyperglycemic hormone. *Gen. comp. Endocr.* **3,** 46-52.

Steele R.W. and Starratt A.N. (1985) *in vitro* inactivation of the insect neuropeptide proctolin in haemolymph. *Insect Biochem.* **15,** 511-519.

Steiner D.F., Docherty K., Chan S.H., Segundo B.S. and Carroll R. (1983) Intracellular proteolytic mechanisms in the biosynthesis of hormones and peptide neurotransmitters. In *Biochemical and Clinical Aspects of Neuropeptides: Synthesis, Processing, and Gene Structure.* (Edited by Koch D. and Richter D.), pp. 3-13, Academic Press, NY.

Stone J.V., Mordue W., Batley K.E. and Morris H.R. (1976) Structure of locust adipokinetic hormone, a neurohormone that regulates lipid utilization during flight. *Nature* **263,** 207-211.

Stone J.V., Mordue W., Broomfield C.E. and Hardy P.M. (1978) Structure-activity relationships for the lipid-mobilising action of locust adipokinetic hormone. Synthesis and activity of a series of hormone analogues. *Eur. J. Biochem.* **89,** 195-202.

Sullivan R.E. and Newcomb R.W. (1982) Structure function analysis of an arthropod peptide hormone: proctolin and synthetic analogues compared on the cockroach hindgut receptor. *Peptides* **3,** 337-344.

Suzuki A. (1986) Prothoracicotropic hormones and neurohormones in *Bombyx mori.* In *Insect Neurochemistry and Neurophysiology 1986* (Edited by Borkovec, A.B. and Gelman, D.B.) pgs. 29-51, Humana Press, Clifton, NJ..

Tager H.S., Markese J., Kramer K.J., Spiers R.D. and Childs C.N. (1976) Glucagon-like and insulin-like hormones in the insect neurosecretory system. *Biochem. J.* **156,** 515-520.

Taylor W.L. and Dixon J. (1978) Characterization of a pyroglutamate aminopeptidase from rat serum that degrades TRH. *J. biol. Chem.* **19,** 6934-6940.

Turner C.D. and Bagnara J.T. (1976). *General Endocrinology,* W.B. Saunders, Philadelphia, pp. 160-166.

Weeda E. (1981) Hormonal regulation of proline synthesis and glucose release in the fat body of the Colorado potato beetle, *Leptinotarsa decemlineata. J. Insect Physiol.* **27,** 411-417.

Weiner R.I. and Ganong W.F. (1978) Role of brain monoamines and histamine in regulation of anterior pituitary secretion. *Physiol. Rev.* **58,** 905-976.

Williams C.M. (1946) Physiology of insect diapause: the role of the brain in the production and termination of pupal dormancy in the giant silkworm, *Platysamia cecropia. Biol. Bull.* **90,** 234-243.

Williams C.M. (1947) Physiology of insect diapause. II. Interaction between the pupal brain and prothoracic glands in the metamorphosis in the giant silkworm, *Platysamia cecropia. Biol. Bull.* **93,** 89-98.

Williams C.M. (1952) Physiology of insect diapause. IV. The brain and prothoracic glands as an endocrine system in the cecropia silkworm. *Biol. Bull.* **103,** 120-138.

Witten J.L., Schaffer M.H., O'Shea M., Cook J.C., Hemling M.E. and Rinehart K.L. (1984) Structures of two cockroach neuropeptides assigned by fast atom bombardment mass spectrometry. *Biochem. biophys. Res. Comm.* **124,** 350-358.

Ziegler R. and Gade G. (1984) Preliminary characterization of glycogen phosphorylase activating hormone and adipokinetic hormone from *Manduca sexta* corpora cardiaca. *Physiol. Entomol.* **9,** 229-236.

Ziegler R., Eckart K., Schwarz H. and Keller R. (1985) Amino acid sequence of *Manduca sexta* adipokinetic hormone elucidated by combined fast atom bombardment (FAB)/tandem mass spectrometry. *Biochem. biophys. Res. Comm.* **133,** 337-342.

Ziegler R., Hoff R. and Rohde M. (1988) Storage site of glycogen phosphorylase activating hormone in larvae of *Manduca sexta*. *J. Insect Physiol.* **34,** 143-150.

Summaries of Recent Research

Neurochemistry

From *Insect Neurochemistry and Neurophysiology • 1989* Edited by A. B. Borkovec and E. P. Masler

OVERVIEW: PROGRESS IN INSECT NEUROCHEMISTRY

Ashok K. Raina

Insect Chemical Ecology Laboratory

Agricultural Research Service, U.S.D.A.

Beltsville, MD 20705

Great progress has been made in the area of neurochemistry since the last International Conference on Insect Neurochemistry and Neurophysiology three years ago. In 1986 only about 10 fully characterized insect peptides were known; today, there are over 40 known structures. New techniques and instrumentation used in the isolation of peptides and for determination of their structure by dedicated scientists were primarily responsible for this progress. David Schooley in his plenary lecture highlighted problems associated with peptide isolation and possible solutions to some of these problems. He emphasized that new, simple and reliable bioassays are an integral part of the isolation process. Bioassays are also very important to establish the biological role of newly discovered neuropeptides. We are now at a transition point where a shift in emphasis from identification of the peptides themselves to more emphasis on studies of the role of peptides in various physiological processes and on the isolation and identification of specific genes coding for these peptides is becoming evident.

Peptide Identification and Biological Activity

Some more structures have been added to the group often referred to as the adipokinetic hormone/red pigment concentrating hormone (AKH/RPCH)

family of peptides since 1986. It has become increasingly evident that the corpora cardiaca (CC) of insects are a rich source of neuropeptides and that pure material can often be obtained with a minimum of purification steps. Of the five peptides isolated and identified from the CC of various Coleoptera and reported at this conference, two represented new structures. From the CC of the American cockroach, three new neuropeptides were reported. One of these, named corazonin, has a very high cardioacceleratory activity wheras the other two resemble the leucosulfakinin (Lom-SK). Two new neuropeptides, neuroparsin A and B, were isolated from the CC of *Locusta migratoria* by a two-step HPLC process and their primary structures elucidated. Neuroparsin B is made up of three polypeptide chains with a total of 78 amino acid residues. A number of varied functions are associated with these two neuropeptides.

Eclosion hormone (EH) is a neuropeptide hormone in insects with an important and well defined function. EH was first identified from *Manduca sexta* in 1987. Now the primary structure of an EH from *Bombyx mori* has been revealed and it was reported to have 82% sequence homology to Mas-EH. The primary structure of a oostatic hormone form a mosquito was reported. This decapeptide is unique in that it has six prolines at the C-terminus and bears a 70% homology with HIV II (human immuno-deficiency virus). Progress was also reported in the isolation of the house fly oostatic hormone as well as a testis ecdysiotropin from the brains of the gypsy moth and the tobacco budworm. Prothoracicotropic activity was reported from pre-hatch eggs of the gypsy moth. A preliminary report on the isolation of a mating factor, presumably of peptidic nature, that appears to interact with the pheromone biosynthesis activating neuropeptide (PBAN) to terminate pheromone production in mated females of *Heliothis zea* was presented. Whereas biology of diuresis in adult *H. zea* was discussed in detail, the actual isolation and characterization of a diuretic hormone (DH) from this species is still in a preliminary stage. Recently, the primary structure of the *M. sexta*- DH was published. A challenge in peptide chemistry came from the report of a peptide with strong diuretic activity isolated from the house cricket. However, when the 16-amino acid residue peptide was synthesized it showed no biological activity.

Radioimmunoassay was used to demonstrate FMRFamide-like peptides in the entire nervous system of the locust. The C-terminus of a fully characterized decapeptide was shown to be homologous to FMRFamide. HPLC in conjunction with immunohistochemistry was used to demonstrate the presence of proctolin in the female reproductive tissue of *Rhodinus* . Hypolipaemic activity in the locust by the ß-chain of insulin provided further evidence of functional similarities, even though the mechanism of action in insects and vertebrates was reported to be different.

Structure Activity and Mode of Action

Several papers dealt with structure/activity studies, binding and degradation, and mode of action of peptides. The importance of side chains and termini and the presence of aromatic amino acids at key positions in the AKH/RPCH peptides, were dicussed. The use of circular dichroism to study secondary structure of these peptides was demonstrated and speculations were made on the importance of ß-turn for biological activity in the decapeptides of this family. The dipteran hypotrehalosemic hormone was shown to preferentially bind to a 100 kD protein in affinity columns. A preliminary study on the metabolism of the dipteran peptides was presented. It was also reported that membranes prepared from the heads of *Drosophila* possess endopeptidases that are capable of degrading peptides having internal phenylalanine residues, as is the case with all known AKH/RPCH peptides. A report that the Δ-11 desaturation step in the pheromone biosynthetic pathway in a moth species is controlled by PBAN was presented. There was speculation that c-AMP may be involved as a second messenger in PBAN-regulated pheromone biosynthesis in *H. armigera.*

The Role of Molecular Biology

So far only 3 genes for insect neuropeptides have been cloned. Complete structure of the gene for *Manduca*-EH was reported and by *in situ* hybridization it was shown that the major sites of expression for this gene are a pair of ventral medial neurosecretory cells in the brain. Larry Keeley in his plenary lecture revealed the structure of the gene for AKH of *M. sexta* . Both of these genes contain a single copy of the hormone. The concept, and pros and cons of using peptides or peptidomimetics for pest control were discussed in plenary lectures by both Larry Keeley and Tomas Kempe. A methodical approach, showing how peptide genes might be engineered into a baculovirus to produce biological pesticides that would not be expected to have the environmental problems often associated with chemical pesticides, concluded the scientific presentations.

Conclusion

Significant progress has been made in the area of peptide identification, but the hormonal nature of most of these peptides has not been conclusively proven. Additional evidence through radioimmunoassays is needed for such studies. In the future, researchers need to increase their efforts on such problems as the mode of action and degradation of peptide hormones. Increased activity in the area of molecular biology involving insect peptides is expected. With the many exciting opportunities before us for increasing our knowledge on insect neuropeptides, we look forward to ICINN '92.

From *Insect Neurochemistry and Neurophysiology • 1989* Edited by A. B. Borkovec and E. P. Masler

BOMBYX ECLOSION HORMONE

Akinori Suzuki, Hiromichi Nagasawa, Takaharu Kono, Bunji Sato, Takehiko Kamito, Hideo Tanaka, Youji Sakagami, Akira Mizoguchi*, Hironori Ishizaki*, and Hajime Fugo**

Department of Agricultural Chemistry, The University of Tokyo, Tokyo 113, *Faculty of Science, Nagoya University, Nagoya 446 and **Faculty of Agriculture, Tokyo University of Agriculture and Technology, Tokyo 183, JAPAN

Adult eclosion of the silkworm, *Bombyx mori*, occurs during a specific period of the day under the control of eclosion hormone (EH). *Bombyx* EH was extracted from the pupal and adult brains (Fugo, et al., 1982), and also the embryos of *B. mori* (Chen, et al., 1986). In 1985, *Bombyx* EH was first isolated from the pharate adult heads of *B. mori* (Nagasawa, et al., 1985). Recently the entire amino acid sequence of *Bombyx* EH was determined as shown in Fig. 1 through the combination of sequencing the isolated *Bombyx* EH (Kono, et al., 1987) and cloning the *Bombyx* EH gene. A monoclonal antibody against the synthetic peptide corresponding to the carboxyl-terminal 13 amino acid residues of *Bombyx* EH was prepared (Kono, et al., in press).

This paper deals briefly with the amino acid sequence of *Bombyx* EH along with some biological properties.

Starting with 770,000 pharate adult heads, two active fractions were isolated with yields of 26.9 μg and 10.3 μg, respectively. When injected into pharate adult *B. mori*, each fraction caused eclosion behavior at a dose of 0.1-0.3 ng/insect (Kono, et al., 1987). *Bombyx* EH was also active in *Manduca sexta* and *Samia cynthia ricini* (Fugo, unpublished).

After reductive alkylation, each fraction was digested with lysyl endopeptidase and V8 protease, and the sequence

H-[1]S-P-A-I-A-S-S-Y-D-[10]A-M-E-I-C-I-E-N-C-A-[20]Q-
[21]C-K-K-M-F-G-P-W-F-[30]E-G-S-L-C-A-E-S-C-I-[40]K-
[41]A-R-G-K-D-I-P-E-C-[50]E-S-F-A-S-I-S-P-F-L-[60]N-
[61]K-L-OH

Fig. 1. Amino acid sequence of Bombyx EH

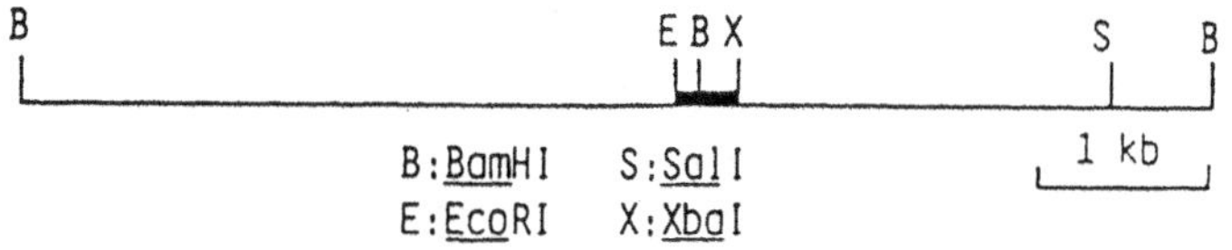


```
                                                   XbaI       9
                                                   CTAGATAAA
10                                                           68
TGTTATTTTCGAACCGAATTTGAATTGGTTCAATATTTTGGAGATTAGAGGGTACAAGC
69                                                          127
GCACAGATAATGTAGTAGTTGAAACTAGTGTCGCGTCGTTGAATGAAAATTTTAATGTG
128             141                                         172
CTTATTTTCACAG   CA.TCC.AGC.TAC.GAC.GCA.ATG.GAA.ATT.TGC.ATC.
               Ala-Ser-Ser-Tyr-Asp-Ala-Met-Glu-Ile-Cys-Ile-
                5                   10                  15
173                                                        217
GAG.AAC.TGC.GCG.CAA.TGC.AAA.AAA.ATG.TTC.GGC.CCT.TGG.TTC.GAG.
Glu-Asn-Cys-Ala-Gln-Cys-Lys-Lys-Met-Phe-Gly-Pro-Trp-Phe-Glu-
                20                  25                  30
218                                                        262
GGA.TCC.CTT.TGC.GCG.GAG.TCC.TGC.ATC.AAA.GCC.AGA.GGC.AAA.GAC.
Gly-Ser-Leu-Cys-Ala-Glu-Ser-Cys-Ile-Lys-Ala-Arg-Gly-Lys-Asp-
                35                  40                  45
263                                                        307
ATT.CCG.GAA.TGC.GAA.AGC.TTC.GCA.TCC.ATA.TCT.CCA.TTC.CTG.AAT.
Ile-Pro-Glu-Cys-Glu-Ser-Phe-Ala-Ser-Ile-Ser-Pro-Phe-Leu-Asn-
                50                  55                  60
308      316                                               362
AAG.CTT.TAA. GACCGATCGTTGATAACTATGATTGATGGACGTAAGTTTATGTAA
Lys-Leu-***
      62
       370
ATGAATTC
   EcoRI
```

Fig. 2. Restriction map and sequence of Bombyx EH gene

analysis of the resulting fragments revealed a probable 61 amino acid sequence from Ser 1 to Lys 61 (Kono, et al., 1987). Amino acid sequence analysis suggested that at least two molecular species of Bombyx EH exist, differing in their amino-terminals; one starts from Ser 1 and another from Ala 3, but the structure of the carboxyl-terminal remained to be clarified (Nagasawa, et al., 1987; Kono, et al., 1987).

To confirm the amino acid sequence and to determine the carboxyl-terminal structure, the gene encoding Bombyx EH was cloned and sequenced. The genomic library containing 6 x 10^6 recombinants was prepared from the genomic DNA of 3rd instar Bombyx larvae and λ EMBL3 vector. Synthetic oligonucleotide probes were designed on the basis of the amino acid sequence of Bombyx EH (probe A, 54mer corresponding to Tyr 8 - Phe 25; probe B, 33mer antisense corresponding to Gln 20 - Glu 30; probe C, 36mer antisense corresponding to Ile 39 - Glu 50; probe D, a mixture of 8 kinds of 17mer corresponding to Tyr 8 - Ile 13). After screening 1.2 x 10^5 clones, fragments of the Bombyx EH gene were subcloned into pUC18 vectors and sequenced to reveal that XbaI-BamHI fragment encoded the amino-terminal from Ala 5 to Gly 31 of Bombyx-EH and BamHI-Eco-RI fragment corresponded to the carboxyl-terminal from Gly 31 to Leu 62, as shown in Fig. 2. A sequence 5' upstream from the nucleotides encoding Ala 5 seems to be an intron, because there is a consensus sequence for an acceptor site of the intron such as (PyPy)nNCAG. The terminal codon of TAA found after the codon of Leu 62 indicates that the carboxyl-terminal of Bombyx EH is Leu-OH. Thus, the amino acid sequence of Bombyx EH was suggested as shown in Fig. 1, showing 80% homology with Manduca EH (Kataoka, et al., 1987; Marti, et al., 1987).

A monoclonal antibody, recognizing specifically both the synthetic peptide and native Bombyx EH, was prepared using a synthetic tridecapeptide, corresponding to Bombyx EH carboxyl-terminal from Cys 49 to Lys 61, as an antigen. Immunohistochemistry using this antibody indicates that Bombyx EH is produced in two pairs of the middorsal-anterior neurosecretory cells of Bombyx brain (Kono, et al., in press).

This work was partly supported by grants from The Ministry of Education, Science, and Culture, Japan.

REFERENCES

Chen, J.H., Fugo, H., Nakajima, M., Nagasawa, H., and A. Suzuki. (1986) The presence of neurohormonal activities in embryos of the silkworm, Bombyx mori. J. Seric. Sci. Jpn., 55: 54-59.

Fugo, H. and Y. Iwata. (1983) Changes of eclosion hormone activity in the brain during pupal and adult development in the silkworm, Bombyx mori. J. Seric. Sci. Jpn., 52: 79-84.

Kataoka, H., Troetschler, R.G., Kramer, S.J., Cesarin, B. and Schooley, D.A. (1987) Isolation and primary structure of the eclosion hormone of the tobacco hornworm, Maduca sexta. Biochem. Biophys. Res. Comm., 146: 746-750.

Kono, T., Nagasawa, H., Isogai, A., Fugo, H. and Suzuki, A. (1987) Amino acid sequence of eclosion hormone of the silkworm. Agric. Biol. Chem., 51: 2307-2308.

Kono, T., Mizoguchi, A., Nagasawa, H., Ishizaki, H., Fugo, H. and Suzuki, A. (1990) A monoclonal antibody against a synthetic carboxyl-fragment of the eclosion hormone of the silkworm, Bombyx mori. Zool. Science, in press.

Marti, T., Takio, K., Walsh, K., Terzi, G. and Truman, J.W. (1987) Microanalysis of the amino acid sequence of the eclosion hormone from the tobacco hornworm, Manduca sexta. FEBS Lett., 219: 415-418.

Nagasawa, H., Kamito, T., Takahashi, S., Isogai, A., Fugo, H. and Suzuki, A. (1985) Eclosion hormone of the silkworm, Bombyx mori: Purification and determination of the N-terminal amino acid sequence. Insect Biochem., 15: 573-578.

Nagasawa, H., Mikogami, T., Kono, Y., Fugo, H., and Suzuki, A. (1987) Molecular heterogeneity of eclosion hormone in adult heads of the silkworm, Bombyx mori. Agric. Biol. Chem., 51: 1741-1743.

From *Insect Neurochemistry and Neurophysiology • 1989* Edited by A. B. Borkovec and E. P. Masler

EXPRESSION OF THE ECLOSION HORMONE GENE IN THE BRAIN OF MANDUCA SEXTA

Frank M. Horodyski, Lynn M. Riddiford,
and James W. Truman
Department of Zoology NJ-15
University of Washington
Seattle, WA 98195

INTRODUCTION

Neuropeptides are an important class of regulatory molecules in insects which affect a number of physiological functions relating to growth, metamorphosis, and homeostasis. One of the better known neuropeptides is eclosion hormone (EH). EH is synthesized in the brain of the insect and released into the hemolymph at a defined developmental time just prior to ecdysis at the end of the molt (Reynolds and Truman, 1983). Once released into the hemolymph, its principal effect is on the nervous system of the insect to activate a program of neural activity which leads to ecdysis. Other effects of EH integrate ecdysis with the physiological events that occur at that time.

Insect neuropeptides have traditionally been difficult to study since they are generally present in very small amounts, and usually synthesized only in a few cells. Yet EH was purified from the tobacco hornworm, *Manduca sexta*, and sequenced (Marti *et al.*, 1987; Kataoka *et al.*, 1987). Using this sequence, we isolated and characterized the EH gene of *Manduca* and begun a study of its developmental expression as summarized below.

RESULTS AND DISCUSSION

From the sequence of the N-terminal 24 amino acids of *Manduca* EH, we designed a unique 72 nucleotide-long synthetic probe to screen the *Manduca* genomic library and obtained clones from three separate genomic regions (Horodyski *et al.*, 1989). A translation of the DNA sequence of one of these clones was identical to the previously determined amino acid sequence of *Manduca* EH from amino acids 5 to 62, and was followed by a termination codon.

To determine the upstream sequence and the organization of the *Manduca* EH gene, we constructed a cDNA library from brain RNA of day 9-12 developing adults and isolated a clone from this library by hybridization to the labelled genomic clone (Horodyski *et al.*, 1989). Sequence analysis of the cDNA clone showed that the EH mRNA has a translated region that codes for a 88 amino acid preEH, which consists of a 26 amino acid signal peptide followed by a single copy of the EH peptide. Using this cDNA as a probe, we showed that the gene is 7.8 kb in length and is composed of 3 exons, which are separated by introns of 3.1 and 4.0 kb (Fig. 1). High stringency hybridization to a genomic Southern blot showed that the *Manduca* EH gene is present in only one copy per haploid genome.

To determine when the EH gene is expressed during adult development, we hybridized the cloned gene to

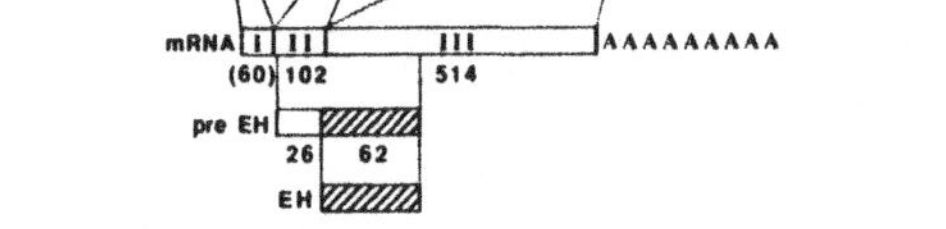

Fig 1. Organization of the *Manduca* EH gene. The top line shows a partial restriction map of the *Manduca* EH genomic clone showing the location of the 3 exons. The size of the exons is shown. The relative location and size of the signal peptide (stippled) and the mature EH (hatched) are shown below. From Horodyski *et al.*, (1989).

Northern blots containing brain RNA isolated from *Manduca* at various stages (Horodyski *et al.*, 1989). A single mRNA of 0.8 kb hybridized to the cloned EH gene. The EH mRNA was present at relatively constant levels throughout adult development and in diapausing pupae. This RNA was also detected in brain of feeding 5th instar larvae, but not in any region of the larval nerve cord or in the epidermis. This result supported the recent findings of Truman and Copenhaver (1989) that in the larva the only EH-immunopositive cells are in the brain. These cells send their axons down the ventral nerve cord to exit in the proctodeal nerve and thus accounts for the earlier findings that EH accumulates in the ventral nervous system prior to each larval and pupal ecdysis (Reynolds and Truman, 1983).

To begin to understand the factors involved in determining the site of EH synthesis, we localized by *in situ* hybridization the cells in the brain which contained the EH mRNA (Horodyski *et al.*, 1989). These studies conclusively showed that the EH mRNA was present only in the 2 pairs of ventromedial neurosecretory cells in feeding 5th instar larvae which are also immunopositive. During adult development, another group of brain cells, the Group 1a neurosecretory cells, also acquire immunoreactive and bioactive EH (Copenhaver and Truman, 1986). These cells project their axons to the corpora cardiaca which is the major site of EH release in the adult. We examined brains from 3 times during adult development using *in situ* hybridization and failed to detect EH message in these lateral cells although it was present abundantly in the ventromedial cells at all times examined (Horodyski *et al.*, 1989). These results suggest that either the level of mRNA in the lateral cells is much lower than in the ventromedial cells, or that the mRNA is only transiently expressed in these cells at a stage not yet examined by *in situ* hybridization.

We are currently examining whether sequences similar to the *Manduca* EH gene can be detected in the genomes of other insects. At low stringency hybridization conditions, a single hybridizing region is detected in the genomes of *Bombyx mori* and *Drosophila melanogaster*. We have isolated a cross-hybridizing *Drosophila* genomic clone and are presently characterizing it. If similar genes could be found in other species, it would support the idea that EH is found universally among insects (Reynolds and Truman, 1983).

The combination of modern molecular approaches with traditional physiological studies are valuable in understanding the structure, regulation and mode of action of neuropeptides. The use of the cloned gene to synthesize large amounts of neuropeptides will provide reagents for physiological and biochemical studies. Also, recombinant technology provides a way for producing structurally modified versions of EH, a task that is impractical by conventional means for this peptide.

REFERENCES

Copenhaver, P.F. and Truman, J.W. (1986) Identification of the cerebral neurosecretory cells that contain eclosion hormone in the moth *Manduca sexta*. *J. Neurosci.* **6**, 1738-1747.

Horodyski, F.M., Riddiford, L.M. and Truman, J.W. (1989) Isolation and expression of the eclosion hormone gene from the tobacco hornworm, *Manduca sexta*. *Proc. Nat. Acad. Sci. USA*, in press.

Kataoka, H., Troetschler, R.G., Kramer, S.J., Cesarin, B.J. and Schooley, D.A. (1987) Isolation and primary structure of the eclosion hormone of the tobacco hormworm, *Manduca sexta*. *Biochem. Biophys. Res. Commun.* **146**, 746-750.

Marti, T., Takio, K., Walsh, K.A., Terzi, G. and Truman, J.W. (1987) Microanalysis of the amino acid sequence of the eclosion hormone from the tobacco hornworm, *Manduca sexta*. *FEBS Lett.* **219**, 415-418.

Reynolds, S.E. and Truman, J.W. (1983) Eclosion hormone. In *Insect Endocrinology* (Edited by Downer, R.G.H. and Laufer, H.), pp. 217-233. A.R. Liss, New York.

Truman, J.W. and Copenhaver, P.F. (1989) The larval eclosion hormone neurones in *Manduca sexta*: Identification of the brain-proctodeal neurosecretory system. *J. Exp. Biol.* in press.

From *Insect Neurochemistry and Neurophysiology • 1989* Edited by A. B. Borkovec and E. P. Masler

SEQUENCING AND BIOLOGICAL ACTIVITIES OF NEUROPARSINS ISOLATED FROM CORPORA CARDIACA OF *LOCUSTA MIGRATORIA*

Josiane Girardie°, Jean-Claude Huet",
Jean-Claude Pernollet and Adrien Girardie
° Lab. Neuroendocrinologie, URA CNRS 1138,
Av.des Facultés, 33405 Talence, France
" Lab. Et. Protéines, INRA Versailles, Fr.

Neuroparsins (NPs) are two distinct and differently charged proteins, present in 70% methanolic extracts of nervous corpora cardiaca (N-CC) of *Locusta migratoria*. Since their production in the central nervous system is restricted to the Al type of neurosecretory cells from the pars intercerebralis (PI), they were named neuroparsins (Girardie *et al*., 1987a ; Bourême *et al*., 1987). Using sodium dodecyl sulfate polyacrylamide gel electrophoresis (SDS-PAGE), the two NPs had the same apparent molecular weight. Reduction by 2 mercaptoethanol demonstrated that both NPs were constituted of two polypeptide chains of the same apparent molecular weight linked by disulfide bridges (Girardie *et al*., 1985). *In vivo* labelling with 35S cysteine revealed that only the most basic NP was present in the PI while both NPs were stored in the N-CC (Girardie *et al*., 1987b). In vitro enzymatic transformation of the PI-NP into NP stored only in the N-CC was obtained following incubation of PI-NP in crude extracts of N-CC (Bourême and Girardie, 1988). Consequently, we designated the parent NP as NPA and the mature NP as NPB.

PURIFICATION

Purification of NPs from 70% methanolic crude extracts of N-CC was performed using only 2 steps of liquid chromatography. Crude extracts of N-CC were first separated on an anion exchange column (Mono Q HR5/5

Pharmacia) using a linear gradient of NaCl. Fifteen to twenty fractions, each corresponding to one or two peaks, were manually collected following absorbance curves simultaneously at 225 and 280 nm. Neuroparsins A and B were identified by their electrophoretic mobility on 7.5% PAGE as the essential protein content of fractions 4 and 6 (Girardie et al., 1989). Each NP was then desalted and fully purified on a reversed-phase column (Pro RPC HR5/10 Pharmacia) using acetonitrile gradient and TFA as ion pair reagent.

CHARACTERIZATION

Neuroparsins were sequenced using an Applied Biosystems gas-phase protein sequencer (Girardie et al., 1989). Native neuroparsins were only partially sequenced. Forty-four residues, with only one residue at every step of Edman degradation, were characterized for NPB. Only twenty-nine residues were obtained for NPA which appeared to be composed of three polypeptide chains : one corresponding to NPB and two others having respectively five and two extra N-terminal residues. Because of the homogeneity of NPB (homodimer), we were first able to analysed its sequence. The entire characterization of NPB (reduced and alkylated) was deduced from sequences of five overlapping peptides obtained following cyanogen bromide cleavage (two peptides) and endoprotease Glu-C digestion (three peptides). Neuroparsin B monomer is composed of seventy-eight amino-acids including twelve cysteines and one methionine. Its calculated molecular weight is 8188 D, in agreement with previous SDS-PAGE estimations of 9000 D (Girardie et al., 1985) and 7000 D (Bourême and Girardie, 1986). The sequence of NPB monomer is :

S C E G A N C V V D L T R C E Y G D V T D F F G R K V C A
K G P G D K C G G P Y E L H G K C G V G M D C R C G L C S
G C S L H N L Q C F F F E G G L P S S C

BIOLOGICAL EFFECTS

Biological activities of neuroparsins were investigated using either pure NPA or NPB and specific immune serum obtained against NPA or NPB (Bourême et al., 1987). Both neuroparsins have similar biological activities. They have anti-juvenile, anti-diuretic hypertrehalosemic and hyperlipemic effects.

Injection of anti-neuroparsin immune serum, initiated in III larval instar until adult life, induced green pigmentation as early as the fourth larval moult, disturbed metamorphosis and advanced sexual maturation in both sexes (Girardie et al., 1987a). These three physiological events are juvenile hormone dependent. Injections of pure NPA or NPB both inhibited sexual maturation (Girardie et al., 1987a) and decreased juvenile hormone levels in hemolymph (unpublished results).

In vitro introduction of NPA or NPB in everted rectum preparations both enhanced, in a dose-response manner, fluid reabsorption (Fournier and Girardie, 1988).

In vivo injection of NPA or NPB into adult male locusts both increased, in a dose-response manner, hemolymph lipid and trehalose levels (Moreau et al., 1988).

In conclusion, a polytropic neurohormone produced in 2 isoforms by the A1 type of the protocerebral median neurosecretory cells has been isolated from Locusta migratoria. Because of the site of production, the two isoforms were named neuroparsins A and B. Only NPB has yet been fully sequenced. It is the largest and most sulfur containing insect neurohormone known to date.

Using immunohistology, neuroparsin-like substances has been demonstrated in the PI of all species tested in Orthoptera, Odonata, Isoptera and Dictyoptera, but not in species of other orders (Tamarelle and Girardie, 1989). Moreover, neuroparsin-like substances of two species of Dictyoptera exhibited antidiuretic activity similarly to neuroparsins (Bourême et al., 1989). Thus, Locusta neuroparsin could constitute the first known member of a new intraphyletic family of insect neurohormones.

REFERENCES

Bourême D. et Girardie J. (1986) Présence de protéines à ponts disulfure dans les corpora cardiaca d'insectes. C.r. Acad. Sci. Paris, **18**, 655-660.

Bourême D. et Girardie J. (1988) Les neuroparsines du criquet (Orthoptera) : microhétérogénéité d'origine enzymatique. Ann. Soc. entomol. Fr., **24**, 390.

Bourême D., Fournier B., Matz G. and Girardie, J. (1989)

Immunological and junctional cross-reactivities between locust neuroparsins and proteins from cockroach corpora cardiaca. J. Insect Physiol., **35**, 265-271.

Bourême D. Tamarelle M. and Girardie J. (1987) Production and characterization of antibodies to neuroparsins A and B isolated from the corpora cardiaca of the locust. Gen. comp. Endocrinol., **67**, 169-177.

Fournier B. and Girardie J. (1988) A new function for the locust neuroparsins : stimulation of water reabsorption. J. Insect Physiol., **34**, 309-313.

Girardie J., Bourême D., Couillaud F., Tamarelle M. and Girardie A. (1987a) Anti-juvenile effect of neuroparsin A, a neuroprotein isolated from locust corpora cardiaca. Insect Biochem., **17**, 977-983.

Girardie J., Bourême D. and Girardie A. (1987b) Production sites of the three neurosecretory proteins characterized in the corpora cardiaca of the migratory locust. Insect Biochem., **17**, 29-36.

Girardie J., Faddoul A. and Girardie A. (1985) Characterization of three neurosecretory proteins from the A median neurosecretory cells of Locusta migratoria by coupled chromatographic, electrophoretic and isoelectrofocusing methods. Insect Biochem., **15**, 85-92.

Girardie J., Girardie A., Huet J.C. and Pernollet J.C. (1989) Amino acid sequence of locust neuroparsins. FEBS Letters, **245**, 4-8.

Moreau R., Gourdoux L. and Girardie J. (1988) Neuroparsin : A new energetic neurohormone in the African locust. Arch. Insect Biochem. Physiol., **8**, 135-145.

Tamarelle M. and Girardie J. (1989) Immunohistochemical investigation of locust neuroparsin-like substances in several insects, in some other invertebrates, and vertebrates. Histochem., **91**, 431-435.

From *Insect Neurochemistry and Neurophysiology • 1989* Edited by A. B. Borkovec and E. P. Masler

ISOLATION AND STRUCTURE OF THREE NEUROPEPTIDES FROM THE CORPORA CARDIACA OF THE AMERICAN COCKROACH

Jan A. Veenstra*

Department of Biological Organic Chemistry,
Centro de Investigación y Desarrollo,
Barcelona, Spain.

In the last years numerous insect neuropeptides have been isolated and sequenced (Holman *et al.*, 1988; Nachman *et al.*, 1986a,b). Here the recent isolation of three new neuropeptides from the corpora cardiaca of the American cockroach, *Periplaneta americana*, is briefly summarized. These peptides are corazonin, a novel cardioacceleratory peptide, and two peptides related to the leucosulfakinins (Veenstra, 1989a,b).

Dissected corpora cardiaca were homogenized in 1 ml Bennet's mixture (1% NaCl, 5% formic acid and 1% TFA in 1 N HCl), centrifuged and the supernatants were loaded on a previously activated and equilibrated C-18 Sep-Pak. After washing the cartridge with 5 ml 0.1% TFA and 5 ml 0.1% TFA in 13% acetonitrile, the peptides were eluted with 5 ml 0.1% TFA in 32% acetonitrile. After reduction of the volume by nitrogen evaporation the peptides were separated by HPLC on 15 x 0.46 cm column and 5 x 0.46 precolumn filled with 5 µm particles Spherisorb ODS II. To purify the peptides, two consecutive gradients between water and acetonitrile, which contained HFBA in the first and TFA in the second gradient as pairing ions (fig. 1), were used.

Determination of the amino acid composition with a Picotag system gave the following results: corazonin: Asx_1, Glx_2, Ser_1, Gly_1, Arg_1, Thr_2, Tyr_1, Phe_1; perisulfakinin: Asx_2, Glx_2, Gly_1, His_1, Arg_1, Tyr_1, Met_1, Phe_2; non-sulfated leucosulfakinin II: Asx_2, Glx_1, Ser_1, Gly_1, His_1, Arg_1, Tyr_1, Met_1, Phe_1. The peptides were present in amounts of ca 2.0, 2.5 an 1.1 pmol per pair CC respectively.

One peptide could be directly sequenced, but the other two had blocked N-terminals, and were sequenced after pyroglutamate amino-

* present address: Department of Entomology, 434 Forbes, bldg # 36, The University of Arizona, Tucson, AZ 85721, USA.

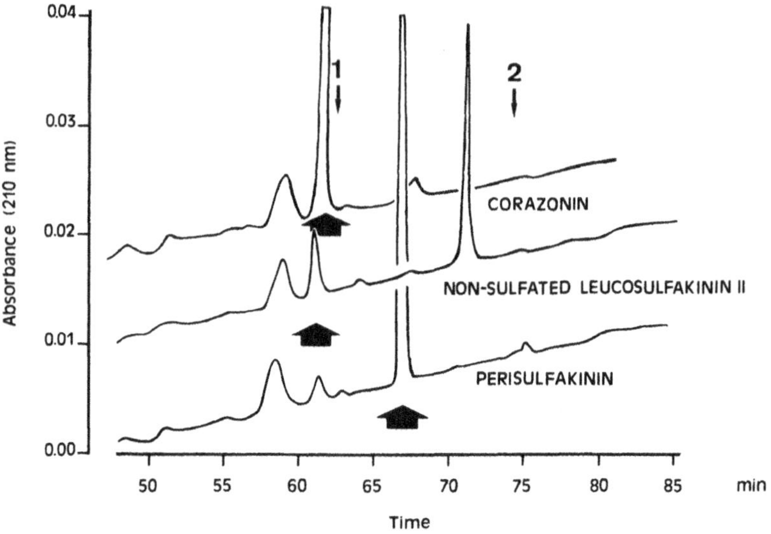

Fig. 1. Final purification of corazonin, non-sulfated leucosulfakinin II and perisulfakinin isolated from the American cockroach. Isocratic elution for 30 min of 13% acetonitrile with 0.1 % TFA, followed by a linear gradient from 13 to 33 % acetonitrile over 60 min, flow rate 1ml/min. Purified peptides indicated by large arrows, small arrows indictate the elution positions of periplanetins CC1 and CC2 respectively.

peptidase digestion. Sequencing of these peptides was performed by "Servicio de sequenciación de Proteinas de la Universidad de Barcelona", using a model 470A protein sequencer of Applied Biosystems. Trypsin digestion was used to determine whether the C-terminals were blocked. The trypsin digests of the leucosulfakinin related peptides were derivatized with phenyl isothiocyanate (PITC), which yielded in both cases PTC-Phe-amide. The C-terminal tetrapeptide fragment from a trypsin digest of corazonin was isolated by HPLC and then submitted to manual Edman degradation for three cycles, but without removing the PTH amino acid residues. After the third cycle the mixture was again reacted with PITC, and then analyzed for the presence of PTC-Asn-amide. A peak with the same retention time as authentic PTC-Asn-amide was found in this mixture and its identity confirmed by acid hydrolysis of the collected PTC-Asn-amide peak, which yielded PTH-Asn and PTC-Asn, the same products that were found after hydrolysis of authentic PTC-Asn-amide. The whole structure of Corazonin was found to be Glp-Thr-Phe-Gln-Tyr-Ser-Arg-Gly-Trp-Thr-Asn-amide. It increased the rate of heart beat at concentrations as low as 0.2 nM (fig. 2), which makes it about 100 times more potent than the periplanetins CC1 and CC2 isolated by Scarborough *et al.* (1984). The two leucosulfakinin-related peptides did not affect the rate of heart beat. It is interesting to note, that although corazonin is not a typical member of the AKH/RPCH-family, it does have some sequence similarity

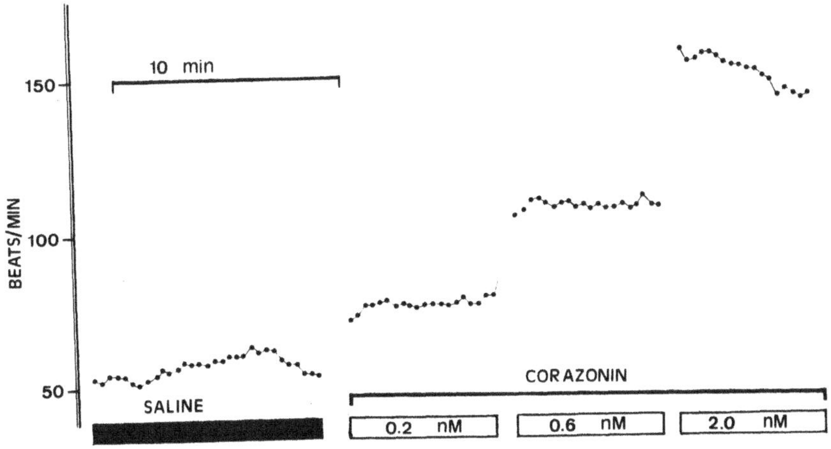

Fig. 2. The changes in the rate of heart beat induced by corazonin.

Glp-Leu-Thr-Phe-Ser-Ser-Gly-Trp-Gly-Asn-NH_2 (1)
Glp-Thr-Phe-Gln-Tyr-Ser-Arg-Gly-Trp-Thr-Asn-NH_2 (2)

Fig. 3. The structures of *Heliothis zea* hypertrehalosemic hormone (1) and corazonin (2).

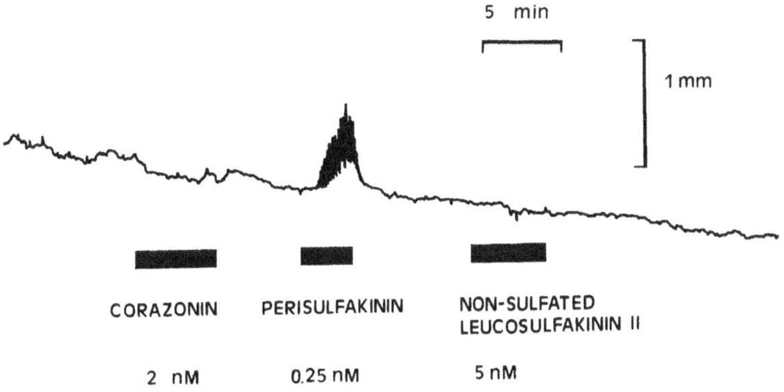

Fig. 4. Effects of corazonin, perisulfakinin and non-sulfated leucosulfakinin II on hindgut contractions in the American cockroach.

to the hypertrehalosemic hormone recently isolated from *Heliothis zea* by Jaffe *et al.* (1988), as shown in fig.3, but the peptide did not affect haemolymph trehalose at physiological doses.

The presence or absence of tyrosine sulfate in the leucosulfakinin-related peptides was determined by PITC derivatization of barium hydroxide hydrolysates.The two peptides related to the leucosulfakinins turned out to be [Asp^4]-leucosulfakinin I, which was called perisulfakinin, and non-sulfated leucosulfakinin II. The first peptide stimulated hindgut contractions in the American cockroach in concentrations as low as 250 pmol, while non-sulfated leucosulfakinin II and corazonin were not active in the concentrations tested (fig. 4).

I thank Henry H. Hagedorn for critically reading the manuscript. This work was supported by a postdoctoral fellowship from the Spanish Ministry of Education and Science and grants from C.A.I.C.Y.T. (84/87) and C.S.I.C. (85/263).

Jaffe, H., Raina, A.K., Riley, C.T., Fraser, B.A., Bird, T.G., Tseng, C.-M., Zhang, Y.-S. and Hayes, D.K. (1988) Isolation and primary structure of a neuropeptide from *Heliothis zea* with hypertrhalosemic and adipokinetic activities. *Biochem. Biophys. Res. Commun.* **155**, 344-350.

Holman, G.M., Wright, M.S. and Nachman, R.J. (1988) Insect neuropeptides: Coming of age. *ISIS Atlas Anim. Plant Sci.* **1**, 129-136.

Nachman, R.J., Holman, G.M., Haddon, W.F. and Ling, N. (1986a) Leucosulfakinin, a sulfated insect neuropeptide with homology to gastrin and cholecystokinin. *Science* **234**, 71-73.

Nachman, R.J., Holman, G.M., Cook, B.J., Haddon, W.F. and Ling, N. (1986b). Leucosulfakinin II, a blocked sulfated insect neuropeptide with homology to cholecystokinin and gastrin. *Biochem. Biophys. Res. Commun.* **140**, 357-364.

Scarborough, R.M., Jamieson, G.C., Kalish, F., Kramer, S.J., McEnroe, G.A., Miller, C.A. and Schooley, D.A. (1984).Isolation and primary structure of two peptides with cardioacceleratory and hyperglycaemic activity from the corpora cardiaca of *Periplaneta americana. Proc. Natl. Acad. Sci. USA* **81**, 5575-5579.

Veenstra, J.A., (1989a) Isolation and structure of two gastrin/CCK-like neuropeptides from the American cockroach homologous to the leucosulfakinins. *Neuropeptides, in press.*

Veenstra, J.A., (1989b) Isolation and structure of corazonin, a cardioactive peptide from the American cockroach. *FEBS LETTERS, in press.*

From *Insect Neurochemistry and Neurophysiology • 1989* Edited by A. B. Borkovec and E. P. Masler

ISOLATION, PURIFICATION AND CHARACTERIZATION OF A DIURETIC PEPTIDE (AP-I) FROM THE HOUSE CRICKET, *ACHETA DOMESTICUS*.

Geoffrey M. Coast, Colin H. Wheeler, Nicholas F. Totty*, Robin J. Philp*, and Graham J. Goldsworthy.
Birkbeck College, Malet Street, London WC1E 7HX. U.K., and *Ludwig Institute for Cancer Research, London, WC1. U.K.

Insect diuretic peptides differ chemically and, even within a single species, several forms of diuretic peptide may be present (Wheeler and Coast, 1989). Thus, in *Locusta migratoria*, diuretic peptides have been described as AVP-like, ACTH-like, and as novel peptides. Only two insect diuretic peptides have been fully characterised, one from *Locusta* (Proux *et al*., 1987), and the other from *Manduca sexta* (Kataoka *et al*., 1989). They have no structural analogy to each other, but each has similarities with some vertebrate peptides. Fractionation of aqueous extracts of the corpora cardiaca of *Acheta domesticus* yields a number of peptides which stimulate fluid secretion by isolated Malpighian tubules (Coast and Wheeler, 1989). One peptide, *Acheta* peptide-I (AP-I), is associated with a major UV absorbing peak, and was selected for sequence analysis.

Freshly dissected corpora cardiaca were sonicated, heated at 100°C for 5 min, and then centrifuged for 2min. at 12000g. AP-I is purified by HPLC on a Lichrosorb RP8 column followed by separation on an Aquapore RP300 column, both eluted in acetonitrile with TFA as the modifier. A single major UV absorbing peak, which has diuretic activity, is obtained and this was used for sequence determination. The purity of such material was validated by reversed-phase (on phenyl and octadecyl columns), and by size exclusion and ion-exchange HPLC. Diuretic activity was always associated with a UV peak identified as AP-I on the Lichrosorb RP8 separation. Purified AP-I when subjected to capillary zonal electrophoresis (20kV; 25mM citrate buffer, pH 2.5) showed only a single peak of UV absorbance at 200nm.

Acheta peptide-I does not absorb strongly at 280nm and, after treatment with dithiothreitol and 4-vinyl pyridine (Tarr, 1986), no absorbance peak at 260nm is detected and biological activity is retained. These data suggest the absence of tyrosinyl, tryptophanyl and cystyl residues Automated Edman-degradation with on-line PTH amino acid analysis was used to determine the primary structure. The intact peptide yields no sequence, but treatment with pyroglutamate aminopeptidase (PGA) gives two products: peak 1 [Arg]^{2}..[Lys]^{16} (see Table 1), and peak 2, a more hydrophobic product, [Arg]^{2}..[Pro]^{15}. Peptide mapping using sequence-grade enzymes and chemical sequencing suggested the entire sequence [pGlu]^{1}..[Lys]^{16} (Table 1).

The amino acid composition of a hydrolysate of AP-I is shown in Table 2. Only minor impurities (leucine and tyrosine) were detected. The molecular weight of AP-I determined by fast atom bombardment mass spectrometry (FAB-MS) was 1925, corresponding to the average molecular weight of the free acid form of [pGlu]^{1}..[Lys]^{16}. AP-I was also isolated on the Lichrosorb RP8 column eluted with acetonitrile/TFA at pH 2.1, and re-chromatographed on the same column, but eluted with acetonitrile at pH 6.8 using ammonium acetate as the modifier. AP-I isolated by this method was biologically active and had an identical molecular weight to the material isolated previously.

Table 1 Sequenced products of peptide mapping studies.

AP-I: Entire sequence	<Q	R	D	I	F	H	A	Q	T	D	I	F	Q	V	P	K-OH
PGA (peak 1)		R	D	I	F	H	A	Q	T	D	I	F	Q	V	P	K
PGA (peak 2)		R	D	I	F	H	A	Q	T	D	I	F	Q	V	P	
TPCK-trypsin			D	I	F	H	A	Q	T	D	I	F	Q	V	P	K
TLCK-chymotrypsin						H	A	Q	T	D	I	F				
Endoprotease V8			D	I	F	H	A	Q	T		I	F	Q	V	P	K

The free acid and amidated forms of [pGlu]^{1}..[Lys]^{16} were synthesised using FMOC chemistry. Neither of these synthetic peptides, however, has diuretic activity when assayed at 10^{-10}M to 10^{-6}M. There are a number of possible reasons why the synthetic peptide failed to elicit biological activity. Natural AP-I (nAP-I) may have been co-purified with an extremely potent minor contaminant which has all of the biological activity. Analysis of the 'purified' peptide by FAB-MS has revealed other minor mass ions and additonal chromatographic procedures are being developed to test the purity of nAP-I.

Table 2 Amino acid composition of a hydrolysate of nAP-I.

Glx	3.34 (3)	Val	1 (1)
Thr	0.90 (1)	Arg	1.03 (1)
Asx	2.25 (2)	His	0.92 (1)
Phe	1.84 (2)	Pro	1.04 (1)
Ala	1.09 (1)	Leu	0.01 (-)
Ile	1.89 (2)	Tyr	0.06 (-)

Figures in parentheses refer to the refer to the number of residues predicted from the sequence data.

Other possible reasons for the lack of activity of the synthetic peptide are that the sequence data for nAP-I is incomplete, or that the synthetic peptide may be an inactive form of the natural product. However, nAP-I used for sequence analysis and for FAB-MS was biologically active. Furthermore, synthetic [pGlu]^{1}..[Lys]^{16}-OH gives the same sequence as nAP-I, has the same molecular weight, and co-elutes with nAP-I. However, one chemical difference between the synthetic and natural peptides has been detected. During treatment with PGA under identical conditions, nAP-I yields two products (see above), whereas the synthetic peptide gives only a single product (= peak 1). A study of the time-course of the reaction with AP-I shows that peak 1 appears first. This product, [Arg]^{2}..[Lys]^{16}, co-elutes with a synthetic peptide having the identical sequence. Peak 2 appears to be derived from peak 1, differing by the absence of the [Lys]^{16} residue. The appearance of two reaction products after treatment with PGA may be due to prolidase activity in the enzyme preparation. However, this fails to explain why our synthetic peptide does not react similarly. This is being investigated further using a combination of sequence analysis and FAB-MS.

Acknowledgements: We are grateful to Professor R. Caprioli (Houston, Texas) and Dr. A. I. Mallet (St. Thomas's Hospital, London) for FAB-MS data, and to Prof. M. Waterfield (Ludwig Institute, London) for access to peptide sequencing and synthesis facilities.

REFERENCES

Coast G.M. and Wheeler C.H. (1989) The distribution and relative potency of diuretic peptides in the house cricket, Acheta domesticus. Physiol. Entomol., In press.

Kataoka H., Troetschler R.G., Li J.P., Kramer S.J., Carney R.L. and Schooley D.A. (1989) Isolation and identification of a diuretic peptide from the tobacco hornworm, Manduca sexta. Proc. Nat. Acad. Sci. USA, 86, 2976-2980.

Proux J. P., Miller .C.A., Li J.P., Carney R.L., Girardie A., Delaage M. and Schooley D.A. (1987) Identification of an arginine vasopressin-like diuretic hormone from Locusta migratoria. Biochem. Biophys. Res. Commun., 149, 180-186.

Tarr, G.E. (1986) Manual Edman sequencing system. In Methods of protein microcharacterization (Ed. by Shively, J.E.), pp. 162-163, Humana press, New York.

Wheeler C.H. and Coast G.M. (1989) Assay and characterisation of diuretic factors in insects. J. Insect Physiol., In press.

From *Insect Neurochemistry and Neurophysiology* • *1989* Edited by A. B. Borkovec and E. P. Masler

MEMBERS OF THE AKH/RPCH-PEPTIDE FAMILY IN COLEOPTERA

Gerd Gäde
Institut für Zoologie, Lehrstuhl für Tierphysiologie, Universität Düsseldorf, D-4000 Düsseldorf 1, F.R.G.

INTRODUCTION

Several peptides localized in the corpora cardiaca (CC) of insects are known to regulate lipid and carbohydrate mobilization. Such adipokinetic and hypertrehalosaemic activities have been found in the neuroendocrine organs from representatives of several insect orders. The primary structures of such neuropeptides are known, however, only from orthopteran species and two species of moths (see, Gäde, 1988). The peptides are octa, nona- or decapeptides belonging to one large peptide family, the so-called adipokinetic hormone/red pigment-concentrating hormone (AKH/RPCH)-family, because of their close structural relationship. Recently, I have become interested in the presence and role of such peptides in another holometabolous order, the Coleoptera. In the present communication I analysed the peptide content of the corpora cardiaca from two members of the family Tenebrionidae (the mealworms T. molitor and Zophobas rugipes), from one member of the family Chrysomelidae (the Colorado potato beetle Leptinotarsa decemlineata), and from four members of the family Scarabaeidae (the African rosechafers Pachnoda marginata and P. sinuata, as well as the May beetles Melolontha hippocastani and M. melolontha.

Methanolic extracts of CC (0.25 to 0.5 gland pairs) from all the investigated beetles but the Melolontha species had a hypertrehalosaemic effect when injected into the American cockroach and caused hyperlipaemia in acceptor migratory locusts. The material from M. melolontha caused a very weak hypertrehalosaemic effect, but was not active in the locust.

Separation of methanolic extracts of CC was performed on a reversed-phase Nucleosil C-18 column (5 µm particle size). The aqueous solvent was 0.11% trifluoroacetic acid (solvent A) and the organic solvent was 0.11% TFA in 60% acetonitrile (solvent B). The solvents were applied as a linear gradient (25 to 80% B in 45 min or 43 to 53% B in 20 min; flow rate 1 ml/min) and the eluent was monitored at 210 or 214 nm. The extract from Z. rugipes showed three main absorbance peaks; only the most hydrophobic one caused hypertrehalosaemia in cockroaches. The extract from the CC of T. molitor had an absorbance peak associated with hypertrehalosaemic activity at the same retention time as the active peak of Z. rugipes. The CC material from L. decemlineata revealed two absorbance peaks with biological activity co-eluting with the hypertrehalosaemic peptides of the American cockroach, Pea-CAH-I and II (nomenclature according to Raina and Gäde, 1988). Extracts of CC from the African rosechafers resulted in a prominent and a minor absorbance peak with adipokinetic activity. An absorbance peak with the same retention time as the prominent peak of the rosechafers was also obtained when CC extract of the two May beetles was chromatographed.

The HPLC-separated material from various species was subjected to amino acid analysis after hydrolysis in hydrochloric and (in some case s) methanesulfonic acid. The materials from Z. rugipes and T. molitor gave the same results, and the data identified the compound as an octapeptide with the following amino acid composition: 2 Asx, 1 Glx, 1 Ser, 1 Pro, 1 Leu, 1 Phe and 1 Trp. The two active fractions from the CC of L. decemlineata had composition data identical to those of the cockraoch hypertrehalosaemic octapeptides; compound I: 2 Asx, 1 Glx, 1 Ser, 1 Pro, 1 Val, 1 Phe and 1 Trp; and compound II: 1 Asx, 1 Glx, 2 Thr, 1 Pro, 1 Leu, 1 Phe and 1 Trp. The prominent HPLC-peak from the CC of P. marginata turned out to be also an octapeptide: 2 Asx, 1 Glx, 1 Ser, 1 Pro, 1 Tyr, 1 Leu and 1 Trp.

The isolated biologically active materials from T. molitor, Z. rugipes and L. decemlineata were enzymatically deblocked using L-pyroglutamate aminopeptidase, and the undigested peptides were separated from the deblocked ones by HPLC. The deblocked peptides were used for pulsed-liquid sequencing. Amidation at the C-terminal was shown by the unability of carboxypeptidase A to cleave the molecules. The sequences for the CC material from both members of the Tenebrionidae were identical; the compound termed Tem-HrTH because of its hypertrehalosaemic effect in T. molitor (Rosinski and Gäde, 1988) has the structure pGlu-Leu-Asn-Phe-Ser-Pro-Asn-TrpNH_2. The two neuropeptides from the Colorado potato beetle are identical to the hypertrehalosaemic peptides of the American cockroach and, therefore, have the same nomenclature: Pea-CAH-I: pGlu-Val-Asn-Phe-Ser-Pro-Asn-TrpNH_2; Pea-CAH-II: pGlu-Leu-Thr-Phe-Thr-Pro-Asn-TrpNH_2. Preliminary experiments have shown that both peptides increase proline synthesis in vitro in the potato beetle (G. Gäde and E. Weeda, unpublished).

In summary, the CC of tenebrionid beetles contain a novel octapeptide member of the AKH/RPCH-family that causes hypertrehalsaemia and phosphorylase-activation in T. molitor and Z. rugipes. Surprisingly, such far-related species as a cockroach (P. americana) and a beetle (L. decemlineata) contain identical neuropeptides. Apparently, the same peptides are involved in different metabolic regulations in these species: carbohydrate in the cockroach and proline in the beetle.

The author thanks Miss B. Schumacher for expert technical help, Dr. R. Kellner (Applied Biosystems GmbH, Weiterstadt, F.R.G.) for pulsed-liquid sequencing, and the Deutsche Forschungsgemeinschaft for financial support (Ga 241/6-4) and the award of a Heisenberg Fellowship (Ga 241/5-2).

REFERENCES

Gäde G. (1988) New structures of insect neuropeptides. In Endocrinological Frontiers in Physiological Insect Ecology (Ed. by Sehnal F., Zabza A. and Denlinger D.L.), pp. 635-650. Technical University Press, Wroclaw.

Raina A.K. and Gäde G. (1988) Insect peptide nomenclature. Insect Biochem. **8**, 785-787.

Rosinski G. and Gäde G. (1988) Hyperglycaemic and myoacitve factors in the corpora cardiaca of the mealworm, Tenebrio molitor. J. Insect Physiol. **34**, 1035-1042.

From *Insect Neurochemistry and Neurophysiology • 1989* Edited by A. B. Borkovec and E. P. Masler

STRUCTURES IN THE AKH FAMILY OF NEUROPEPTIDES

Colin H. Wheeler, Alex F. Drake, Carrie M. Wilmot, Janet M. Thornton, Gerd Gade* and Graham J. Goldsworthy.

Birkbeck College, Malet Street, London, WC1E 7HX., U.K. and *Universitat Dusseldorf, Dusseldorf, D-4000, F.R.G.

In locusts, adipokinetic hormones have well defined actions: they mobilise diacylglycerols from the fat body and stimulate oxidation of lipids by the flight muscles (Goldsworthy and Wheeler, 1989). Their actions presumably depend on interactions with specific receptors in the tissues, but there have been no direct studies of AKH receptors. Some information about receptors can, however, be inferred from biological assay data. For most actions, AKH-I and to a lesser extent AKH-II are both effective agonists, but AKH-II stimulates cAMP accumulation in the fat body more than AKH-I (Goldsworthy et al., 1986), and AKH-II releases carbohydrate from the fat body, whereas AKH-I does not (Loughton and Orchard, 1981). These differences in actions suggest that AKH-I and -II may act via different receptors at the fat body, although their high degree of cross reactivity in other assays indicates that they may also stimulate the same receptors. The availability of synthetic naturally occurring analogues of adipokinetic peptides has prompted an examination of their structure/activity relationships. The remarkably high degree of conservatism displayed by this family of peptides indicates the importance of specific residues within the peptide. All have $[pGlu]^1$, an amidated C-terminal, $[Phe]^4$, $[Trp]^8$ or conservative substitutions of $[Leu]^2$ or $[Val]^2$, $[Asp]^3$ or $[Thr]^3$, and $[Thr]^5$ or $[Ser]^5$. Despite the great similarities between these peptides, they exhibit a range of activities when

Table 1. PRIMARY STRUCTURES AND PREDICTED TYPE-I β-TURNS (UNDERLINED) FOR MEMBERS OF THE AKH FAMILY

Name	Sequence and turn prediction	CD Evidence
Locust AKH-I	<QLNFTPNWGT$_{NH2}$	Yes
Carausius-HTH2	<QLTFTPNWGT$_{NH2}$	Yes
Nauphoeta/Blaberus HTH	<QVNFSPGWGT$_{NH2}$	Yes
Romalea-I AKH	<QVNFTPNWGT$_{NH2}$	Yes
Manduca/Heliothis AKH	<QLTFTSSWG$_{NH2}$	No
Crustacean RPCH	<QLNFSPGW$_{NH2}$	Yes
Locusta AKH-II	<QLNFSAGW$_{NH2}$	No
Schistocerca AKH-II	<QLNFSTGW$_{NH2}$	No
Periplaneta M-I	<QVNFSPNW$_{NH2}$	Yes
Periplaneta M-II	<QLTFTPNW$_{NH2}$	Yes
Gryllus/Romalea-II/ *Acheta* AKH	<QVNFSTGW$_{NH2}$	No

A type-I β-turn between [Ser]^6..[Gly]^9 is also predicted for Manduca-AKH.

assayed for hyperlipaemia in locusts, with ED_{50} ranging from 1 to 10pmol, ED_{max} between 3 and 40pmol, and maximum responses between 45 and 100% that to *Locusta* AKH-I (Goldsworthy and Wheeler, 1989). In many respects the value of these studies is limited because the effects of individual substitutions are often impossible to interpret. Nevertheless, the availability of a range of natural peptides provides a means of assessing methods to determine secondary structure in peptides: interpretation of structure/activity studies is limited without some concept of the shape of the molecules. Stone *et al.* (1978) predicted that AKH-I would adopt a β-turn and this would allow interaction between residues at each end of the molecule to aid stability. We have used a β-turn prediction algorithm (Wilmot and Thornton, 1988) to compute the probability of a β-turn occurring, and its type and position. Type-I β-turns were consistently predicted in all peptides listed in Table I except *Locusta* and *Schistocerca* AKH-IIs, and *Romalea*-II. Such predictions relate only to statistical probabilities and we have therefore tried to correlate these with actual measurements, using circular dichroism (CD) spectroscopy.

Fig. 1. Stereoscopic skeletal model of locust AKH-I suggesting a possible conformation for receptor binding

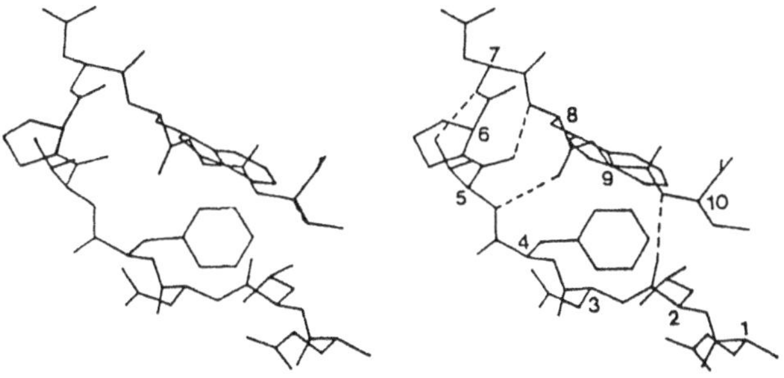

Preliminary CD spectroscopy of AKH-I suggested that it had no identifiable secondary structure in solution. However, secondary structures of peptides can be influenced by low temperatures or the presence of SDS micelles. The CD spectrum of AKH-I was therefore monitored after addition of SDS at 15°C and in a solution of ethanediol:water (2:1) at -110°C. Under such conditions spectra were obtained which were characteristic of a type-I and type-II β-turn respectively. Thus, at room temperature, AKH-I normally exists in equilibrium between a type-I and a type-II β-turn, giving an averaged CD spectrum. Only those peptides which possess proline adopt a β-turn in the presence of SDS (Table 1). Romalea-II and the locust AKH-IIs had no identifiable structure, but contrary to prediction, Manduca-AKH did not adopt a turn. On the basis that AKH-I can adopt a type-I β-turn, we have modelled AKH-I (Fig. 1) using HYDRA graphics on an E & S PS300 graphics system linked to a VAX/750. Note that $[pGlu]^1$ and $[Thr]^{10}$ have been modelled using standard amino acids. A type-I β-turn incorporating residues 5-8 was built using standard torsions and all other angles were set to an ideal β-conformation. This resulted in orientation of $[Leu]^2$, $[Phe]^4$ and $[Trp]^8$ side-chains on one side of the molecule.

The aromatic rings of [Phe]^4 and [Trp]^8 were orientated using analysis of ring-ring interactions (Singh and Thornton, 1985) and the model energy minimised (600 cycles) using GROMOS (Van Gunsteren and Berendsen, 1987). The model suggests an amphipathic molecule with a distinct hydrophobic cluster consisting of [Leu]^2, [Phe]^4 and [Trp]^8 side-chains on one face of the molecule which is likely to be important for interaction with receptors. This model provides a useful aid for the rational design of analogues and for the interpretation of structure activity studies of AKH-like peptides.

REFERENCES

Goldsworthy G.J. and Wheeler C.H. (1989) Physiological and structural aspects of adipokinetic hormone function in locusts. Pestic. Sci. 25, 85-95.

Goldsworthy G.J., Mallison K. and Wheeler C.H. (1986) The relative potencies of two known locust adipokinetic hormones. J. Insect Physiol. 32, 95-101.

Loughton B.L. and Orchard I. (1981) The nature of the hyperglycaemic factor from the glandular lobes of the corpus cardiacum of Locusta migratora. J. Insect Physiol. 27, 383-385.

Singh J. and Thornton J.M. (1985) The interaction between phenylalanine rings in proteins. FEBS Letters 191, 1-6.

Stone J.V., Mordue W., Broomfield C.E. and Hardy P.M. (1978) Structure-activity relationships for the lipid mobilising action of locust adipokinetic hormone. Eur. J. Biochem. 89, 195-202.

Van Gunsteren W.F. and Berendsen H.J.C. (1987) Groningen molecular simulation (GROMOS) Library manual, BIOMOS, Mijenborgh 16, Groningen, The Netherlands, pp. 1-29.

Wilmot C.M. and Thornton J.M. (1988) Analysis and prediction of the different types of β-turn in proteins. J. Mol. Biol. 203, 221-232.

From *Insect Neurochemistry and Neurophysiology • 1989* Edited by A. B. Borkovec and E. P. Masler

QUATERNARY STRUCTURE AND HYDROPHOBIC INTERACTIONS OF *DROSOPHILA* ACETYLCHOLINESTERASE IN WILD TYPE FLIES AND IN MUTANTS OF THE *Ace* LOCUS.

Jean-Pierre Toutant[1] and Martine Arpagaus[2]

(1) Pharmacology, Case Western Reserve University, Cleveland, Ohio
(2) Pharmacology, University of Michigan, Ann Arbor.

Adult *Drosophila* acetylcholinesterase (AChE) is predominantly a disulfide-linked dimer of globular subunits (Toutant *et al.*, 1988). Each catalytic subunit possesses a C-terminal phosphatidylinositol that anchors the molecule to the membrane (Gnagey *et al.*, 1987; Fournier *et al.*, 1988a). Phosphatidylinositol-specific phospholipase C (PI-PLC) releases the terminal glycolipid and converts the membrane-bound AChE to a hydrophilic form which no longer interacts with nondenaturing detergents. Thus, the original dimer is referred to as an **amphiphilic** molecule. We present here additional structural data on the dimeric AChE of adult wild type *Drosophila.* We also report preliminary results on the structure of AChE in certain mutants of the *Ace* locus (Hall *et al.*, 1980; Greenspan *et al.*, 1980).

A scheme of the structure of the C-terminal region of *Drosophila* AChE is presented in figure 1 (see Toutant, 1989). It contains the C-terminal cysteine of the protein, a complex glycan and a terminal phosphatidylinositol. The cleavage sites of proteinase K (PK), PI-PLC (PLC), anchor specific phospholipase D (PLD) and hydroxylamine (NH_2OH) at alkaline pH are indicated.

Sensitivity of *Drosophila* AChE to PI-PLC and PLD

Nondenaturing gel electrophoresis was used to unambiguously resolve amphiphilic and hydrophilic AChE forms (Arpagaus and Toutant, 1985). Figure 2 shows that PI-PLC from *Bacillus thuringiensis* converted the amphiphilic G_2 form (lane a, band 1) to a fast migrating hydrophilic G_2 form (lane b, band 2). An incomplete conversion was achieved with rabbit serum PLD (lane c). The additional band observed in the sample treated with

PLD (*) was a cholinesterase from the rabbit serum (see figure 6, lane d). In figure 2, lane d, the amphiphilic dimer in lane a was treated with 10^{-2} M dithiothreitol, for one hour at 37°C at pH 8.6 and was converted to an amphiphilic monomer (Band 1').

Effect of Alkaline Hydroxylamine

The glycolipid anchors of membrane proteins investigated thus far fall into two structural types: diacylglycerol (type 1) as in the trypanosome variant surface proteins or alkylacylglycerol (type 2) as in mammalian AChE (Rosenberry *et al.*, 1989 for a review). Hydroxylamine at pH>10.5 cleaves the ester bonds releasing the fatty acyl chains but not the ether-linked alkyl chain (Toutant *et al.*, 1989). Hydroxylamine therefore converts glycolipid-anchored proteins of type 1 to hydrophilic counterparts (complete deacylation). Type 2 proteins retain their hydrophobic interactions after an alkaline hydroxylamine treatment because the remaining alkyl chain is still able to bind detergent. We treated *Drosophila* AChE by hydroxylamine at pH 10.6 for 1 hour at 37°C in the presence of edrophonium chloride (1mM) for the protection of enzymatic activity. Figure 3 shows that the migration of the amphiphilic G_2 AChE (lane a) was not modified by NH_2OH treatment (lane c) indicating that the hydrophobic interactions of the molecule were not altered. Thus, the hydrophobic domain of the glycoinositol phospholipid of *Drosophila* AChE is likely an alkylacylglycerol (see figure 1).

Effect of Proteinase K Treatment

Mild treatments with proteases convert the glycolipid-anchored dimers of torpedo and Human AChEs to hydrophilic dimers. Proteinase K cleaves in the peptidic sequence, removing at least the last C-terminal amino acid (a Cys residue at position 615 in *Drosophila* AChE, Haas *et al.*, 1988). It has been proposed that this Cys residue is also involved in the interchain disulfide bond in the dimer (Fournier *et al.*, 1988b). In this hypothesis, proteinase K treatment would release the disulfide bond as well as the hydrophobic domain and produce hydrophilic **monomers.** Figure 4 shows that the treatment produced essentially hydrophilic **dimers**. Lanes b, c and d correspond to 15, 30 and 60 min. digestions with 0.1ug/ul of PK at 37°C. The results suggest that the interchain link does not involve the C-terminal Cys but rather the other free Cys residue at position 328 (Hall and Spierer, 1986).

Mutants of the *Ace* Locus

The following *Ace* mutants in *Drosophila melanogaster:* j19, j29, j40, j50 and lm38 were a gift of Dr. J.C. Hall (Brandeis University). Homozygous *Ace* j29/*Ace* j29 are cold-sensitive but survive at 25-29°C. *Ace* j40/*Ace* j40 are heat-sensitive and survive only at 18°C. Homozygous J40 and J29 flies raised at permissive temperatures had AChE activity respectively 30 and 40% that of wild types. The mutations j19, j50 and

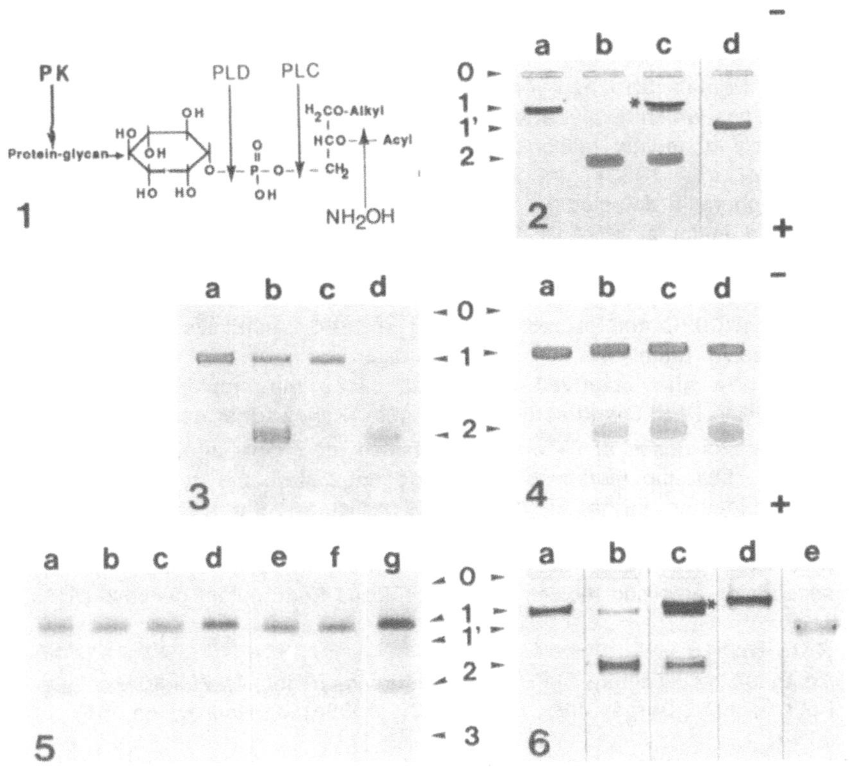

Figure 1: scheme of the glycolipid anchor. **Figures 2-4:** nondenaturing gel electrophoresis of wild type *Drosophila* AChE treated by PI-PLC and serum PLD (2), hydroxylamine (3) and proteinase K (4). In figure 3, lane a: control; lane b: PI-PLC-treated; lane c: hydroxylamine-treated; lane d: hydroxylamine then PI-PLC. **Figure 5:** molecular forms of AChE in j19/j29 or j29/j50 flies raised at 29°C (permissive temp.) and shifted to 18 or 29°C after eclosion; lanes a, d and g: j19/+, j29/+, j50/+; lanes b and c: j19/j29 at 18 and 29°C respectively; lanes e and f: j29/j50 at 18 and 29°C respectively. **Figure 6:** sensitivity of AChE (a) to PI-PLC (b) and serum PLD (c) in J29/j29 mutants raised at 29°C and maintained for 7 days at 18°C (restrictive temp.); lane d: PLD alone showing the endogenous rabbit cholinesterase (*); lane e: disulfide bond reduction.
O: origin of migrations; **1** and **1'**: amphiphilic dimer and monomer; **2** and **3**: hydrophilic dimer and monomer.

lm38 are nonconditional lethal. Different heteroallelic combinations such as j19/j29, j29/j50 or j29/lm38 developed well at 29°C and the adults were placed for several days at the restrictive or permissive temperatures. These animals also had reduced AChE activities. We tested whether the mutations, which likely affect the AChE protein, corresponded to faulty assembly of mutant monomers or to defects in the membrane interactions (Hall *et al.*, 1980). In all mutants examined, nondenaturing gel electrophoresis detected the same molecular forms of AChE as in the wild types: a major amphiphilic dimer (1) with minor amounts of an amphiphilic monomer (1') and hydrophilic dimer (2) and monomer (3). Figure 5 shows this in the case of the heteroallelic combinations j19/j29 and j29/j50 raised at 29°C and placed thereafter at 18°C (restrictive temp.) or 29°C (permissive temp.).

We also observed that in all cases the amphiphilic dimer was sensitive to PI-PLC and serum PLD. Figure 6 shows the results obtained with J29/j29 flies raised at 29°c and maintained for 7 days at 18°C.

Thus, the mutations studied do not disturb the overall structure of the molecular forms of AChE; in particular, the association of the amphiphilic dimer with the membrane by a glycolipid anchor appears unaffected. We are currently localizing directly the sites of the mutations by sequencing amplified regions of the *Ace* locus from mutant genomic DNAs.

References

Arpagaus, M., Toutant, J.P. (1985) Neurochem. Int. **7**, 793-804.

Fournier, D., Bergé, J.B., Bordier, C. (1988a) J. Neurochem. **50**, 1158-1163.

Fournier, D., Bride, J.M., Karch, F., Bergé, J.B. (1988b) FEBS Lett. **238**, 333-337.

Gnagey, A.L., Forte, M., Rosenberry, T.L. (1987) J. Biol. Chem. **262**, 13290-13298.

Greenspan, R.J., Finn, J.A., Hall, J.C. (1980) J. Comp. Neurol. **189**, 741-774.

Haas, R., Marshall, T.L., Rosenberry, T.L. (1988) Biochemistry **27**, 6453-6457.

Hall, J.C., Alahiotis, S.N., Strumpf, D.A., White, K. (1980) Genetics **96**, 939-965.

Hall, L.C.M., Spierer, P. (1986) EMBO J. **5**, 2949-2954.

Rosenberry, T.L., Toutant, J.P., Haas, R., Roberts, W.L., (1989) Meth. Cell Biol. **32**, 231-255.

Toutant, J.P. (1989) Prog. Neurobiol. **32**, 423-446.

Toutant, J.P., Arpagaus, M., Fournier, D. (1988) J. Neurochem. **50**, 209-218.

Toutant, J.P., Roberts, W.L., Murray N.R., Rosenberry, T.L. (1989) Eur. J. Biochem. **180**, 305-310.

From *Insect Neurochemistry and Neurophysiology* • *1989* Edited by A. B. Borkovec and E. P. Masler

INSECT HYPERTREHALOSEMIC HORMONE: STRUCTURE-ACTIVITY STUDIES IN *PERIPLANETA AMERICANA*

Gerd Gäde[1] and Timothy K. Hayes[2]
[1]Institut für Zoologie, Lehrstuhl für Tierphysiologie, Heinrich-Heine-Universität, D-400 Düsseldorf 1, F.R.G.
[2]Laboratories for Invertebrate Neuroendocrine Research Department of Entomology, Texas A&M University College Station, Texas 77843, U.S.A.

The hypertrehalosemic hormone (HTH, Bld-HrTH) is a decapeptide that has been isolated from both *Blaberus discoidalis* (Hayes et. al., 1986) and *Nauphoeta cinerea* (Gäde and Rinehart, 1986) cockroaches. The major function of the peptide in both insects is the elevation of hemolymph carbohydrates. MI (Pea-CAHI) is the endogenous hypertrehalosemic peptide in *Periplaneta americana*. Both MI and HTH are members of the of the AKH-RPCH family of neuropeptides. Both peptides are active hypertrehalosemic factors when injected into either *P. americana* or *B. discoidalis*.

Analogs to HTH were originally prepared to determine what amino acid residues and termini are important for the hyperglycemic response in *B. discoidalis* (Ford et al., 1988). These studies have been repeated in *P. americana* to test for changes in how the two animals read the biological information coded in HTH. This manuscript compares the results between the two insect species and demonstrates that neuropeptide analogs can be designed to control species selectivity.

Analogs to HTH were prepared by solid-phase peptide synthesis, purified by RP-HPLC , and their identity tested through analytical HPLC and amino acid analysis (Hayes et. al., 1986). Solvent was removed prior to reconstitution of each analog sample in water and aliquots (10 µl) were injected into adult male *P. americana*. Before and 2 hr after sample injection, a sample (1 µl) of hemolymph was drawn from each test animal. Carbohydrate concentration of each extraction sample was determined by a modified anthrone procedure. Results are expressed as percentage of the change of the hemolymph carbohydrate concentration elicited by the injection of 5 pmoles of MI.

The information contained in the structure of a peptide like HTH is translated into biological action through reception at the surface membrane of the target cell. There are two steps involved in the interaction of a neurohormone at the target cell receptor: 1) binding (recognition), and 2) an induced conformational change by the receptor that stimulates the cellular response (transduction). Information about the structural elements of the peptide responsible for these two events can be inferred from structure-activity studies where full dose-response curves have been tested. The potency (ED_{50}) of an analog relative to the natural hormone gives information about how structural changes influence that part of the hormone responsible for receptor recognition (the "address"). And, changes in the maximum activity found in a dose-response curve give information on the part of the neurohormone that induced the transduction (the "message"). Address and message can exist on separate or overlapping portions of the hormone sequence. We have used full dose-response curves to calculate the bioassay parameters that suggest information about the "address" (ED_{50}) and "message" (Maximum Activity) of HTH in *P. americana.*

Several amino acids in the HTH sequence appear important for the "address" as read by *P. americana* for the hyperglycemic response (Table 1). For convenience we list the sequence of HTH in three letter amino acid code:

1 2 3 4 5 6 7 8 9 10
pGlu-Val-Asn-Phe-Ser-Pro-Gly-Trp-Gly-Thr-NH_2

The aromatic residues at positions 4 and 8 are critical to the recognition of the peptide in this cockroach. Both [Ala^4]-HTH and [Leu^8]-HTH are inactive. These analogs could not block the action of MI when coinjected into assay animals. Neither of these analogs have any interaction with the MI receptor. Analogs to positions 2,3 and 6 show the strongest potency decreases without complete inactivation.

The conformation of HTH also appears to be important to a functional address for the peptide in *P. americana.* A β-turn is predicted in HTH at residues 5-8. The relative potencies of single amino acid replacement analogs and the frequency of amino acids in the first three positions of β-turns show a direct correlation for positions 5-7 as was previously observed for *B. discoidalis* (see Hayes et al., this volume).

Depression of the saturable maximum activity by several analogs in the *P. americana* hyperglycemic assay suggest that several side chains contribute to the "message" in HTH (Table 1). Side chains at positions 3, 5 and 6 all fall into this category. The aromatic side chain of Phe at position 4 might be important to the message as well. However, the Phe is so important to the address that effects on message can not be ascertained, since recognition (address) is required prior to transduction (message) of the response.

Table 1. MI and HTH analog sequences and bioassay parameters for hyperglycemia in *P. americana.*

Peptide Sequence	ED_{50} (pmole)	Maximum Activity (% Max MI Response)
MI:		
pQVNFSPNW-NH_2	1	100
HTH:		
pQVNFSPGWGT-NH_2	3	102
Ac-AVNFSPGWGT-NH_2	55	105
pQANFSPGWGT-NH_2	130	97
pQVAFSPGWGT-NH_2	200	39*
pQVNASPGWGT-NH_2	NA	NA
pQVNFGPGWGT-NH_2	50	59*
pQVNFSGGWGT-NH_2	155	54*
pQVNFSPNWGT-NH_2	3	93
pQVNFSPNWGT-NH_2	4	95
pQVNFSPGLGT-NH_2	NA	NA
pQVNFSPGWAT-NH_2	5	100
pQVNFSPGWAT-NH_2	35	100
pQVNFSPGWGA-NH_2	3	106
pQVNFSPGWGT-OH	125	78*

•Single underlined amino acid=single L-amino acid replacement
•Double underlined amino acid=single D-amino acid replacement
*Significant depression of maximum activity
•NA = No activity detected

Several of these analogs were tested in time course experiments to determine if degradation or elimination functions were interfering with the dose-response information obtained in our studies. All analogs tested, paralleled the time course of MI. These results are consistent with MI and the HTH analogs undergoing similar degradation and elimination processes. Thus, our observations should reflect peptide-receptor interactions in *P. americana*.

A comparison of an identical analog series in two separate insects suggests that slight differences in receptor binding sites exist as illustrated in Figure 1. These differences can be exploited to control the species selectivity of analogs to insect neuropeptides. Analogs can be specific to one insect. For example, [N-acetyl-Ala^1]-HTH is a strong agonist in *P.*

americana and inactive in *B. discoidalis*. Alternatively, several analogs including both endogenous hormones are strong hyperglycemic peptides in both species.

	1	2	3	4	5	6	7	8	9	10
	pGlu-	Val-	Asn-	Phe-	Ser-	Pro-	Gly-	Trp-	Gly-	Thr-NH_2
Blaberus:	A	-	a	A	M	a	-	A	a	-
Periplaneta:	-	a	aM	A	M	aM	-	A	-	-

Figure 1. Comparison of the suggested locations of "address" and "message" functions on the sequence of HTH in *B. discoidalis* and *P. americana*. A=critical influence on Address. a=strong but non critical influence on address. M=observable influence on Message. (-)= only a small or no influence on address or message.

ACKNOWLEDGEMENTS

The authors with to thank the Deutsche Forschung-shemeinschaft (Ga 241/6-2, 6-3, 6-4 to G.G.) and the Heisenberg Fellowship (Ga 241/5-2 to G.G.), the U.S. Public Health Service [Grant TMP(AHR)R01 NS20137 to TKH] and the U.S. Department of Agriculture [Grant 86-CRCR-1-2008 to TKH] for support of this research. We also wish to thank the Texas Agriculture Experiment Station's Biotechnology Support Laboratory for suppling the capability for peptide synthesis, sequencing, and amino acid analysis necessary for these studies. We would also like to thank Beate Schumacher, Allison Strey, Virginia Johnson, Mary Ford and Robert Jamroz for their various degrees of technical assistance.

LITERATURE CITED

Ford M. M., Hayes T. K. and Keeley L. L. (1988) Structure-activity relationships for insect hypertrehalosemic hormone: the importance of side chains and termini. In *Peptides: Chemistry and Biology* (Edited by Marshall, G. R.), pp. 653-655, ESCOM, Leiden.

Gade G., Rinehart K.L., (1986) Amino acid sequence of a hypertrehalosemic neuropeptide from the corpus cardiacum of the cockroach, Nauphoeta cinerea. Biochem. Biophys. Res.Comm. 141:774-781

Hayes T.K., Keeley L.L. and Knight D.W. (1986) Insect hypertrehalosemic hormone: isolation and primary structure from *Blaberus discoidalis* cockroaches. *Biochem. biophys. Res. Comm.* **140,** 674-678.

From *Insect Neurochemistry and Neurophysiology* • *1989* Edited by A. B. Borkovec and E. P. Masler

INSECT HYPERTREHALOSEMIC HORMONE: STRUCTURE-ACTIVITY STUDIES IN *BLABERUS DISCOIDALIS*

Timothy K. Hayes, M.M. Ford and L.L. Keeley
Laboratories for Invertebrate Neuroendocrine Research
Department of Entomology
Texas A&M University
College Station, TX 77843

Hypertrehalosemic hormone (HTH, Bld-HrTH) is a decapeptide member of the AKH-RPCH family of neuropeptides as indicated by its structure.

1 2 3 4 5 6 7 8 9 10
pGlu-Val-Asn-Phe-Ser-Pro-Gly-Trp-Gly-Thr-NH_2

The peptide was isolated from *Blaberus discoidalis* (Hayes et al., 1986) where it influences the elevation of hemolymph trehalose concentrations (Hayes and Keeley, 1985). The results presented in this manuscript represent an extension of our initial studies into how biological information is encoded by the structure of HTH.

Analogs to HTH were prepared by solid-phase peptide synthesis, purified by RP-HPLC , and their identity tested through analytical HPLC and amino acid analysis (Hayes et. al., 1986). Solvent was removed prior to reconstitution of each analog sample in an appropriate concentration of Ringer solution. Samples were injected into test animals (24 hr decapitated, adult, male *B. discoidalis*) in 10 µl of Ringer. Before and 2 hr after sample injection, a sample (5 µl) of hemolymph was drawn from each test animal. Free carbohydrate was extracted from whole hemolymph samples with 95µl of Ringer. Carbohydrate concentration of each extraction sample was determined by a modified anthrone procedure.

The information contained in the structure of a peptide like HTH is translated into biological action through reception at the surface membrane of the target cell. There are two steps involved in the interaction of a neurohormone at the target cell receptor: 1) binding (recognition), and 2) an induced conformational change by the receptor that stimulates the cellular response (transduction). Information about the structural elements of the peptide responsible for these two events can be inferred from structure-activity studies where full dose-response curves have been tested. The potency (ED_{50}) of an analog relative to the natural hormone gives information about how structural changes influence that part of the hormone responsible for receptor recognition (the "address"). And, changes in the maximum activity found in a dose-response curve give information on the part of the neurohormone that induced the transduction (the "message"). Address and message can exist on separate or overlapping portions of the hormone sequence. We have used full dose-response curves to calculate the bioassay parameters that suggest information about the "address" (ED_{50}) and "message" (Maximum Activity) of HTH.

Our initial investigations into this subject focused on single amino acid replacement analogs (Ford et. al., 1988) of HTH to help determine the importance of each amino acid side chain and termini for hyperglycemic activity. Each analog had one amino acid replacement and the hyperglycemic activity of each analog was compared to HTH. Positions 1, 4 and 8 are critical to the HTH "address". Positions 3, 6, and 9 are important but less critical to the "address". Removal of the side chain at position 5 results in an analog with reduced maximum activity which suggests that the Ser^5 side chain is important to the HTH "message". Positions 2, 7, and 10 tolerate a change in amino acid without much change in the bioassay parameters of the analog when compared to HTH.

Table 1. A comparison of turn frequency coefficients and the potency of HTH analogs at positions 5-7.

Sequence Position	HTH Residue	Analog Residue	Frequency Coefficient	Potency (ED_{50})
5(f_i)	Ser	Gly	0.120(0.102)	1(9)
6(f_i+1)	Pro	Gly	0.850(0.301)	1(200)
7(f_i+2)	Gly	Asn	0.191(0.190)	1(1)

Values in parentheses are for HTH single amino acid replacement analogs.
Values outside parentheses are for unmodified HTH.
f_i=the N-terminal amino acid position in a 4 amino acid turn.

Other information has suggested that a folded conformation for HTH may be important for the "address" of this peptide and its action in *B. discoidalis*. First, Chou-Fasman calculations (Chou and Fasman, 1978) predict that residues 5-8 form a β-turn. Also, Wheeler and coworkers (see article in this volume) have observed through CD analysis, evidence that HTH forms a β-turn in SDS-micelles. Further, we have compared HTH to single amino acid replacement analogs at positions 5-7 and have found correlations between the amino acid frequency coefficients and relative analog potency (Table 1). For example, Gly is found in the f_i position (HTH position 5) of β-turns slightly less often than Ser. [Gly^5]-HTH is only slightly less potent than HTH. On the other hand, Gly is found much less often than Pro at the f_{i+1} β-turn position, and [Gly^6]-HTH is much less potent than HTH. Gly and Asn are found in almost identical frequency at the f_{i+2} position. Likewise, [Asn^7]-HTH has identical potency values as HTH. Finally, in other studies we have found that position 7 of HTH tolerates replacement by D-amino acids much better than the rest of the molecule. Tolerance of D-amino acids has been an important clue to the presence and position of a β-turn in other small neuropeptides like HTH, for example LHRH (Rose et al., 1985).

Table 2. Bioassay parameters for select aromatic analogs of the hypertrehalosemic hormone in *Blaberus discoidalis*.

Peptide	ED_{50} (pmole)	% of HTH Max. Act. (μg/μl)
HTH	1	100
[Leu^4]-HTH	NA	NA
[Tyr^4]-HTH	2	100
[4-nitro-Phe]-HTH	15	98
[Leu^8]-HTH	NA	NA
[Phe^8]-HTH	200	SNR
[2-Naphthyl-Ala^8]-HTH	23	100

NA=No activity
SNR=Saturation not reached in the assay

We have looked in more detail at the aromatic residues at HTH positions 4 and 8, since these are critical to "address" function (Table 2). Aromaticity and not just hydrophobicity is required at both positions since replacement of either aromatic residue with Leu results in an inactive analog. A decrease in electron density of the aromatic ring at position 4 has a minor influence on the potency of the PNP-analog. A bicyclic aromatic ring is required at position 8 since [Phe^8]-HTH loses much more activity than the bicyclic [2-Naphthyl-Ala^8]-HTH. The positioning of the two aromatic rings on either end of the proposed turn sequence could be important. Computer models of HTH with a turn at the proposed 5-8 position, spatially locate the Trp^8 side chain under Gly^9 where the distal portion of the bicyclic indole ring can interact with the Phe^4 aromatic ring to stabilize the a conformation or perhaps coordinate the two rings to interact with a receptor aromatic residue.

ACKNOWLEDGEMENTS

The authors wish to thank the U.S. Public Health Service [Grant TMP(AHR)R01 NS20137 to TKH] and the U.S. Department of Agriculture [Grant 86-CRCR-1-2008 to TKH] for support of this research. We also wish to thank the Texas Agriculture Experiment Station's Biotechnology Support Laboratory for suppling the capability for peptide synthesis, sequencing, and amino acid analysis necessary for these studies. We also wish to thank Robert Jamroz, Virginia Johnson and Allison Strey for technical assistance.

LITERATURE CITED

Chou P.Y., Fasman G.D. (1978) Empirical predictions of protein conformation. *Ann. Rev. Biochem.* **47**, 251-276.

Ford M. M., Hayes T. K. and Keeley L. L. (1988) Structure-activity relationships for insect hypertrehalosemic hormone: the importance of side chains and termini. In *Peptides: Chemistry and Biology* (Edited by Marshall, G. R.), pp. 653-655, ESCOM, Leiden.

Hayes T.K. and Keeley L.L. (1985) Properties of an in vitro bioassay for hypertrehalosemic hormone of *Blaberus discoidalis* cockroaches. *Gen. comp. Endocr.* **57**, 246-256.

Hayes T.K., Keeley L.L. and Knight D.W. (1986) Insect hypertrehalosemic hormone: isolation and primary structure from *Blaberus discoidalis* cockroaches. *Biochem. biophys. Res. Comm.* **140**, 674-678.

Rose G.D., Gierasch L.M., Smith J.A. (1985) Turns in peptides and proteins. *Advances in Protein Chemistry* **37**, 1-108.

From *Insect Neurochemistry and Neurophysiology • 1989* Edited by A. B. Borkovec and E. P. Masler

THE ROLE OF MOSQUITO OOSTATIC HORMONE IN THE REGULATION OF MIDGUT SERINE PROTEASES

Dov Borovsky and David A. Carlson

University of Florida- IFAS, Florida Medical Entomology Laboratory, 200 9th Street S.E. Vero Beach, Florida 32962

U. S. Department of Agriculture, Agric. Research Service, Insects Affecting Man and Animals Research Laboratory, Gainesville, Florida 32604

Antigonadotropins or hormones that inhibit egg development (oostatic hormones) have been found in the cockroach, eye gnat, crustaceans, house fly, and mosquitoes (Iwanov and Mescherskaya, 1935; Carlisle and Knowles, 1959; Adams et al., 1968; Meola and Lea, 1972; Else and Judson, 1972; Kelly et al., 1984; Borovsky, 1985; Kelly et al. 1986). Kelly et al. (1984, 1986) reported that crude extract of oostatic hormone from *Musca domestica* inhibited both egg development and the biosynthesis of ecdysteroids in the autogenous mosquito *Aedes atropalpus* which led them to believe that the role of oostatic hormone is to inhibit ecdysteroid biosynthesis in the ovary. On the other hand, Davey (1978) and Davey and Kunster (1981) suggested that in *Rhodnius prolixus* oostatic hormone inhibits JH stimulation of the ovary, thus blocking the transport of vitellogenin from the hemolymph into the ovary. Borovsky (1985) reported that a 7,000-fold purified mosquito oostatic hormone that was injected into female *Aedes aegypti* inhibited egg development and the biosynthesis of vitellogenin. Egg development in anautogenous mosquitoes depends on blood digestion (Gooding, 1966) and the release of egg development neurosecretory hormone (EDNH) by the blood meal (Lea, 1967). Borovsky has shown recently (1985) that oostatic hormone does not block the release of EDNH from the brain. An attempt was made to study the possibility that oostatic hormone inhibits the biosynthesis of trypsinlike enzymes in the midgut of female mosquitoes. This report demonstrates that both native oostatic hormone and synthetic analogs modulate the biosynthesis of serine proteases in the midgut of female mosquito.

Mosquito oostatic hormone has been purified from the ovaries of 30,000 female A. aegypti using low and high pressure liquid chromatography. The hormone was eluted as a single peak after C_{18} reversed phase HPLC. The hormone was then subjected to amino acid analysis and sequenced using Tandem Quadrupole Mass Spectrometry with Ion Cyclotron Resonance. The hormone is a decapeptide of M_r 1,047 which exhibited 70% homology to HIV II X-ORF protein. Two decapeptide, an octapeptide and a pentapeptide analogs were synthesized using solid phase peptide synthesizer.

Injections of the natural hormone into female A. aegypti at concentrations of 240 pmol caused 23% inhibition of protease activity in the midgut and 62% inhibition in yolk deposition 24 h after the blood meal. Injections of larger concentrations of oostatic hormone 1.5-2.9 nmol caused 30%-91% inhibition of proteolytic enzyme activity in the gut and 56%-98% inhibition of yolk deposition, respectively. Similar results were obtained with synthetic oostatic hormone and a decapeptide analog in which the positions of amino acids 1 and 2 were switched. An octapeptide (amino acids 3-10) poorly inhibited egg development and proteolytic enzymes biosynthesis indicating that amino acid 1 and 2 may be involved in binding the hormone to cell specific receptors. On the other hand, a pentapeptide analog (amino acids 1-5) was very effective in inhibiting egg development and proteolytic enzymes biosynthesis 50%-100% at concentrations of 8-15 nmol, respectively.

Oostatic hormone is not species specific (Borovsky, 1985), when several concentrations of A. aegypti synthetic oostatic hormone (0.29-10 nmol) were injected into A. aegypti, C. variipennis, A. quadrimaculatus, Stomoxys calcitrans, Ctenocephalides felis (Bouche) and Lutzomyia anthophora (addis) synthesis of trypsinlike enzymes was reduced by 50% to 90% as compared with controls.

Collating these observations, a model for the control of trypsin biosynthesis in the mosquito is proposed. After the blood meal the midgut epithelial cells start to synthesize trypsin in concert with the neuroendocrine system that synthesizes oostatic hormone, as was proposed for Rhodnius (Davey and Kunster, 1981), which accumulates in the developing ovary as a prohormone. The active hormone is released into the hemolymph 24 h later, and signals the midgut epithelial cells to stop the synthesis of trypsinlike enzymes which ceases 55 h later.

ACKNOWLEDGEMENTS

This work was partially supported by USDA research grant CRC-1-2394 to Dr. Dov Borovsky.

REFERENCES

Adams, T. S., Hintz, A. M. and Pomonis, J. G. (1968) Oostatic hormone production in houseflies, *Musca domestica* with developing ovaries. *J. Insect Physiol.* *14*, 983-993.

Borovsky, D. (1985) Isolation and characterization of highly purified mosquito oostatic hormone. *Arch. Insect Biochem. Physiol.* *2*, 333-349.

Carlisle, D. B. and Knowles, F. (1959) Endocrine control in crustacean. *Cambridge Monog. Exp. Biol.* *10*, 1-120.

Davey, K. G. (1978) Hormonal stimulation and inhibition in the ovary of an insect *Rhodnius prolixus*. In *Comparative Endocrinology* (Ed. by Gaillard, P. J. and Boer, H. H.), pp. 13-16. Elsevier/North Holland, Amsterdam.

Davey, K. G. and Kunster, J. F. (1981) The source of an antigondotropin in the female of *Rhodnius prolixus* Stal. *Cand. J. Zool.* *59*, 761-764.

Else, J. G. and Judson, C. L. (1972) Enforced egg-retention and its effects on vitellogenesis in the mosquito *Aedes aegypti*. *J. Med. Entomol.* *9*, 527-530.

Gooding, R. H. (1966) Physiological aspects of digestion of the blood meal by Aedes aegypti (L.) and Culex fatigans Wiedemann. J. Med. Entomol. 3, 53-60.

Iwanov, P. P. and Mescherskaya, K. A. (1935) Die physiologischen Besonderheiten der geschlechtlich unreifen Insektenovarien und die zylischen Veranderungen ihrer Eigenschaften. Zool. Jb. (Physiol) 55, 281-348.

Kelly, T. J., Birnbaum, M. J., Woods, C. W. and Borkovec, A. B. (1984) Effects of housefly oostatic hormone on egg development neurosecretory hormone action in Aedes atropalpus. J. Exp. Zool. 229, 491-496.

Kelly, T. J., Masler, P. E., Schwartz, M. B. and Haught, S. B. (1986) Inhibitory effects of oostatic hormone on ovarian maturation and ecdysteroid production in diptera. Insect Biochem. 16, 273-279.

Lea, A. O. (1967) The medial neurosecretory cells and egg maturation in mosquitoes. J. Insect Physiol. 13, 419-429.

Meola, R. and Lea, A. O. (1972) Humoral inhibition of egg development in mosquitoes. J. Med. Entomol. 9, 99-103.

From *Insect Neurochemistry and Neurophysiology • 1989* Edited by A. B. Borkovec and E. P. Masler

RECENT DEVELOPMENTS IN THE USE OF HPLC FOR THE PURIFICATION OF HOUSE FLY OOSTATIC HORMONE

DeMilo A.B., Kelly T.J., Masler E.P. and Bell R.A.

Insect Reproduction Laboratory, USDA, ARS, Beltsville, MD, USA 20705

House fly oostatic hormone (OH) was first reported 21 years ago and postulated to be involved in a humoral feedback mechanism controlling cyclical production of eggs (Adams *et al*., 1968; Adams, 1981). Active OH extracts have been prepared from mature adult ovaries, abdomens, or whole bodies (Adams, *et al*., 1968; Adams 1977; Kelly *et al*., 1984a). While extensive studies to elucidate oostatic mechanisms have been conducted in *Rhodnius prolixus*, *Diploptera punctata*, *Aedes aegypti*, *Aedes atropalpus* and *Musca domestica*, information on the chemical nature of these factors has been reported in only two of these species (*R. prolixus* Liu and Davey, 1974; and *A. aegypti* Borovsky, 1985).

Despite the long history of *M. domestica* OH, little has been reported on a systematic chromatographic approach to its purification. Reports on the preparation of semi-purified extracts included techniques such as dialysis and acetone precipitation (Adams *et al*., 1970) and solvent partitioning and adsorption chromatography (Adams *et al*., 1977). A scheme involving reverse-phase (RP) HPLC was recently used to further characterize OH (Kelly *et al*., 1984b). We have now added size-exclusion (SE) and normal-phase (NP) HPLC to RP-HPLC methods in our attempts to purify this hormone. Crude extracts and HPLC fractions were tested for activity using an *in vivo* *A. atropalpus* ovarian maturation assay (Kelly *et al*., 1984b).

An extract, prepared by homogenizing ovaries in 25% MeCN/0.1% TFA, was processed by SE-HPLC (Protein Pak 125,

25% MeCN/0.1% TFA, 0.75ml/min). Several consecutive fractions with OH activity were collected at elution times that corresponded to low molecular weight components ($\leq$2.5 KD). Fractions corresponding to higher molecular weight components were inactive. Active SE fractions were combined, lyophilyzed and reconstituted in 0.1%TFA for processing by RP-HPLC (Fig. 1). Three consecutive fractions with activity eluted near the solvent front (2-5 min). A single fraction (24 min) was marginally inhibitory but this result was not reproducible. Despite the fact that a different method was used to extract ovaries, these RP-HPLC results were strikingly similar to those reported earlier (Kelly et al., 1984b).

Different RP columns and/or conditions offered marginal improvements in peak resolution in the 2-5-min polar region (Fig. 1), but these changes never affected baseline separation. Low retentivity of active fractions from the C-18 column suggested that OH is very polar and that HPLC columns stressing adsorption rather than absorption characteristics might be more suitable for its purification. Accordingly, we found that a standard 5µ silica gel column gave excellent separation of one of the RP-HPLC fractions that eluted at 3-4 min (Fig. 2). Four NP-HPLC fractions, numbered 1-4, were collected and assayed. Fractions 2 and 4 were moderately active but fraction 3 (broad peak eluting at 9-13 min) was the most active. Further processing of fraction 3 by NP-HPLC revealed the presence of at least two UV detectable materials, one of which was volatile. Chromatographic analysis and bioassay results indicated that the volatile component in that fraction contained all of the inhibitory activity.

In summary, NP-HPLC is a useful technique to resolve OH-active peaks that elute in the highly polar region of the RP chromatogram. At least 3 regions of the NP chromatogram elicited oostatic activity in the A. atropalpus ovarian maturation assay. Since these regions resulted from NP processing of only one of the three highly polar RP fractions, it is entirely possible that other inhibitors may yet be found. Partially-purified oostatic factors are highly polar and have apparent molecular weights of $\leq$2.5 KD. RP-HPLC characteristics of house fly oostatic factors suggest these factors will have markedly different chemistries from that of mosquito oostatic hormone.

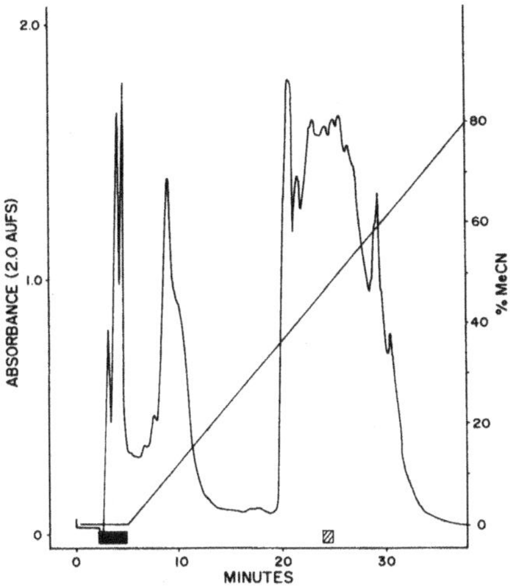

Fig. 1. RP-HPLC of pooled active fractions. Shandon Hypersil-ODS (4.6x250 mm, 5μ); UV Trace 214 nm, 37 ovary equivalents. Flow 1 ml/min; isocratic 5 min then gradient 35 min. Fractions (1 min intervals) were freeze-dried and redissolved in Aedes saline for bioassay. (Solid bar, strong OH response; cross-hatched bar, weak OH response).

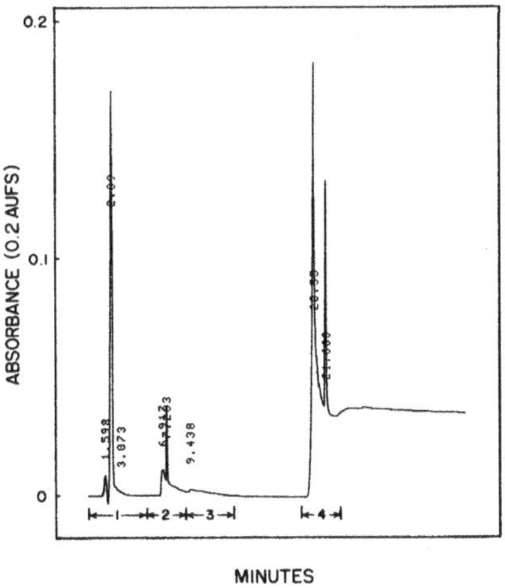

Fig. 2. NP-HPLC of fraction 4 from RP separation. Shandon Hypersil (3.9x300 mm.). UV Trace 214 nm, 3.8 ovary equivalents. Isocratic: 90% MeCN/0.1% TFA (0-16 min, 1 ml/min) then 50% MeCN/0.1% TFA (16-43 min, 1.5 ml/min).

REFERENCES

Adams T.S. (1977) Effect of the oostatic hormone on egg development and neurosecretion in the house fly, Musca domestica. In Advances in Invertebrate Reproduction, Vol. 1 (Ed. by Adiyodi K.G. and Adiyodi R.G.) pp 380-387. Peralam-Kenoth, Kerala, India.

Adams T.S. (1981) The role of ovarian hormones in maintaining cyclical egg production in insects. In Advances in Invertebrate Reproduction. (Ed. by Clark W.M., Jr. and Adams T.S.), pp 109-125. Elsevier/North Holland Biomedical Press, Amsterdam.

Adams T.S., Hintz A.M. and Pomonis J.T. (1968) Oostatic hormone production in house flies, Musca domestica, with developing ovaries. J.Insect Physiol. 14, 983-993.

Adams T.S., Johnson J.D., Fatland C.L. and Olstad G. (1970) Preparation of a semipurified extract of the oostatic hormone and its effects on egg maturation in the house fly. Ann. Entomol. Soc. Amer. 63, 1565-1569.

Borovsky D. (1985) Isolation and characterization of highly purified mosquito oostatic hormone. Arch. Insect Biochem. Physiol. 2, 333-349.

Kelly T.J., Birnbaum M.J., Woods C.W. and Borkovec A.B. (1984a). Effects of house fly oostatic hormone on egg development-neurosecretory hormone action in Aedes atropalpus. J. Exp. Zool., 229, 491-496.

Kelly T.J., Woods C.W., Birnbaum M.J. and Borkovec A.B. (1984b). Oostatic hormone and biogenic amines as inhibitors of ovarian maturation in house flies and mosquitoes. In Insect Neurochemistry and Neurophysiology. (Ed. by Borkovec A.B. and Kelly T.J.), pp 401-403. Plenum Press, New York.

Liu T.P. and Davey K.G. (1974) Partial characterization of a proposed antigonadotropin from the ovaries of the insect, Rhodnius prolixus, Stal. Gen. Comp. Endocrinol. 24, 405-408.

From *Insect Neurochemistry and Neurophysiology • 1989* Edited by A. B. Borkovec and E. P. Masler

CHARACTERIZATION OF TESTIS ECDYSIOTROPIN (TE) FROM BRAINS OF LEPIDOPTERA

Marcia J. Loeb, Albert B. DeMilo and Carol A. Sheppard

Insect Reproduction Laboratory, ARS, U.S. Department of Agriculture, Beltsville, MD 20705, USA

Testis ecdysiotropin (TE) induces synthesis of ecdysteroids by testes of Heliothis virescens and Lymantria dispar. TE was detected in extracts of male brains, at times corresponding to fusion of paired testes (Loeb et al., 1987, 1988) and development of the genital tract (Loeb, unpublished) and did not necessarily correspond to major hemolymph ecdysteroid peaks (Loeb et al, 1987, 1988; Schnee et al., 1984). TE may be comparable to the neuropeptides prothoracicotropic hormone (PTTH) and egg development neurosecretory hormone (EDNH).

Brains of 4-6 day H. virescens and 6-7 day L. dispar pupae were each pooled in methanol: water: acetic acid: thiodiglycol (90:9:0.9:0.1), and stored at -20° C. Batches of 200 H. virescens or 2000 L. dispar brains were sonicated (Heat Systems) and centrifuged at 1470 x g for 15 min. Supernatants were subjected to size-exclusion (SE) high precision liquid chromatography (HPLC) (Waters Protein Pak I-125), and reverse phase (RP) HPLC (WP-C4 Shandon Hypersil) using acetonitrile-0.1 % trifluoroacetic acid (MeCN-TFA) or acetonitrile-0.1 % heptafluorobutyric acid (MeCN-HFBA) as cosolvents for gradient runs. Peaks were identified by UV absorption at 214 nm. Each fraction was dried (Savant Speed Vac) at room temperature and dissolved in Ringer medium prior to bioassay with testes from day 1 or 2 fifth instar L. dispar larvae (Loeb et al 1987). Data are expressed as ug ecdysteroid synthesized, as determined by radioimmunoassay (RIA) (Borst and O'Connor, 1974).

SE-HPLC of supernatant yielded 4 bioactive fractions of approximately 1-2, 3, 7, and greater than 29 KD for H. virescens (Fig. 1A) and 1-2, 4, 9, and greater than 29 KD for L. dispar (Fig. 1B). The 3, 4, 7, and 9 KD fractions those determined for small PTTH (Bollenbacher et al., 1984) and EDNH (Hagedorn et al., 1979). The highest molecular weight fractions with TE activity are in the range of big PTTH (Bollenbacher et al., 1984, Suzuki,1986). However, ecdysiotropins of 1-2 KD have not yet been ascribed functions other than testis activation. The TE-active fractions in the middle molecular weight ranges in brain extracts from A. virescens and L. dispar are not the same. Surprisingly, the activity of crude TE is greater in cross species bioassay than in the same species (Loeb et al., 1988). Suzuki (1986) discusses PTTHs extracted from brains of larval and pupal Bombyx mori and Samia cynthia which all induce molting, although not necessarily in the same species; differences in the content of big and little PTTH in the extracts are responsible. Non-identical TE-active molecules in crude brain extracts may similarly affect activity in different lepidopterans.

The 1-2 KD fraction, common to extracts of both H. virescens and L. dispar brains, was subjected to RP-HPLC (10-60% MeCN-TFA). Several TE-active peaks were detected. The most active fraction of L. dispar origin, eluting close to an insulin standard, was rechromatographed

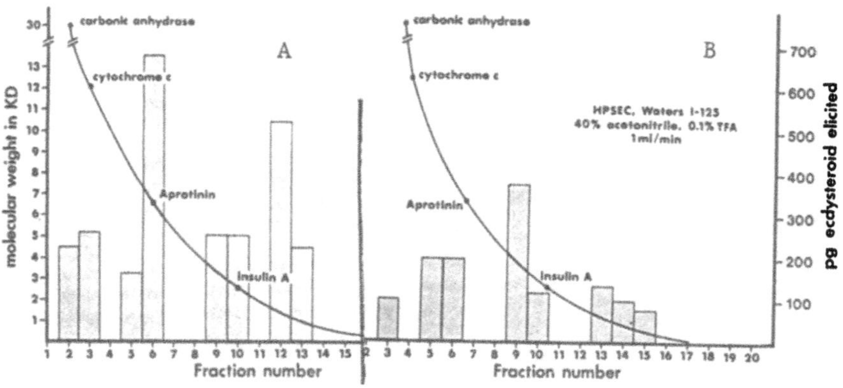

Figure 1. Approximate molecular weights of active TE fractions by SE-HPLC. Bars: TE activity. Curve: molecular markers. A: H. virescens; B: L. dispar.

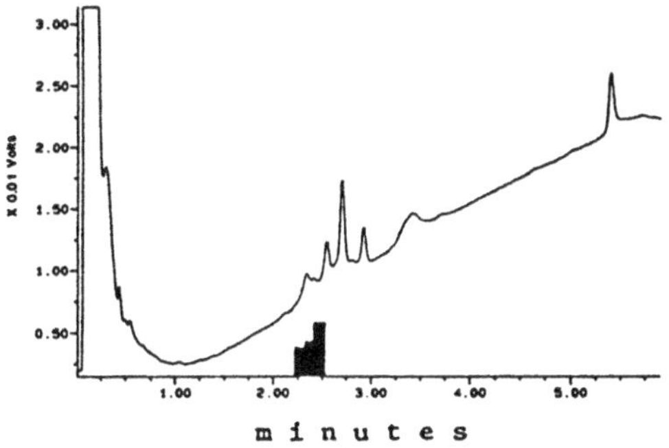

Figure 2: RP-HPLC with MeCN-HFBA. Bars show TE activity in 2 small peaks. 1v = 1 AU.

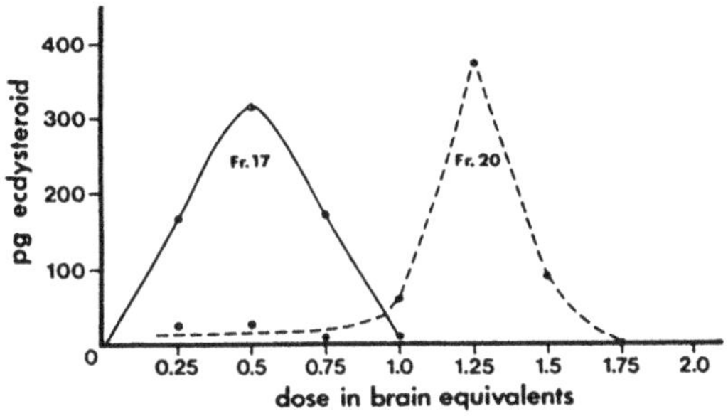

Figure 3: Dose responses of 2 TE active fractions from MeCN-TFA RP-HPLC.

(RP-HPLC, 25-60% MeCN-TFA). The most active of the resulting peaks was again subjected to RPHPLC (20-50% MeCN-HFBA). Two discrete peaks (22-24 min) were detected (Fig.2). Further study of this material is in progress.

Response to TE increases as the dose rises to a maximum point; further increase in TE titer induces progressively less ecdysteroid secretion (Loeb et al.,1987, 1988). Such a dose response occurs as mosquito ovaries synthesize ecdysone in response to EDNH (Hagedorn et al., 1979) and Leydig cells (vertebrate testis) synthesize testosterone in response to luteinizing hormone (Catt et al., 1980). Dose responses of two active fractions from MeCN-TFA separation of 1-2 KD material are similar (Fig. 3), suggesting that the receptor response to partially purified TE is similar to the response to multimolecular forms of TE in crude extracts. Although 45% of the activity of crude TE was lost by boiling for two min, partially purified TE was stable to heating. Trypsin destroyed activity of partially purified TE, as in the crude material. Therefore, 1-2 KD TE is probably a small peptide.

REFERENCES

Bollenbacher, W.E., Katahira, G.J., O;Brien, M., Gilbert, L.I., Thomas, M.K., Agui, N. and Baumhover, A.H. (1984). Insect prothoracicotropic hormone: evidence for two molecular forms. Science 224: 1243-1245.

Borst, D.W. and O'Connor, J.D. (1974) Trace analysis of ecdysone by gas-liquid chromatography, radioimmunoassay and bioassay. Steroids 24: 637-645.

Catt, K.J., Harwood, J.P., Clayton, R.N., Davies, T.F., Chan, V., Katikinen, M., Nozu, K. and Dufau, M.L.(1980). Regulation of peptide hormone receptors and gonadal steroidogenesis. Recent Prog. Horm. Res. 36: 557-622.

Hagedorn, H.H., Shapiro, J.P., Hanaoka, K. (1979) Ovarian ecdysone secretion is controlled by a brain hormone in an adult mosquito. Nature 282: 92-94.

Loeb, M.J., Brandt, E.P., Woods, C.W. and Borkovec, A.B. (1987). An ecdysiotropic factor from brains of Heliothis virescens induces testes to produce immunodetectable ecdysteroid in vitro. J. Exp. Zool. 243: 275-282.

Loeb, M.J., Brandt, E.P., Woods, C.W. and Bell, R.A.(1988) Secretion of ecdysteroid by sheaths of testes of the gypsy moth, Lymantria dispar, and its regulation by testis ecdysiotropin. J. Exp. Zool. 248: 94-100.

Schnee, M.E., Ma, M., and Kelly, T.J. (1984) Hormonal basis of adult eclosion in the gypsy moth. J. Insect Physiol. 30: 351-356.

Suzuki, A. (1986) Prothoracicotropic hormones and neurohormones in Bombyx mori. In Insect Neurochemistry and Neurophysiology, eds. Borkovec, A.B. and Gelman, D.B. The Humana Press, Clifton, N.J. pp. 29-51.

From *Insect Neurochemistry and Neurophysiology • 1989* Edited by A. B. Borkovec and E. P. Masler

BINDING OF SYNTHETIC DIPTERAN NEUROPEPTIDE PREPARATIONS TO HEAD AND ABDOMINAL PROTEINS OF 3-DAY OLD FACE FLIES

Dora K. Hayes

Livestock Insects Lab., L.P.S.I., Agr. Res. Serv., U. S. Dept. Agr., Beltsville, MD 20705

Receptors for hormones function in both time and space. In order to learn how best to use neuropeptides in insect population management programs, determination of when most receptor activity is found in the insect and the nature of that receptor are legitimate avenues of investigation. Keeley and Hayes (1987) state that "practical disruption of hormone-receptor interactions depends on the future development of ... hormone analogs designed from information of the biologically active structures of specific hormones. The structures of biological interest are those formed by peptide chains when they bind and activate target cell receptors." I here report on proteins which bind to a neuropeptide, but which have not been specifically identified as receptors. The neuropeptide chosen for study was a synthetic molecule with an amino acid sequence identical to that for an adipokinetic-type hormone isolated from the ring glands of tabanids by Jaffe et al. (1988). Because vertebrate receptors have been shown to change with time over a 24 h span (Scheving et al. 1989, for example), binding of proteins from extracts prepared at different times after lights on (HALO) was determined.

Face flies, (*Musca autumnalis* de Geer), from the Beltsville colony were reared under a regimen of LD 12:12 until pupation. Pupae were placed in a LD 12:12 chamber, lights on 6:00 am, 28 degrees C and those adults emerging over a one hour time span were selected for the study.

Flies were held for 18 hours and the sampling was begun at 0715 hours Eastern Standard time. Samples were prepared by freezing about 75 to eighty flies in dry ice at each time point. The heads and abdomens were then removed without thawing the flies. The pooled abdomens or heads were weighed and ground in a mortar that had been chilled in dry ice. The frozen powder was suspended in a solution of phosphate-buffered saline (PBS) which contained 100 μl of saturated phenylthiourea, 20 μl aprotinin (13 units/cc), 10 μl phenylmethylsulfonyl-fluoride in DMSO (200 mM), 10 μl pepstatin (0.7 μg/μl) and 10 μl leupeptin (0.5 μg/μl) for every 1000 μl of PBS. The mixture was well agitated and then centrifuged at 4500 X gravity for 15 minutes. To the pellet was added 4 ml of the PBS-inhibitor mixture which had been made 0.5% in n-octyl-β-D-glucopyranoside. The suspension was allowed to stand at room temperature for one hour and then centrifuged. The supernantant was diluted to four times its volume and then poured first through an albumin affinity column and then through an affinity column prepared with the crude synthetic dipteran peptide, *Tabanus atratus* hypotrehalosemic hormone (Taa-HoTH or DCC-II) (pGlu-Leu-Thr-Phe-Thr-Pro-Gly-Trp-Gly-Tyr-NH) as described by Hayes et al. (1986). The column was washed with 10 times the column volume with PBS and then eluted sequentially with (a) 5 ml of 2M NaCl in PBS, (b) 2 M diethylamine and (c) 50 mM PBS, pH 3.0. Samples were concentrated in Amicon Centricon tubes with a molecular cutoff of 10,000, subjected to polyacrylamide gel electrophoresis in SDS (PAGE-SDS) and silver stained.

Differences in the bands prepared from heads or from abdomens at different times of day were observed. Fig. 1 is a plot of semiquantitative evaluations of these differences on one band selected from electrophoretic runs of head preparations and two bands selected from runs prepared from abdomens.

These data suggest that some proteins which bind to the peptide, Taa-HoTH, and which might be receptors vary in apparent levels of concentration during a 24 h time span. This may reflect a real difference in concentrations of these proteins or it may be a result of slight conformational or other changes making the proteins unavailable for extraction and binding by the techniques employed. The data suggest that there may be a need to consider the time during a photoperiodic regimen that an extraction is performed; the 50 kd protein extracted from abdomens was hardly detectable in the samples prepared at 7:15 p.m. EST and found in high concentration at 11:15 in the morning.

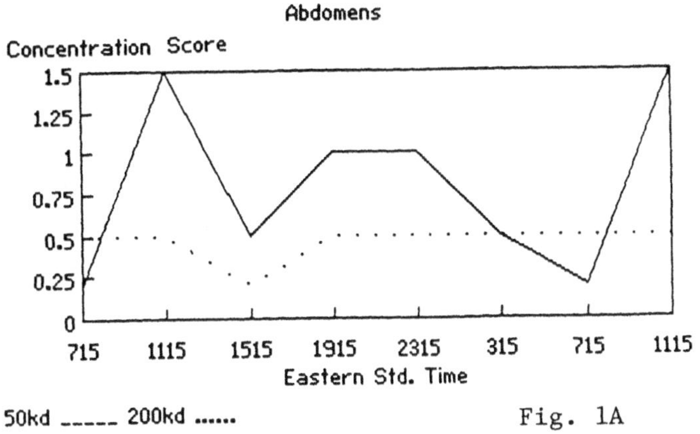

Fig. 1A

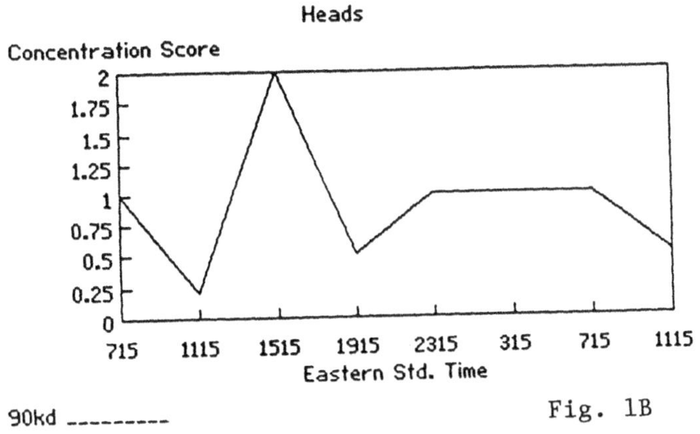

Fig. 1B

Figure 1. Relative concentrations of protein in bands from SDS-PAGE electrophoresis of eluates from Taa-HoTH affinity columns. Fig. 1A 50Kd and 200Kd bands, diethyl amine eluate, membrane extract from fly abdomens. Fig. 1B 90Kd band, 2 M NaCl in PBS eluate from face fly heads.

REFERENCES

Hayes D. K., Jaffe H., Morgan N. O. and Redfern R. E. (1986) Binding of synthetic insect peptide analogs to components of insect membrane preparations. In Insect Neurochemistry and Neurophysiology (Ed. by Borkovec A. B. and Gelman D. B.), pp. 247-250. The Humana Press, Clifton, N.J.

Jaffe H., Raina A. K., Fraser B. A., Keim P., Rao K. R., Zhang Y. S., Lancaster J. L. and Hayes D. K. (1988) Isolation of two neuropeptides in the AKH/RPCH family from horseflies (Diptera). Biochem. Biophys. Res. Communs. 151, 656-663.

Keeley L. L. and Hayes T. K. (1987) Speculations on biotechnology applications for insect neuroendocrine research. Insect Biochem. 17, 639-651.

Scheving L. A., Tsai T. H., Cornett L. E. Feuers R. J. and Scheving L. E. (1989). Circadian variation of epidermal growth factor receptor in mouse liver. Chronobiologia 16, 177-178.

From *Insect Neurochemistry and Neurophysiology* • *1989* Edited by A. B. Borkovec and E. P. Masler

NEUROPEPTIDE METABOLISM: PROPERTIES OF AN ENDOPEPTIDASE FROM HEADS OF DROSOPHILA MELANOGASTER

R. Elwyn Isaac and Richard M. Priestly

Department of Pure and Applied Biology,
University of Leeds, Leeds LS2 9JT, U.K.

INTRODUCTION

The hydrolysis of neuropeptides by ectoenzymes positioned on the surface of plasma membranes is now recognised to be an important step in the inactivation of many peptide messengers (Turner et al., 1987). In vertebrates relatively few membrane-bound enzymes are responsible for the metabolism of a wide range of neuropeptide transmitters and hormones, in a variety of tissues. For example, endopeptidase-24:11, an enzyme that appears to have a general role in the metabolism of a number of mammalian neuropeptides and can initiate the breakdown of peptides with blocked N- and C- termini and internal hydro-phobic residue (Turner, 1987). Recently, we have shown that a similar enzyme is present in membranes prepared from nervous tissue, fat-body, hindgut and Malpighian tubules of the locust, *Schistocerca gregaria*, and is enriched in a synaptic membrane preparation (Isaac, 1988). The enzyme can initiate the hydrolysis of the blocked peptide AKH I by cleaving the Asn-Phe bond. Although the substrate specificity of this enzyme has not yet been defined, it may have an important role in the metabolism of many insect peptides possessing an internal phenylalanine residue. In the present paper we report the occurrence of a similar endopeptidase in membranes prepared from the heads of *Drosophila melanogaster*. Our considerable knowledge of the genetics of *D. melanogaster* make this insect a particularly convenient model for the molecular analysis of neuropeptide action.

Membranes were prepared from heads of flies by centrifugation of the homogenate at 30,000g for 20 min and were subsequently washed with high salt buffer. Two peptide substrates, locust AKH I and [D-Ala[2],Leu[5]] enkephalin, were used to measure endopeptidase activity by established assays using hplc (Isaac, 1988; Matsas et al., 1983).

Drosophila membranes hydrolysed AKH I generating pGlu-Leu-Asn and Phe-Thr-Pro-Asn-Trp-Gly-ThrNH_2. The latter fragment was protected from further degradation by the presence of an aminopeptidase inhibitor, amastatin, and was quantified by hplc (280nm). This endopeptidase displayed optimal activity (89±4 pmol/min/mg of protein; S.E.M.;n=3) between pH7 and 8.5 and was inhibited by 1mM-EDTA and 1mM-phenanthroline (75 and 91% inhibition, respectively), indicating a requirement for a divalent metal ion. The bulk of the activity was inhibited by 10uM-phosphoramidon (90% inhibition) and 100uM-retrothiorphan (70% inhibition), both of which are inhibitors of Zn-metallopeptidases.

[D-Ala^2,Leu^5]Enkephalin is often used to assay the phosphoramidon-sensitive endopeptidase 24:11 or 'enkephalinase' in mammalian tissues. Hydrolysis at the Gly-Phe bond releases Tyr-D-Ala-Gly which is protected from further metabolism and can be quantified by hplc 214nm. However, the use of this substrate to measure peptidase activity from D. melanogaster was often limited by the presence of pigments that interfered with the hplc analysis. More importantly, this activity was only partially inhibited (54%) by 10uM-phosphoramidon suggesting that other enkephalin degrading enzymes were present in the crude membrane preparations.

The membrane-bound, AKH-degrading endopeptidase was readily solubilised by treatment with Triton X100 resulting in 75% of the activity being recovered in the soluble fraction after centrifugation at 40,000g for 3h. Solubilised enzyme was chromatographed by gel-filtration FPLC and the endopeptidase activity eluted as a single peak corresponding to a protein-detergent complex with a Mr of around 300,000. Chromatography of solubilised enzyme from synaptic membranes of S.gregaria also resulted in the elution of a single AKH I-degrading activity with similar mass. The partially purified soluble endopeptidases from fly and locust could not be distinguished by their response to phosphoramidon (IC_{50},0.3uM). Gel-filtration chromatography resolved two

[D-Ala^2,Leu^5]enkephalin degrading activities, one corresponding to the enzyme-detergent complex at Mr 300,000 which was inhibited by phosphoramidon and a second activity eluting in a lower molecular weight range (Mr 300,000) and insensitive to the same inhibitor. The latter enzyme activity generated Tyr-D-Ala-Gly but not Phe-Leu. These results explain the partial inhibition by phosphoramidon of the metabolism of the enkephalin analogue by crude membranes and demonstrates the usefulness of AKH I as a peptide substrate for measuring membrane-bound endopeptidase activity in insects. The Mr of the solubilised insect enzyme is similar to that of the Triton X100-protein complex of solubilised endopeptidase-24:11 from pig kidney and brain which comprises a protein dimer and a detergent micelle (Relton et al., 1983).

We have shown that membranes prepared from heads of D. melanogaster contain a neutral metallo-endopeptidase that is sensitive to phosphoramidon. This activity cannot be distinguished from that of an endopeptidase found in locust synaptic membranes and both insect enzymes appear to have some properties in common with those of mammalian endopeptidase-24:11.

REFERENCES

Isaac, R.E. (1988) Neuropeptide-degrading endopeptidase activity of locust (Schistocerca gregaria) synaptic membranes. Biochem. J. 25, 843-847.

Matsas, R., Fulcher, I.S., Kenny, A.J. & Turner, A.J. (1983) Substance P and [Leu]enkephalin are hydrolysed by an enzyme in pig caudate synaptic membranes that is identical with the endopeptidase of kidney microvilli. Proc. Natl. Acad. Sci. USA. 80, 3111-3115.

Relton, J.M., Gee, N.S., Matsas, R., Turner, A.J. and Kenny, A.J. (1983) Purification of endopeptidase-24:11 from pig brain by immunoadsorbant chromatography. Biochem.J. 215, 519-523.

Turner, A.J. (1987) Endopeptidase-24:11 and neuropeptide metabolism, in Neuropeptides and their Peptidases (Turner A.J., ed) pp. 183-201, Ellis Horwood, Chichester.

Turner, A.J., Hooper, N.M. & Kenny, A.J. (1987) Metabolism of neuropeptides in Mammalian Ectoenzymes (Kenny A.J. & Turner A.J. eds) pp. 211-248, Elsevier Science Publishers B.V.

From *Insect Neurochemistry and Neurophysiology • 1989* Edited by A. B. Borkovec and E. P. Masler

DEGRADATION OF DIPTERAN NEUROPEPTIDES BY MUSCA AUTUMNALIS

R. M. Wagner, T. G. Bird, and J. Kochanksy

USDA,ARS, VTERL/LIL/IHL

Route 5, Box 810, College Station, TX 77840

A large number of insect neuropeptides have been purified and sequenced in the past ten years, although the physiological functions of some have not been fully established. In the action of a peptide, there is a series of events which are crucial to the initiation, regulation, and termination of its activities. These consist of biosynthesis and processing of precursors to the active form of the peptide, release of the active form at or near the site of action, binding of the peptide to a specific receptor, induction of the physiological response(s), and elimination or degradation of the peptide for termination of the biological response. The design of peptide analogues as possible control agents for insects, therefore, must target one or several of these events in order to provide a rational mode of inhibition.

In the field of medicine, a number of peptide and non-peptide analogues have been designed and tested that are metabolically stable. These are either super-agonists, which mimic the action of the natural product, or antagonists, which inhibit or block the action of their naturally-occurring counterparts. For example, H-Tyr-D-Ala-Gly-me-Phe-Met(O)-ol is a highly potent analogue of enkephalins, with extended analgesic activity (Roemer and Pless, 1979) because of its relative resistance to metabolic degradation; and as such, it has undergone extensive clinical evaluation for its uses as a therapeutic drug.

Before such analogues may be designed to overstimulate or inhibit a given physiological response which is mediated by a peptide(s) in insects, the nature of the metabolism of the native peptide within the insect must be determined. As the two neuropeptides (DCCI/Taa-AKH (Raina and Gade, 1988) and DCCII/Taa-HoTH) isolated from horsefly corpora cardiaca (Jaffe et al., 1988) were shown to have adipokinetic and hypotrehalosemic activity, respectively, in Musca autumnalis (Jaffe et al., submitted), the metabolism of these peptides was investigated in the face fly.

Taa-AKH and Taa-HoTH were synthesized by solid-phase techniques on a Milligen-Biosearch Model 9600 Peptide Synthesizer, using t-butyloxycarbonyl-amino acids and coupling by diisopropylcarbodiimide activation in dimethylformamide/dichloromethane, with hydroxybenzotriazole as necessary. Peptides were removed from the resins with HF/p-cresol/thiocresol and purified on a Vydac preparative C-18 column, using a Waters HPLC system. Appropriate peaks were collected and the structure confirmed by protein sequencing and bioassay.

Each of the two peptides was incubated with haemolymph of the face fly in 0.1 M TRIS·HCl buffer at pH 7.5. At appropriate time intervals, aliquots were diluted with distilled water and centrifuged through a filter unit with a molecular exclusion limit of 10,000. Samples were analyzed by high pressure liquid chromatography on a Vydac C-18 column, under the conditions of Jaffe et al.(1988). Peaks were collected and the amino acid content of each determined. Controls consisted of 60 min incubations of peptide only, haemolymph only, or peptide with haemolymph which had been molecular weight-filtered prior to incubation. Taa-AKH was hydrolyzed slowly by face fly haemolymph (product at 16.4 min), but Taa-HoTH was metabolized more rapidly to one major product (39.4 min) which was further degraded, but more slowly (products at 15.8 min and 19.4 min) (Figs. 1 and 2).

The identification of the sites of cleavage, together with the amino acids required for receptor binding and/or biological activity, will enable rational design of metabolically-stable analogues for trial as possible insect control agents.

Jaffe, H., Raina, A. K., Fraser, B. A., Keim, P., Rao, K. R., Zhang, Y.-S., Lancaster, J. L. and Hayes, D. K. (1988) Isolation of two neuropeptides in the AKH/RPCH-family from horseflies (Diptera). Biochem. Biophys. Res. Commun. 151:656-663.

Jaffe, H., Raina, A. K., Riley, C. T., Fraser, B. A., Nachman, R. J., Vogel, V. W., Zhang, Y.-S. and Hayes, D. K. (submitted for publication) Primary structure of two neuropeptide hormones with adipokinetic and hypotrehalosemic activity isolated from the corpora cardiaca of horseflies (Diptera). Proc. Natl. Acad. Sci., USA.

Raina, A. and Gade, G. (1988) Insect peptide nomenclature. Insect Biochem. 18:785-787.

Roemer, D. and Pless, J. (1979) Structure-activity relationship of orally active enkephalin analogues as analgesics. Life Sciences 24:621-624.

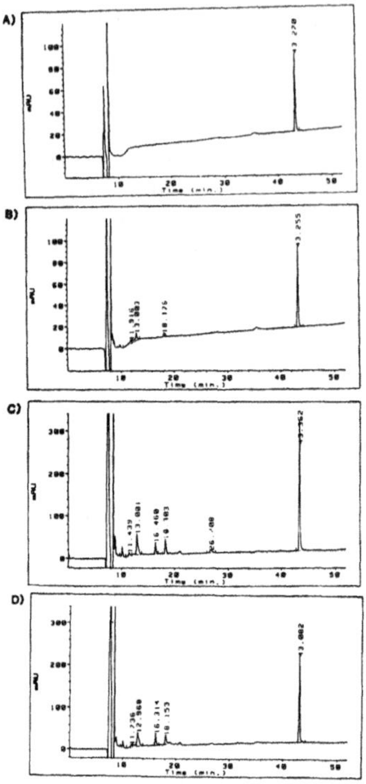

Figure 1

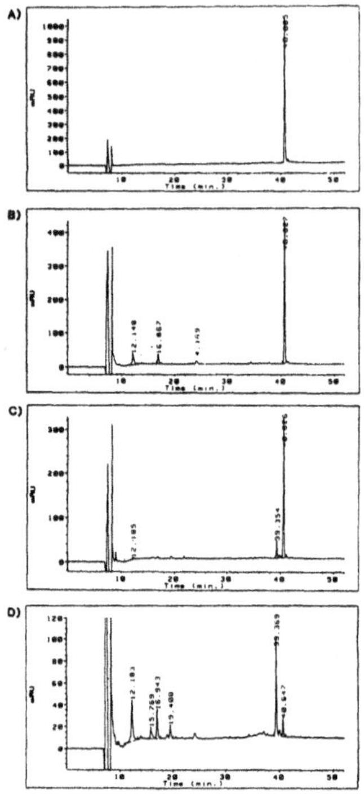

Figure 2

Fig. 1. Degradation of Taa-AKH by haemolymph of <u>Musca autumnalis</u> in 0.1 M TRIS·HCl, pH 7.5. Each chromatogram represents the equivalent of 2 nmol peptide and 1.5 ul haemolymph. A) 1 nmol Taa-AKH, B) MW-filtered haemolymph + 1 nmol Taa-AKH, 60 min, C) complete haemolymph + Taa-AKH, 30 min incubation, D) 60 min incubation.

Fig. 2. Degradation of Taa-HoTH by haemolymph of <u>Musca autumnalis</u> in 0.1 M TRIS·HCl, pH 7.5. Each chromatogram represents the equivalent of 1 nmol peptide and 1.0 ul haemolymph. A) 4 nmol Taa-HoTH, B) MW-filtered haemolymph + Taa-HoTH, 60 min, C) complete haemolymph + Taa-HoTH, 15 min incubation, D) 60 min incubation.

From *Insect Neurochemistry and Neurophysiology • 1989* Edited by A. B. Borkovec and E. P. Masler

METABOLISM OF INOSITOLPHOSPHATES IN THE CEREBRAL GANGLIA OF THE AMERICAN COCKROACH, *PERIPLANETA AMERICANA* L.

Peter Verhaert*, Hilde Walgraeve and Roger Downer.

University of Waterloo, Department of Biology, WATERLOO, Ontario, N2L 3G1 Canada
*present address: K.U.Leuven, Zoological Institute, Naamsestraat 59, B-3000 LEUVEN, Belgium

INTRODUCTION

Extracellular signals generated by "primary" messenger molecules (hormones, neurotransmitters, neuromodulators,...) are converted by responsive "target" cells into intracellular signalling processes through mediation of so-called "second" messengers, such as cyclic AMP (cAMP) and calcium ions. Synthesis of cAMP is catalyzed by cell membrane associated adenylate cyclases (AC) while Ca^{2+} mobilization appears to be linked to the phosphoinositide (PI) cycle (see Altman, 1988).

It is well documented that the insect nervous tissue contains a wide variety of AC (see Bodnaryk, 1983). Several of these agonist [(biogenic) amine/ (neuro)peptide] sensitive AC have been pharmacologically characterized in the central nervous system and/or retrocerebral complex of the American cockroach and other insects (see e.g. Orr et al., 1987; Gole et al., 1987). Few studies, however, have looked at potential interactions between agonists and the Ca^{2+}-inositolphosphate (IP) signalling pathway in insect neural tissues. The present work investigates PI metabolism in the cephalic neuroendocrine system of the cockroach, *Periplaneta americana*.

MEASUREMENTS OF IPS IN COCKROACH BRAINS

Male *P. americana* (1-3 months after the final molt) were taken from a colony maintained in the laboratory under standard conditions (Matthews and Downer, 1973). About 16 hours prior to rapid removal (2 min) of the brain/suboesophageal ganglion (SOG)/corpora cardiaca (CC)- allata (CA)-complex, insects were individually set aside in separate petri-dishes. The dissections were performed in ice cold *P. americana* Ringer (Becht et al., 1960). Tissues were collected on ice in flat-bottomed vials (dia. 1.5 cm) containing Ringer solution.

Prior to incubation with radiolabel, the Ringer was replaced by Hepes buffer (HB: 5 mM Hepes, 140 mM NaCl, 10 mM KCl, 2 mM $CaCl_2$, 3 mM $MgCl_2$, pH 7.0; 50 µl/brain). Following addition of 2.5 µCi *myo*-[^{3}H]-inositol (Ins), tissues were "labelled" for 60 min under gentle agitation at 25°C ("group-labelling"). After a triple rinse with 1 ml cold HB, tissues were carefully transferred into other vials (5 cerebral complexes per experimental group), to which 250 µl HB containing agonist was added. LiCl (10 mM) was included for its inhibiting effect on the hydrolysis of Ins-monophosphate (IP_1) to Ins (Berridge et al., 1982). Incubation in the presence of agonist was terminated by adding 1.0 ml of $CHCl_3$:CH_3OH (1:2, v/v). The incubation fluid was prelevated and tissues were homogenized (potter) in 250 µl 0.5 M HCl. Subsequently, equal amounts of H_2O and $CHCl_3$ were added. This homogenate was combined with the above "incubation fluid", and the resulting solution was vortexed for 1 minute. Aqueous and organic phase were separated by low speed centrifugation. The upper layer, containing the water-soluble IPs, was transferred into 13x10 mm glass tubes and dried under nitrogen. The chloroform layer (containing phosphoinositides PIP_1 and PIP_2) was directly pipetted into scintillation vials, to count the lipid-associated label after evaporation of the solvent. IPs were separated by standard anion exchange chromatography according to Berridge and co-workers (1983): the dried aqueous phase was redissolved in 1.0 ml of distiled water and applied to a small column containing 0.5 ml AG1-X8 resin (100-200 Mesh, formate form, Bio-Rad Labs. Canada Ltd., Missisauga, Ont.). After sample application, the column was washed with 10.0 ml of distiled water to elute all free Ins. The different IPs were then selectively eluted by use of experimentally standardized volumes of eluentia of increasing ionic strength: 5.0 ml 5 mM disodiumtetraborate/60 mM sodium formate (glycero-IP), 7.0 ml 0.1 M formic acid/0.2 M ammonium formate (IP_1), 5.0 ml 0.1 M formic acid/0.4 M ammonium formate (Ins-bisphosphate; IP_2), 5.0 ml 0.1 M formic acid/1.0 M ammonium formate (Ins-trisphosphate, IP_3). One ml aliquots of each of the eluate fractions were counted in duplicate upon addition of 7.5 ml Ecolume^R, in a Beckman LS 1701 liquid scintillation counter (dpm, desintegrations per minute).

OBSERVATIONS AND DISCUSSION

Myo-[^{3}H]-Ins incubated brains exhibit a relatively consistent pattern of labelled phosphatidylinositol metabolites IP_1, IP_2 & IP_3. This indicates that the fraction of Ins-label incorporated by the tissue enters the PI cycle. When lithium ions were added to the medium, basal IP levels were significantly elevated. As expected Li^+ amplifies/stabilizes the obtained signal due to accumulation of IPs (see above). Similar data were described in the locust metathoracic ganglion, suggesting a high rate in IP turnover caused by the release of endogenous transmitters (Trimmer and Berridge, 1985).

Incubation of brain/SOG/CC-CA complexes in the presence of both Li^+ and several "hormonal" (ant)agonists (different times varying between 30 sec and 120 min; various concentrations of octopamine (OA), serotonin (5-HT), carbachol, leucine-enkephalin; singly or in combination) did not result in a significant change in levels of IPs. These results do not exclude the possibility that some of the agonists affect PI metabolism at certain cerebral "target" sites. Such effect(s) may be obscured by variability in the labelling of different experimental brain groups, or by too high an activation of the PI system by endogenous agonists (see above). To investigate these hypotheses, we designed several experiments, but none could unequivocally demonstrate a clear agonist-dependent effect. These tests included work on aliquots (5 brain-equivalents) of a large uniformly labelled brain homogenate; inclusion of a one hour "calm down" period before the labelling step (to lower stress-induced effects on endogenous transmitter activity); application of atropine, a muscarinic antagonist, throughout the labelling period (to block acetylcholine-induced effects); and pretreatment of animals with reserpine (24 hrs before the experiment) to deplete brains of aromatic biogenic amines (5-HT, dopamine, OA), as confirmed by HPLC with electrochemical detection (G. Milton, P. Verhaert and R. Downer, unpublished). The two latter treatments appeared to result in a slight decrease in the uptake of Ins label, but did not allow us to measure any significant agonist-induced alteration in the ratios of PI metabolites. Our findings confirm the report of Trimmer and Berridge (1985) who were unable to demonstrate amine-induced enhancement of PI metabolism in the locust nervous system. However, it should be stressed that the results do not eliminate the possibility that some aminergic responses are mediated through this pathway; for example, the effect may be restricted to a small percentage of the total nervous tissue and, as a result, amine-induced changes are not detected in whole brains.

Table 1. Effect of CC-extract (30 min incubation) on the distribution of radioactivity incorporated into PI metabolites of the cockroach brain.

Fraction	Control (dpm)	CC (dpm)
Organic layer	86054	71517
Aqueous total	78927	94858
Free inositol	44356	50050
Inositol-1-P	9266	14594
Inositol-1,4-P	4209	5507
Inositol-1,4,5-P	4540	4639

However, after incubation of brain-SOG complexes (deprived of retrocerebral complex) with crude extracts of CC (1 equivalent per brain), IP levels were significantly elevated (Table 1), suggesting that CC contain material(s) that may act on brain receptors coupled to PI metabolism. Values indicated in the table represent the mean of two separate experiments, run in duplicate and on different days. Upon stimulation, the amount of radioactivity, associated with the aqueous fraction clearly increases, whereas the lipid associated label decreases. In particular, a significant increase in the IP_1 fraction is observed (see also Fig. 1).

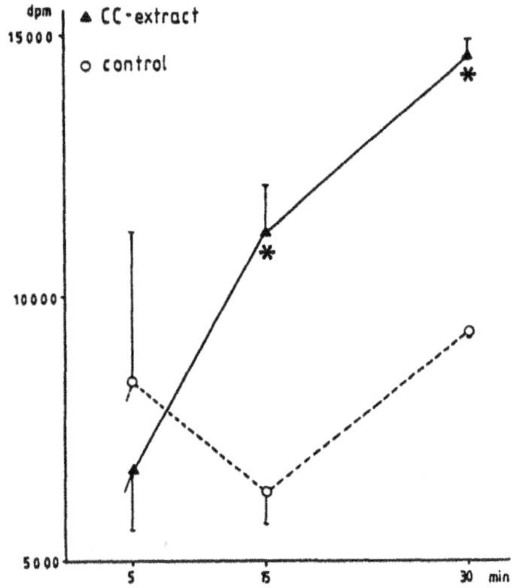

Fig. 1. Time dependent production of radiolabelled IP_1 in cockroach brain stimulated with CC-extract (1 equivalent/ brain). Statistical significance (p = 0.014) was determined by use of the Mann-Whitney-U-test.

A considerable loss in activity was noticed after storing the extract overnight at 4°C, suggesting that the bioactive material may be rapidly degraded. However, the CC-extract remained active in elevating IP levels after boiling (1 min), indicating that, at least, part of the responsible factors are heat-stable (small peptides?). In this respect it is interesting to note that HT_1 & HT_2, two small hypertrehalosemic neuropeptides from *P. americana* CC, appear to act through the PI pathway in the cockroach fat body (J. Steele, University of Western Ontario, personal communication). Finally, it should be mentioned that the same experimental conditions that allowed us to demonstrate this dramatic CC effect on IP hydrolysis in the cockroach brain, indicate a clear stimulatory effect evoked by OA (10 μM) on PI breakdown of a lepidopteran haemocyte cell line (our unpublished observations).

ACKNOWLEDGEMENTS

The authors thank Drs. I. Orchard and A. Lange (University of Toronto) for an introduction to practical aspects of IP determinations. This study was financially supported by NATO, the Belgian NFWO, and NSERC. P.V. was a 1987-88 fellow of World University Service Canada.

REFERENCES

Altman J. (1988) Inositol phosphates. Ins and outs of cell signalling. *Nature* **331**, 119-120

Becht G., Hoyle G., Usherwood P.N.R. (1960) Neuromuscular transmission in the coxal muscles of the cockroach. *J. Insect Physiol.* **4**: 191-201

Berridge M.J., Downes C.P., Hanley M.R. (1982) Lithium amplifies agonist-dependent phosphatidylinositol responses in brain and salivary glands. *Biochem. J.* **206**, 587-595

Berridge M.J., Dawson R.M.C., Downes C.P., Heslop J.P., Irvine R.F. (1983) Changes in the levels of inositol phosphates after agonist-dependent hydrolysis of membrane phosphoinositides. *Biochem. J.* **212**, 473-482

Bodnaryk R.P. (1983) Cyclic nucleotides. In *Endocrinology of Insects* (Ed. by Downer R.G.H., Laufer H), pp. 567-614

Gole J.W.D., Orr G.L., Downer R.G.H. (1987) Pharmacology of octopamine-, dopamnie-, and 5-hydroxytryptamine-stimulated cyclic AMP accumulation in the corpus cardiacum of the American cockroach, *Periplaneta americana* L.. *Arch. Insect Biochem. Physiol.* **5**: 119-128

Matthews J.R., Downer R.G.H. (1973) Hyperglycaemia induced by anaesthesia in the American cockroach, *Periplaneta americana*. *Can. J. Zool.* **54**, 395-397

Orr G.L., Gole J.W.D., Notman H.J., Downer R.G.H. (1987) Pharmacological characterisation of the dopamine-sensitive adenylate cyclase in cockroach brain: Evidence for a distinct dopamine receptor. *Life Sci.* **41**: 2705-2715

Trimmer B.A., Berridge M.J. (1985) Inositol phosphates in the insect nervous system. *Insect Biochem.* **15**, 811-815

From *Insect Neurochemistry and Neurophysiology • 1989* Edited by A. B. Borkovec and E. P. Masler

TYRAMINE-β- HYDROXYLASE ACTIVITY IN THE SYNGANGLION OF THE CATTLE TICK BOOPHILUS MICROPLUS (ACARI: IXODIDAE)

*Luke R C Kempton, +R John Willis, +John B Pillmoor and *R Elwyn Isaac

*Department of Pure and Applied Biology, University of Leeds, Leeds LS2 9JT, U.K.
+Chesterford Park Research Station, Agrochemicals Division of Schering AG, Berlin, W. Germany

INTRODUCTION

Studies on insects and crustacea have established octopamine as an important chemical messenger with multifunctional roles in arthropods (Evans, 1985). However there is relatively little information describing the octopaminergic system of ticks, an economically and medically important group of ectoparasites. The acaricidal properties of the formamidines, a class of chemical believed to act as octopaminergic agonists (Hollingworth, 1976), suggest that octopamine is an important neurotransmitter/neurohormone in ticks and the demonstration of an octopamine-sensitive adenylate cyclase in the CNS of Boophilus microplus (Morton, 1984) is consistent with such a role. The aim of our work is the characterisation of the enzymes involved in the synthesis and metabolism of octopamine in the cattle tick, B. microplus. In this report properties of tyramine-β-hydroxylase (T-β-H) from the synganglion of B. microplus are described and compared to the well-characterised mammalian enzyme, dopamine -β-hydroxylase (D-β-H).

Crude enzyme was prepared by homogenising synganglia from engorged female ticks in 0.05M PIPES buffer (pH 6.6). The supernatant recovered from a low speed spin (200xg, 10 mins) was used as the source of enzyme activity. In routine assays enzyme (1.8ug protein) was incubated with ring - [3H]tyramine (1.5 kBq,

30uM), ascorbate (20mM), catalase (650 units, Boehringer) and pargyline (3uM) in a total volume of 15 ul of 0.05M PIPES buffer (pH 6.6) at 20°C. After 4 hours the conversion of [3H]tyramine to [3H]octopamine was determined by analysis of the reaction mixture by ion-pair reversed phase hplc (Column: 3u ODS ultrasphere 4.6 x 75mm, eluent: 75mM NaH_2PO_4, 20mM TCA, 1mM sodium dodecyl sulphate (SDS), pH 3.1 in methanol, increasing from 2.5 to 60% methanol in 15 mins.) and liquid scintillation counting of hplc fractions.

T-β-H activity was dependant upon the presence of ascorbate, which at a concentration of 20mM resulted in maximal octopamine synthesis. The conversion of tyramine to octopamine never exceeded 15% and the reaction was linear with time for up to 4 hours with a pH optimum of 6.6. Pargyline was also added to the reaction mixture to minimise the metabolism of tyramine by monamine oxidase enzymes.

1mM-EDTA and 10uM-diethyldithiocarbamate (DDC) strongly inhibited tick T-β-H (96% and 98% inhibition respectively) demonstrating a requirement for divalent metal ions. The sulphydryl reagents cysteine (0.5mM), glutathione (1mM), and mercaptoethanol (1mM) also inhibited the enzyme activity (56%, 94%, and 92% respectively). Enzyme activity lost through chelation with DDC was recovered by the addition of Cu^{2+} but not other divalent metal ions which is consistent with Cu^{2+} being the metal ion within the active site of the enzyme. Activity of the untreated enzyme was stimulated by the addition of Cu^{2+} with a 7-fold activation at 10^{-4}m. At higher concentrations Cu^{2+} inhibited the β-hydroxylase activity.

Mammalian D-β-H is recognised as a potential target for the treatment of cardiovascular disorders and a number of multisubstrate inhibitors have been synthesised (Kruse et al., 1987). Two benzylimidazole-2-thione compounds were tested and found to be potent inhibitors of the tick T-β-H enzyme with similar IC_{50} values to those reported for bovine D-β-H (Table 1).

Table 1.

Inhibitor	IC_{50} (uM) Tick T- -H	Bovine D- -H[a]
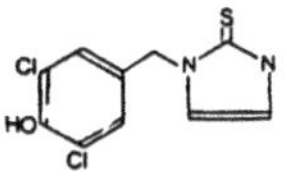	0.1	0.68
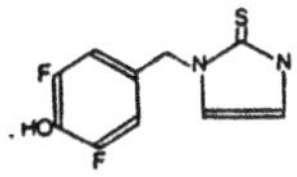	0.05	0.07

[a]- data from Kruse et.al., 1987

β-hydroxylation of tyramine was inhibited by the addition of dopamine. Hydroxylation rates determined at different tyramine concentrations, in the absence and presence of dopamine (200uM), gave an apparent Km for tyramine of around 230uM and established dopamine as a competitive inhibitor of the reaction (Ki, 240uM). This demonstrates that dopamine can bind to the active site of T-β-H and may serve as an alternative substrate for this enzyme.

In an attempt to localise the β-hydroxylase enzyme within the synganglion, antiserum to bovine D-β-H was obtained for immunocytochemical studies (Eugene Tech International, Inc., NJ.). It was first established that the antiserum cross-reacted with tick enzyme by immunoprecipitation of the enzyme activity with antibodies conjugated to protein A/sepharose. Immunocytochemical staining in whole mount preparations of synganglia was most frequently seen in fibres within the pedal nerves and on some occasions cells within the synganglia itself were also immunoreactive. However consistent staining was not always obtained despite attempts to optimise the fixation and processing protocols.

We have demonstrated that the tick synganglion possesses a tyramine-β-hydroxylase activity which is similar to the vertebrate D-β-H enzyme in its requirements for ascorbate, Cu^{2+}, and susceptibility to multisubstrate inhibitors. In addition the tick enzyme has antigenic epitopes common to bovine D-β-H. Competitive inhibition of tyramine-β-hydroxylation by dopamine and the existence of dopamine and its β-hydroxylated derivative, noradrenaline, in the synganglion of ticks (Megaw and Robertson, 1974) suggests a further possible role for this enzyme in catecholamine metabolism. The cellular localisation of T-β-H and potential amine substrates within the tick CNS, and the purification of the β-hydroxylase activity, will help to define the physiological role of this enzyme.

Acknowledgements. This work was supported by a SERC/CASE studentship (no. 86518793) to LRCK.

REFERENCES

Evans, P.D. (1985) Octopamine. In: Comprehensive Insect Physiology, Biochemistry and Pharmacology. Eds. G.A. Kerkut & L.I. Gilbert, Pergamon Press, pp. 499-527.

Hollingworth, R.M. (1976) Chemistry, biological activity and uses of formamidine pesticides. Envir. Hlth. Persp. 14, 57-69.

Kruse, L.I., Kaiser, C., DeWolf, W.E., Frazee, J.S., Ross, S.T., Wawro, J., Wise, M., Flaim, K.E., Sawyer, J.L., Erickson, R.W., Ezekiel, M., Ohlstein, E.H., Berkowitz, B.A. (1987). Multisubstrate inhibitors of dopamine- - hydroxylase. Structure-activity relationships at the phenethylamine binding site. J. Med. Chem. 30, 486-494.

Megaw, M.W.J. and Roberston, H.A. (1974). Dopamine and noradrenaline in the salivary glands and brain of the tick, Boophilus microplus: effect of reserpine. Experientia 30, 1261-1262.

Morton, D.B. (1984). Pharmacology of the octopamine-stimulated adenylate cyclase of the locust and tick CNS. Comp. Biochem. Physiol. 78C, 153-158.

From *Insect Neurochemistry and Neurophysiology • 1989* Edited by A. B. Borkovec and E. P. Masler

AN ECOLOGICALLY ACCEPTABLE STRATEGY FOR THE USE OF GENETICALLY ENGINEERED BACULOVIRUS PESTICIDES

H. Alan Wood, Patrick R. Hughes,
Nikolai van Beek*, and Martha Hamblin

Boyce Thompson Institute
Tower Road, Ithaca, NY 14850, USA
*Present address Sandoz Crop Protection,
975 California Avenue, Palo Alto, CA 94304

The basis for the first U. S. Environmental Protection Agency approval for the field release of a genetically-engineered virus has been to take advantage of the biological properties of baculovirus in such a way that an engineered virus could possess enhanced pesticidal properties but, at the same time, would pose no environmental or health hazards. The ultimate goal of these investigations is to reduce the agricultural requirement for synthetic chemical pesticides through the development of viral pesticides with enhanced pesticidal properties.

Over 500 naturally-occurring baculoviruses have been described. They all infect invertebrates, and 90% infect lepidopterous insects, many of which are major agricultural pests. Viral pesticides have very limited host ranges and do no harm to beneficial insects such as honey bees, parasitic wasps or other natural predators. Following extensive evaluations, the U. S. Environmental Protection Agency has granted registration of the cotton bollworm, douglas fir tussock moth, pine sawfly and gypsy moth baculoviruses for use as pesticides.

Despite the environmental advantages of viral pesticides, the above-mentioned baculoviruses have not been commercially successful. The most significant problems have been that viral pesticides are more expensive and may take from 2 to 10 days to kill an insect pest. Accordingly, under present conditions, it is cheaper and more effective to use synthetic pesticides than viral pesticides.

An obvious solution to these problems is the genetic improvement of viral pesticides. This can be achieved easily by using the baculovirus expression vector (BEV) system (Smith et al. 1983). The basis of this system is the production of a protein product (the polyhedrin protein) that is nonessential and highly expressed by baculoviruses, such as the Autographa californica nuclear polyhedrosis virus (AcNMPV). The polyhedrin gene sequences from the viral genome can be replaced with the coding region of a foreign gene which is under the control of the strong polyhedrin gene promoter. The BEV system has been used extensively in basic research programs to produce large amounts of proteins that are otherwise difficult to obtain.

This approach can also be used to insert into baculoviruses new genes whose products would have high pesticidal activity. The virus would provide a specific delivery system while the new gene product would quickly stop insect pests from feeding and/or kill them.

A major problem in this approach has been that, in the absence of polyhedrin protein (poly-minus), the baculovirus particles are not occluded within polyhedra and, therefore, are rapidly inactivated in the environment. This problem was overcome by co-infection of cells with the wild type AcMNPV and an altered, poly-minus AcMNPV isolate. Under these conditions the wild type virus provides the polyhedrin protein which occludes both the wild type and poly-minus virus particles. This co-occlusion process provides a way of delivering the engineered virus to the field in an infective form.

The field release of a genetically-engineered baculovirus pesticide raises several environmental concerns. Among these concerns are persistence of the engineered virus and the possibility that engineered virus might displace natural virus populations. Using the co-infection/co-occlusion strategy, the persistence of the engineered virus in a virus population is determined by the probability of co-infection of individual larvae and cells with both virus types as the virus is passed from insect to insect. We have studied the dynamics of this process under laboratory conditions. Experiments were conducted to evaluate the amount of occluded poly-minus virus produced following larval infections with different virus inoculum dosages and with inocula containing varying ratios of co-occluded poly-minus and wild type virions. All of the laboratory data indicate that a poly-minus virus could not persist in a virus population under natural conditions.

In order to evaluate this process in nature, a small field trial is being performed in which the rate of loss of a poly-minus virus under natural conditions will be determined. This study is using co-occluded wild type AcMNPV and an engineered AcMNPV which lacks its polyhedrin gene but **has no foreign gene insert**.

Recent investigations have concentrated on identifying foreign genes which, when expressed in virus-infected insects, would have significant pesticidal activities, leading to cessation of feeding and/or quick death of larvae. Among the most promising foreign pesticidal genes are insect genes associated with development. Expression of such genes, under the regulation of the virus rather than an insect regulatory system, is expected to achieve ecologically acceptable, fast and effective control of insect pests.

The insertion of pesticidal genes into baculoviruses will necessitate addressing many additional concerns. The release of a genetically-enhanced viral pesticide will have to satisfy an even wider range of environmental and health issues. In addition, such a virus will have to satisfy commercial requirements relating to efficacy and cost/benefit ratios relative to available chemical pesticides. We feel that the co-infection/co-occlusion approach presents an ideal strategy by which a genetically enhanced viral pesticide can be developed in a manner consistent with environmental and health concerns.

Acknowledgements

Although the research described in this article has been supported by the U.S. Environmental Protection Agency through Assistance Agreement No. CR-814530 to H. A. Wood, it has not been subjected to Agency review and, therefore, does not necessarily reflect the views of the Agency and no official endorsement should be inferred.

Reference

Smith, G.E., Summers, M.D. and Faser, M.J. (1983) Production of human beta interferon in insect cells infected with a baculovirus expression vector. Molecular and Cellular Biology 3, 3156-2165.

Neurophysiology

From *Insect Neurochemistry and Neurophysiology • 1989* Edited by A. B. Borkovec and E. P. Masler

OVERVIEW: PROGRESS IN INSECT NEUROPHYSIOLOGY

Chih-Ming Yin, Dept. of Entomology and
Program in Molecular and Cellular Biology,
Univ. of Mass., Amherst, MA 01003, USA

ICINN89, following the tradition of ICINN83 and ICINN86, continued to attract outstanding international participants whose research interests in insect neurophysiology focus on the neuroendocrine regulatory mechanisms governing insect development, reproduction, metabolism, and behavior. During the past three years, there has been exciting progress in our discipline. For instance, a new putative juvenile hormone, methyl-6,7;10,11-bisepoxy-3,7,11-trimethyl-(2E)-dodecenoate (JHB_3) had been previously reported from studies of *in vitro* production of juvenile hormone (JH) by the ring glands of third instar *Drosophila melanogaster*. At this meeting, the control of JHB_3 synthesis, *in vitro*, during the reproductive development of female *D. melanogaster* was reported (Altaratz *et al.*). Further, previous studies on the structure and function of the eclosion hormone (EH) in *Manduca sexta* have led to the identification of sources and pathways of release of EH, as well as its amino acid sequence. Oligonucleotide probes were then constructed, according to the identified amino acid sequence, and used in the cloning of the EH gene. Expression of this gene in the brain of *M. sexta* was described (Horodyski *et al.*). These studies represent the trend of insect neurophysiological research in which researchers will employ the latest conceptual and technical advances in chemistry, biochemistry, immunology and molecular biology to solve their problems.

Electro- and neurophysiology. Work on the wing stretch-receptor (SR) of the locust has suggested that the SR activity was facilitated by octopamine following the onset of flight. SR controlled the wingbeat frequency, and the octopamine modulation of the SR might influence the motor pattern generation in flying locusts (Ramirez et al.). Baines and Downer reported that the mandibular adductor muscles received innervations from at least 20 neurons in the suboesophageal ganglion. Some of these neurons were found immunoreactive to serotonin antibody. Perfusion of serotonin over the adductor changed the nature of its contraction. Using the stable fly, Stomoxys calcitrans, Cook and Wagner found that octopamine (10^{-7} M) produced strong contraction in the ovarian sheath while proctolin (10^{-9} M) increased both contraction frequency and amplitude. In contrast, glutamic acid (10^{-4} M) inhibited the contraction of the ovarian sheath. Work by McQuiston et al. revealed the presence of voltage-dependent Na^+ and Ca^{++} channels in the cell membrane of corpus allatum (CA) cells of Diploptera punctata.

Regulation of pheromone production. Four papers dealt with mechanisms regulating pheromone production. Raina found that, in the male accessory glands of Heliothis zea, a proteinaceous material, when transferred to the female during mating, may terminate her mating receptivity and reduce her sex pheromone production. Alstein and Dunkelblum reported the presence of a brain neurohormone which activated pheromone biosynthesis in the tomato looper, Chrysodeixis chalcites. Extracts of the brain-suboesophageal ganglion complex of Heliothis armigera stimulated the pheromone biosynthesis, in vitro, by isolated pheromone glands. This stimulation was enhanced by the presence of intact terminal abdominal ganglia in the medium during incubation. Glands stimulated by brain extracts showed increases in cAMP levels. Forskolin, IBMX, and 8-bromo-cAMP also stimulated the pheromone glands in vitro (Rafaeli et al.). In the gypsy moth, Lymantria dispar, it has been shown that the presence of viable sperms in the spermatheca of a female was essential for the termination of pheromone production, although an initial rapid decrease in pheromone production did not require the transfer of the spermatophore. Pheromone titers in females from which the spermatheca was removed or in those mated to sterile males subsequently increased back to virgin-like levels (Giebultowicz et al.).

Control of reproduction. Yin *et al*. reported that precocene II treatment suppressed syntheses of JH, ecdysteroids, and vitellogenin (Vg) in *Phormia regina*. Results suggested that ecdysteroids may regulate Vg synthesis while JH controls Vg uptake. Activation of CA in *Blattella germanica* was linked to increases in the volume of CA cells (Chiang *et al*.). Further, during short-term incubations, dissociated cells of active CA synthesized JH III *in vitro* at rates similar to those of the intact glands. CA from adults, ovariectomized as nymphs, produced JH at higher rates than controls. Thus, it appeared that ovaries in *B. germanica* are not required for CA activation but may be important to CA inhibition (Gadot *et al*.). Davis *et al*. found that in *L. dispar*, Vg synthesis in the last instar larvae was blocked by either head ligation, starvation, or ZR-515 treatment. They also isolated female specific cDNA clones. Sevala and Davey described that the activation of protein kinase C in follicle cells is important in the activation of JH-dependent membrane Na^+, K^+-ATPase which in turn played a key role in reducing the volume of follicle cells to cause patency in *Rhodnius prolixus*. Gerst and Adams showed that 6-hydroxydopamine retarded ovarian development of *M. domestica*. Ecdysteroid titers of treated flies were unaffected, while Vg synthesis was at slower rate but Ug kept accumulating.

Peptidal hormones. Six papers dealt with the activity of prothoracicotropic hormone (PTTH) in *L. dispar* from embryo to adult (Bell *et al*., Fescemyer *et al*.; Kelly *et al*.; Masler *et al*.; Thyagaraja *et al*. a,b). Ecdysteroid titers and PTTH activities were estimated for diapausing and non-diapausing embryos, larvae and pupae. Different stabilities were noticed in the embryonic and post-embryonic prothoracictropic factors of the gypsy moth. Factors from boiled brain extracts of *Ostrinia nubilalis* promoted the production of ecdysone and 3-dehydroecdysone, *in vitro*, by the prothoracic glands of gypsy moth (Gelman *et al*.). Pener *et al*. reported that injection of azadirachtin (AZA), a substance from neem tree, caused fully grown last instar larvae of *M. sexta* to pupate within 1 to 2 days. But these pupae showed retarded adult development. They found that the AZA treatment did not affect PTTH synthesis and PTTH content in the brain. However, it greatly reduced the hemolymph ecdysteroid titer. The presence of a human atrial natriuretic peptide-like (ANP-like) substance in *S. calcitrans* was described by

Chen. Six hours after a blood meal, this ANP-like material from the whole body extract increased at least five-fold from 25 pg/fly prior to the blood meal. Isolation of this fly ANP and its physiological function were reported.

Metabolism of nervous tissues. Two papers dealt with metabolism in nervous tissues. After the incorporation of ^{3}H-*myo*-inositol, by isolated brains of *Periplaneta americana*, into cell membrane-bound inositol phosphoinositides , five prominent radiolabelled metabolites; inositol, inositol 1-monophosphate, inositol 1,4-bisphosphate inositol, 1,4,5-trisphosphate and glycerophosphoinositol were found. Aqueous corpora cardiaca extracts increased the inositol phosphate formation *in vitro*, while addition of serotonin, octopamine, carbachol and atropine into the incubation medium had little effect (Verhaert *et al*.). A sublethal dose of endosulfan increased glycogenolysis and total protein and reduced glutathione content in the nerve cord of treated 5th instar larvae of *Bombyx mori*. Later these larvae spun cocoons weighing significantly more than those of the untreated larvae (Jadhav and Kallapur).

Arthur Kornberg, one of the most prominent biochemists in the world, recently commented in his book "For the Love of Enzymes" (Harvard Univ. Press, 1989) that the medical science in the 20th century has seen a succession of hunters going after microbes, vitamins, enzymes, and genes. He feels that , with some technical advancements, neurobiologists- who he called "head hunters"- may dominate the last part of the century. Works presented in ICINN89 clearly indicate that insect neurobiologists will not be left out of the hunt.

From *Insect Neurochemistry and Neurophysiology* • *1989* Edited by A. B. Borkovec and E. P. Masler

GENE EXPRESSION DURING PUPAL-ADULT NEUROGENESIS IN *MANDUCA SEXTA*

M.-T. B. Davis, *R. S. Hakim, and J. A. Svoboda

USDA-ARS, B-467, BARC-East, Beltsville, MD 20705 &

*Dept. of Anatomy, Howard Univ., Wash., D.C. 20059

Different from hemimetabolous insects in which the adult central nervous system (CNS) is essentially a replica of the embryonic CNS, most of the adult brain in holometabolous insects develops during postembryonic neurogenesis (Panov, 1963). Postembryonic neurogenesis has been well-described in various insects (Panov, 1963; Technau and Heisenberg, 1982; Booker and Truman, 1987). Since postembryonic neurogenesis in Lepidoptera is a critical period of terminal neuroblast differentiation, neuronal cell death and synaptogenesis for transformation of larval neurons into adult neurons, we have been interested in identifying proteins and genes that influence this developmental process in *M. sexta*. In this paper, we review our previous work concerning: 1) changes in expression of *Manduca* intracellular brain polypeptides, and of *Manduca* brain polypeptides that are secreted into an incubation medium at different stages of post-embryonic neurogenesis (Davis, submitted), and 2) cloning, from a *Manduca* brain cDNA library, genes that are activated during metamorphosis (Davis and Svoboda, 1989). Since it is not known how many of the secreted polypeptides are actual neurosecretory polypeptides, the secreted polypeptides will be referred to as putative neurosecretory polypeptides.

I. Changes in Expression of Intracellular and Putative Neurosecretory Polypeptides During Postembryonic Neurogenesis

A. Intracellular Brain Polypeptides

Both the steady-state level and *in vitro* synthesis of *Manduca* cellular brain polypeptides were characterized in late 5th instar larvae, pupae and pharate adults, all active periods of postembryonic neurogenesis. By improving the O'Farrell (1975) two-dimensional gel electrophoresis, we were able to resolve, in a single gel, complex mixtures of *Manduca* brain polypeptides in which molecular weights range from less than 14,000 daltons to more than 95,000 daltons. Whereas the *in vitro* steady-state levels of many polypeptides, which reflect both the rate of protein synthesis and degradation, remained fairly constant during postembryonic neurogenesis, *in vitro* protein synthesis declined as metamorphosis progressed. In addition, many other changes in concentrations of 21 synthesized brain polypeptides were also detected as related to brain development: 1) Of the 70% methanol-soluble polypeptides, one was expressed only in 0-12-h-old pupae; and one occurred in medium abundance in larvae and 0-12-h-old pupae, but was not seen in 7-8-day-old pharate adults and was barely detectable in 11-12-day-old pharate adults; and 2) Nineteen polypeptides, insoluble in 70% methanol, showed distinct changes in concentration during development. Of these 19 polypeptides, six had different concentrations at different developmental stages; twelve were specific to larvae and/or 0-12-h-old pupae; six were newly synthesized starting at 0-12 h and 7-8 days after pupal ecdysis, and continued to be present in older pharate adults; and four had higher concentrations in pharate adults than in 0-12-h-old pupae. Some of the above polypeptides belonged to more than one category of expression.

B. Putative Neurosecretory Polypeptides

Four distinct patterns of differential expression of *in vitro* synthesized putative neurosecretory polypeptides were observed during brain development from 5th instar larvae to pharate adult: 1) 24 polypeptides were present at different concentrations in different developmental stages examined; 2) 24 polypeptides were stage-specific; 3) six polypeptides were newly synthesized beginning at 0-12 h, 7-8 days and 11-12 days after pupal ecdysis, with the synthesis of most of those polypeptides in 0-12-h-old pupae and 7-8-day-old pharate adults continuing in older pharate adults; and 4) 11 polypeptides were more abundant in pharate adults than in 0-12-h-old pupae and/or larvae.

II. Cloning Genes that are Activated During Postembryonic Neurogenesis

We have constructed a brain cDNA library from 0-12-h-old *Manduca* pupae in lambda Zap phage, and using a differential screening method, cloned novel genes that are activated during metamorphosis. The isolated clones, three of which are brain-specific, represent six different patterns of gene expression. The intensity of gene expression was quantified by counting the radioactivity of dot-blotted RNA which was obtained from *Manduca* brains at different developmental stages, and was hybridized to the isolated cDNAs. In 0-12-h-old pupae, genes #1 and 2 which had single transcripts at 2.5 kb and 1.45 kb, respectively, were expressed eight and four times as strongly as in 7-8-day-old pharate adults. Gene #3 had multiple transcripts and was expressed two and three times as strongly in 7-8-day-old and 11-12-day-old pharate adults, respectively, as in 0-12-h-old pupae. Genes #5 and #6, which had single transcripts at 0.90 kb and 1.75 kb respectively, had peak expression at 7-8 days and 11-12 days after pupal ecdysis. Gene #4, which had multiple transcripts, had a relatively constant level of expression throughout development.

Of the hundreds of brain polypeptides analyzed, only 3 intracellular and 3 putative neurosecretory polypeptides were detected to have the same patterns of differential expression during brain development as the cloned genes #3, 5 and 6. That is, two cellular and two putative neurosecretory polypeptides, corresponding to clones #5 and 3, had a peak expression at 7-8 days and 11-12 days respectively, after pupal ecdysis. One cellular and one putative neurosecretory polypeptide, corresponding to clone #6, were equally abundant in 7-8-day-old and 11-12-day-old pharate adult, and least abundant in 0-12-h-old pupae.

A ^{32}P-labeled cDNA probe from clone #4 was hybridized to frozen sections of *Manduca* brains which were fixed with 4% paraformaldehyde. Preliminary *in-situ* hybridization data showed that gene #4 was expressed in defined peripheral areas of the brain and of subesophageal ganglia and in one area next to the oesophageal foramen (Fig. 1).

In conclusion, using the improved high-resolution O'Farrell two-dimensional gel electrophoresis, we have identified 21 cellular brain polypeptides and 65 putative neurosecretory polypeptides that change in concentrations during adult brain development in *M. sexta*. We have also cloned six genes that are differentially expressed during postembry-

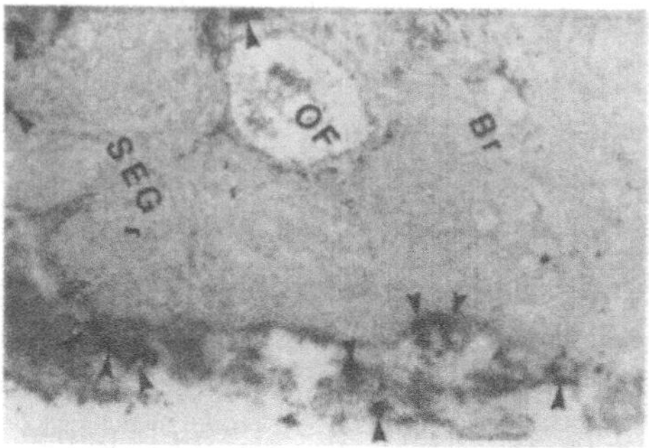

Fig. 1. Localization of expression of gene #4 in the brain and suboesophageal ganglion of a *Manduca* 7-8-day-old pharate adult (arrows). Br, brain; OF, oesophageal foramen; SEG, suboesophageal ganglion.

onic neurogenesis. Activation of the cloned genes during this period suggests that the controlled expression of these genes is important for postembryonic neurogenesis and brain differentiation.

REFERENCES

Booker, R. and Truman, J. W. (1987). Postembryonic neurogenesis in the CNS of the tobacco hornworm, *Manduca sexta*. I. Neuroblast arrays and the fate of their progeny during metamorphosis. *J. Comp. Neurol.* 255:548-559.

Davis, M.-T. B. and Svoboda, J. A. (1989). Genes differentially expressed during pupal/adult neurogenesis in *Manduca sexta*. *Insect Biochem.* Accepted.

O'Farrell, P.H. (1975). High resolution two-dimensional electrophoresis of proteins. *J. Biol. Chem.* 250:4007-4021.

Panov, A. A. (1963). The origin and fate of neuroblasts, neurons, and neuroglial cells in the central nervous system of the Chinese silkmoth, *Antheraea pernyi guer.* (Lepidoptera, Attacidae). *Rev. Entomol. URSS* 42:337-350.

Technau, G. and Heisenberg, M. (1982). Neural reorganization during metamorphosis of the corpora pedunculata in *Drosophila melanogaster*. *Nature* 295:405-407.

From *Insect Neurochemistry and Neurophysiology • 1989* Edited by A. B. Borkovec and E. P. Masler

RECEPTIVITY TERMINATING FACTOR IN *HELIOTHIS ZEA* AND ITS INTERACTION WITH PBAN

A. K. Raina[1], T. G. Bird[2] and J.M. Giebultowicz[3]

[1]Insect Chemical Ecology Lab. and [2]Livestock Insects Lab., Agric. Res. Serv., USDA, Beltsville, MD 20705, USA. [3]Dept. of Zool., Univ. of Maryland, College Park, MD 20742, USA.

Mating in most species of Lepidoptera is mediated by the production and release of sex pheromones. In the corn earworm, *Heliothis zea*, virgin females, while releasing the pheromone, are receptive to approaching males. After mating there is a switch from the virgin to the mated behavior expressed by termination of pheromone production and loss of receptivity to further mating (Raina et al., 1986; Raina, 1989). This switch is caused by a putative receptivity terminating factor (RTF), transferred by the male to the female at the time of mating (Raina, 1989). The testes or the sperm do not contribute to the presence or transfer of this factor, and it was postulated that the source of RTF is the male accessory glands (AG). The material, after transfer along with the spermatophore, is released into the hemolymph and, in an as yet unknown way, blocks the release and/or action of the pheromone biosynthesis activating neuropeptide (PBAN). Similar observations have been reported in the house fly, *Musca domestica,* by Leopold et al. (1971), who noted that radiolabelled male ejaculatory duct material, after transfer to the female, was released into the hemolymph and carried to the head region. The factor caused termination of receptivity in the female house flies. Among the Lepidoptera, Riddiford and Ashenhurst (1973) speculated that receptivity in *Hyalophora cecropia* was terminated by the release of a substance from either the spermatheca or bursa of a mated female and its passage *via* hemolymph to a hypothetical neural center in the brain.

In order to determine if the AG are the source of RTF in *H. zea*, one-day-old virgin females were injected with either 0.25 AG or 0.25 Malpighian tubule (MT) equivalents in 20 µl sucrose-phosphate buffer. A third group of females was injected with 20 µl buffer. Pheromone was extracted 2 h after injection and quantified by capillary gas chromatography (Raina et al., 1986). The experiment was conducted during scoptophase and replicated 15 times. Whereas injection of MT caused only 7.5% reduction in the pheromone titer compared to buffer-injected females, there was 94.5% reduction caused by the AG homogenate. In a second experiment the AG, duplex and *ductus ejaculatorius simplex* (DES) complex (Fig. 1) was dissected from three males; each complex was divided into three respective sections, which were individually homogenized in 60 µl buffer and, centrifuged at 10,000 x g for 1 min. The supernatant was injected at a dose of 0.25 equivalents into virgin females and pheromone titer was determined as described earlier. Once again, percent reduction in pheromone titer in relation to buffer-injected control females was calculated. The pheromone titers were reduced by 80.8 ± 5.0, 28.8 ± 18.1 and 46.9 ± 16.0% in response to AG, duplex and DES, respectively, (mean ± SD, N=15). These observations indicate that AGs are the source of RTF in *H. zea*. A dose response was studied with crude AG homogenates. Fig. 2 shows that reduction in pheromone titer at 0.25 AG equivalent was not significantly different from that at 1 AG equivalent.

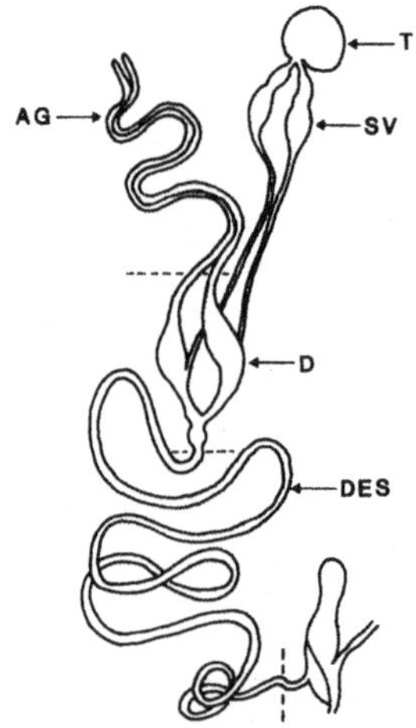

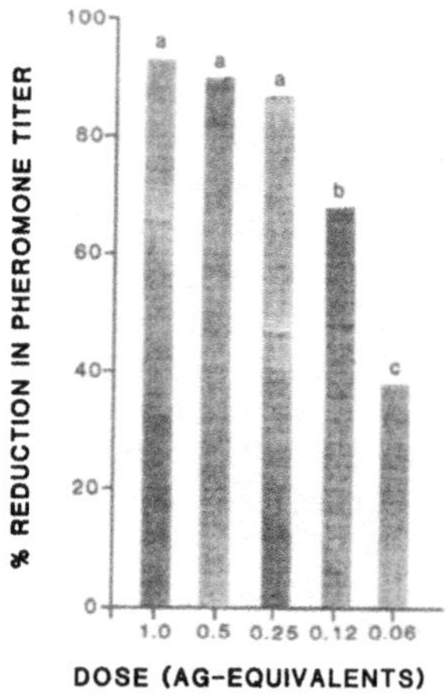

Fig. 1. Reproductive organs of male *H. zea*. Accessory glands (AG), duplex (D), *ductus ejaculatorius simplex* (DES), seminal vesicle (SV) and testes (T).

Fig. 2. Dose x response of crude accessory gland homogenate. Bars with common letters indicate values that are not significantly different at P>0.05 (N=5).

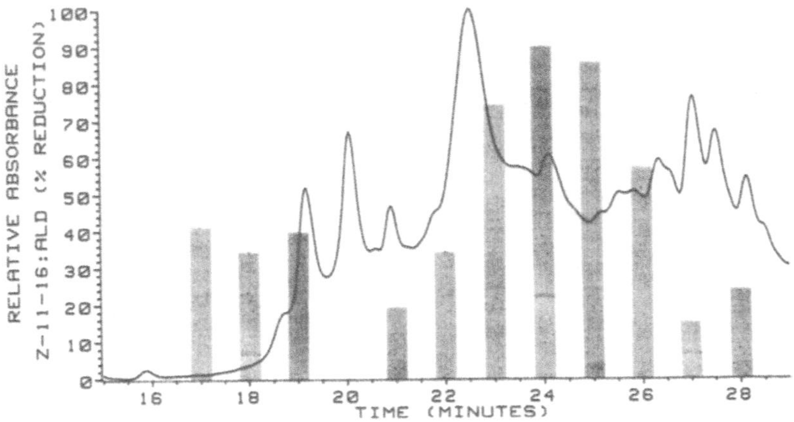

Fig. 3. Biological activity [suppression of pheromone ((z)-11-hexadecenal) production, indicated by bars] associated with HPLC fractions of accessory glands. Fractions 16 and 20 caused a slight increase in the pheromone titer compared to buffer injected control females.

Initial purification of RTF was carried out with AGs removed from 180 *H. zea* males. The tissue was homogenized in Bennett's buffer without NaCl, and centrifuged for 15 min at 15,000 x g at -4°C. The process was repeated with the pellet, and the combined supernatant was chromatographed on a Waters Model 840 HPLC fitted with a Vydac C-18 semi-preparative column using method C described by Jaffe et al. (1988). Fractions were collected over 1-min intervals, and the solvent was removed by Speed-Vac. Five AG equivalents from each fraction were used for the bioassay. Each portion was dissolved in 50 µl buffer and five two-day-old virgin females were each injected with 5 µl (0.5 AG equivalent). Pheromone titers were determined 2 hr after injection, as described earlier. Highest pheromone suppression activity was associated with 23-25 min fractions (Fig. 3). Crude AG extract at a similar concentration caused a 96% reduction in the pheromone titer.

It was reported that termination of pheromone production in mated females could be reversed by the injection of PBAN into such females (Raina, 1989). Based on those results it was speculated that RTF affects the release or action of PBAN. In a recent experiment, virgin *H. zea* females were ligated and, after 17 h, injected with 5 pmoles of synthetic Hez-PBAN (Raina et al., 1989). After 1 h, half of the group was injected with AG extract and the other half with 5 μl buffer. All the females were extracted for pheromone 2 h after the second injection. Pheromone titer of AG-injected females ranged from 1-5 ng/female, compared to 130-160 ng for the control females. These preliminary results indicate that RTF may be blocking the action of PBAN and stimulating the metabolism of existing pheromone.

REFERENCES

Jaffe H., Raina A.K., Fraser B.A., Keim P., Ranga Rao K., Zang Y-S., Lancaster J.L. and Hayes D.K. (1988) Isolation of two neuropeptides in the AKH/RPCH-family from horse flies (Diptera). *Biochem. Biophys. Res. Commun.* **151**, 656-663

Leopold R.A., Teranova A.C., Thorson B.J. and Degrugillier, M.E. (1971). The biosynthesis of the male housefly accessory secretion and its fate in the mated female. *J. Insect Physiol.* **17**, 987-1003.

Raina A.K. (1989) Male-induced termination of sex pheromone production and receptivity in mated females of *Heliothis zea*. *J. Insect Physiol.* (in press).

Raina A.K., Klun J.A. and Stadelbacher E.A. (1986) Diel periodicity and effect of age and mating on female sex pheromone titer in *Heliothis zea* (Lepidoptera: Noctuidae). *Ann. ent. Soc. Am.* **79**, 128-131.

Raina A.K., Jaffe H., Kempe T.G., Keim P., Blacher R.W., Fales H.M., Riley C.T., Klun J.A., Ridgway R.L. and Hayes D.K. (1989) Identification of a neuropeptide hormone that regulates sex pheromone production in female moths. *Science* **244**, 796-798.

Riddiford L.M. and Ashenhurst J.B. (1973) The switchover from virgin to mated behavior in female Cecropia moths: the role of the bursa copulatrix. *Biol. Bull.* **144**, 162-171.

From *Insect Neurochemistry and Neurophysiology • 1989* Edited by A. B. Borkovec and E. P. Masler

REGULATION OF SEX PHEROMONE BIOSYNTHESIS IN CHRYSODEIXIS CHALCITES AND HELIOTHIS PELTIGERA BY PBAN.

Miriam Altstein, Yoav Gazit and Ezra Dunkelblum

Institute of Plant Protection, ARO,

The Volcani Center, Bet Dagan, Israel.

INTRODUCTION

The purpose of our study was to characterize the biological activity of PBAN (Pheromone Biosynthesis Activating Neuropeptide, Raina and Klun, 1984; Raina et al., 1989), and to study its mode of action on the biosynthetic process in the pheromone gland of two moths species: Heliothis peltigera and Chrysodeixis chalcites.

The involvement of PBAN in sex pheromone biosynthesis was determined by comparing pheromone gland content in untreated females, in females ligated between the head and the thorax, and in ligated females injected with head ganglia extracts. Analysis of pheromone gland content was performed by capillary gas chromatography (CGC). The most abundant components of the pheromone blend in each moth, namely: (Z)-11-hexadecenal (Z11-16:Ald) (Dunkelblum and Kehat, 1989), and (Z)-7-dodecenyl acetate (Z7-12:Ac) (Dunkelblum et al., 1987) were chosen to monitor sex pheromone content in H. peltigera and C. chalcites, respectively.

Ligation of 3-day-old H. peltigera females revealed complete depletion of pheromone content (Table 1). Injection of 1/5 equivalent of head ganglia extracts from 3-day-old donors to the ligated females resulted in a recovery of pheromone content, which increased from <0.5 to

Table 1. Pheromone content in untreated and in ligated *Heliothis peltigera* before and after injection of female and male head ganglia extracts

	Heliothis peltigera (3-day-old)	(7-day-old)
	Z11-16:Ald (ng/gland ± SEM)	
Control	105.9±17.3(a) (n=10)	109.3±23.6(a) (n=7)
Ligated	<0.5(c) (n=12)	<0.5(c) (n=10)
Ligated & injected:		
Female extract	56.6±11.3(b) (n=12)	49.6±10.3(b) (n=9)
Male extract	79.6±16.3(ab) (n=8)	n.t.

56.6 ng/gland. Similar results were obtained when 7-day-old females were ligated and injected with the same amount of head ganglia extracts from 7-day-old donors (Table 1). Content analysis of 3-day-old untreated females showed that only 50% of the females produce pheromone in amounts comparable to those found in 7-day-old untreated females. No pheromone could be detected in the other 50% prior to injection, probably as a result of immaturity. Injection of ganglia extracts to the 3-day-old females elevated pheromone content to the same level in all moths, indicating that the lack in biosynthetic capacity in the immature females may have resulted from a deficiency of PBAN content or release.

Similar results were obtained when the bioassay was applied to *C. chalcites*. Ligation resulted in a marked decrease in pheromone content (from 560 ng/gland to 264 ng/gland in untreated and ligated females, respectively). Injection of 1/5 equivalent of head ganglia extract, from 2.5-day-old donors, resulted in a full recovery of sex pheromone content (Altstein et al., 1989).

Recently it has been reported that male moths also produce sex pheromone-like components (Teal and Tumlinson, 1989). The function of the sex pheromones in males is not fully understood at present and the mechanism that regulates sex pheromone biosynthesis in males is unknown. The presence of PBAN in male moths has been reported in several studies (Raina and Klun, 1984; Ando et al., 1988). We have confirmed the presence of a PBAN like activity in male head ganglia extracts of H. peltigera and C. chalcites. Injection of 1/5 equivalent of H. peltigera head ganglia extract elevated the pheromone content from <0.5 to 79.6 ng/gland, indistinguishable from pheromone levels in untreated control moths (Table 1). Similar results were obtained with C. chalcites (Altstein et al., 1989).

Pheromone biosynthesis is affected by a variety of environmental factors, in particular photoperiodicity. Moths are nocturnal animals and sex pheromone biosynthesis occurs mainly during the scotophase. The mechanism by which light switches off pheromone biosynthesis is not understood. Since PBAN controls sex pheromone production, it is very likely that light may have an inhibitory effect on the biosynthesis of PBAN, its release into the hemolymph, or alternatively on its action on the target organ. The effect of light on PBAN activity in H. peltigera was analyzed by comparing the ability of PBAN to stimulate sex pheromone biosynthesis during scotophase and photophase, and by comparing the activity of PBAN extracted from head ganglia in the two photoperiods (Table 2).

Injection of PBAN (1/5 head ganglia equivalent) into ligated females during the scotophase elevated pheromone levels from <0.5 to 56.6 ng/gland. Injection of PBAN to untreated females during the photophase elevated pheromone levels from 0.5 to 63.3 ng/gland, indicating that light does not inhibit PBAN activity or the pheromone biosynthetic pathway in the gland. Injection of head ganglia extracts from donors at photophase raised the pheromone content to 62.7 and 92.3 ng/gland, in scotophase and photophase recipients, respectively, indicating that PBAN content in the neurosecretory cells is not depleted during the light period. On the basis of these results, it is very likely that light has an inhibitory effect on the release of PBAN into the hemolymph.

Table 2. Pheromone content during scotophase and photophase in the presence and absence of PBAN in Heliothis peltigera

Recipient (3-day-old)	Donor (3-day-old)	Treatment	Z11-16:Ald content (ng/gland ± SEM)	
scotophase	-	untreated	105.9±17.3(a)	(n=10)
scotophase	-	ligated	<0.5(c)	(n=12)
scotophase	scotophase	ligated & inj.	56.6±11.3(b)	(n=12)
scotophase	photophase	ligated & inj.	62.7±27.9(ab)	(n=7)
photophase	-	untreated	0.5±0.5(c)	(n=6)
photophase	scotophase	injected	63.3±28.1(ab)	(n=7)
photophase	photophase	injected	92.3±16.2(ab)	(n=10)

Studies on the effect of PBAN on the biosynthetic pathway of sex pheromones was performed by analyzing the content of the pheromone biosynthetic precursors hexadecanoate, (Z)-11-hexadecenoate, (Z)-9-tetradecenoate and (Z)-7-dodecenoate, and the most abundant pheromone components (Z)-7-dodecenyl acetate and (Z)-9-tetradecenyl acetate in the presence and absence of PBAN, namely, in ligated, nonligated, and ligated plus injected C. chalcites females.

CGC analysis of the methyl ester and alcohol derivatives of these components (generated by a basic methanolysis of gland extracts) revealed that the content of all three unsaturated fatty acyls as well as the two major pheromone components decreased significantly in the absence of PBAN, whereas the content of palmitic acid was not affected (Altstein et al., 1989). Since the 11 desaturase is considered to be involved in sex pheromone biosynthesis in C. chalcites (Roelofs and Wolf, 1988), generating Z11-16:Acyl from hexadecanoic acid, our results indicate that this step is affected by the neuroendocrine factor.

ACKNOWLEDGEMENT

This research was supported by the Fund for Basic Research administered by the Israel Academy of Science and Humanities.

REFERENCES

Altstein, M., Harel, M., and Dunkelblum, E. (1989) Effect of a neuroendocrine factor on sex pheromone biosynthesis in the tomato looper, Chrysodeixis chalcites (Lepidoptera: Noctuidae). Insect Biochem. 19, In press.

Ando, T., Arima, R., Uchiyama, M., Nagasawa, H., Inoue, T. and Suzuki, A. (1988) Pheromone biosynthesis activating neuropeptide hormone in heads of the silkworm moth. Agric. biol. Chem. 52, 881-883.

Dunkelblum, E. and Kehat, M. (1989) Female sex pheromone components of Heliothis peltigera (Lepidoptera: Noctuidae): Chemical identification from gland extracts and male response. J. chem. Ecol., 15, 2233-2245.

Dunkelblum, E. Snir, R. Gothilf, S. and Harpaz, I. (1987) Identification of sex pheromone components from pheromone gland volatiles of the tomato looper, Plusia chalcites (Esp.). J. chem. Ecol., 13, 991-1003.

Raina, A.K., Jaffe, H., Kempe, T.G., Keim, P., Blacher, R.W., Fales, H.M., Riley, C.T., Klun, J.A., Ridgway, R.L. and Hayes, D.K. (1989) Identification of a neuropeptide hormone that regulates sex pheromone production in female moths. Science 244, 796-798.

Raina, A.K. and Klun, J.A. (1984) Brain factor control of sex pheromone production in the female corn earworm moth. Science 225, 531-533.

Roelofs, W.L. and Wolf, W.A. (1988) Pheromone biosynthesis in Lepidoptera. J. chem. Ecol. 14, 2019-2031.

Teal, P.E.A. and Tumlinson, J.H. (1989) Isolation, identification and biosynthesis of compounds produced by male hair-pencil gland of Heliothis virescens (F.) (Lepidoptera: Noctuidae). J. chem. Ecol. 15, 413-427.

From *Insect Neurochemistry and Neurophysiology • 1989* Edited by A. B. Borkovec and E. P. Masler

STIMULATION OF DE NOVO PHEROMONE BIOSYNTHESIS BY IN VITRO PHEROMONE GLANDS OF HELIOTHIS SPP.

Rafaeli, A.[1], Soroker, V.[1]
Klun, J.A.[2] and Raina, A.K.[2]

[1] Department of Entomology, Faculty of Agriculture, The Hebrew University, Rehovot 76-100. [2] Insect Chemical Ecology Lab., USDA, Beltsville, Maryland 20705, USA.

We have recently demonstrated (Soroker & Rafaeli, 1989) that isolated *in vitro* pheromone glands of *Heliothis armigera* incorporate radioactivity from ^{14}C sodium acetate and that this total incorporation is stimulated by a pheromonotropic factor present in brain-complex extracts. This stimulation was found to be mediated by cyclic-AMP (Rafaeli & Soroker, 1989b). Subsequent TLC and HPLC separations of the radioactive products extracted by hexane revealed co-elution with Z (11)-hexadecenal (Z-11-16:ald). Furthermore, capillary GC analysis of the hexane soluble products from such incubations, in the absence of ^{14}C sodium acetate, revealed a stimulation of the concentration of Z-11-16:ald by brain extracts. However, levels of pheromone attained were found to be lower than those attained by live virgin females (Rafaeli & Soroker, 1989a).

In vitro stimulation of acetate incorporation suggests that the pheromonotropic factor acts directly on the pheromone glands. However, nerve endings from terminal abdominal ganglia are present in the *in vitro* preparation. The implication that terminal abdominal ganglia (TAG) are involved in pheromone production was made by Tang *et al*. (1987) who showed that transection of the ventral nerve cord anterior to the TAG in pupae affected the normal development of the TAG during metamorphosis and subsequently affected the pheromone emission as well as calling behavior by females of Lymantria dispar.

This report presents verification of the stimulation of Z-11-16:ald production by in vitro incubations of pheromone glands of H. armigera and H. zea using a preparative radio-GLC separation technique. In addition, the involvement of the TAG on de novo biosynthesis is investigated.

In vitro incubations of pheromone glands were performed on 2-3 day old virgin females during the first-third hour of scotophase as reported previously (Soroker & Rafaeli, 1989). For incubations containing TAG, abdominal tips were dissected containing the 5th, 6th, 7th , 8th and 9th abdominal segments. These were cut lengthwise and extraneous tissues were removed thus exposing the TAG situated between the 5th and 6th abdominal segments. This complex was washed several times in TC-199 (pH 7.2) (Sigma, USA) incubated as reported for abdominal tips containing pheromone glands alone (Soroker & Rafaeli, 1989).

Preparative radio-GLC was performed using a 4% Carbowax 20M packed column at 180°C isothermal conditions. Effluent collections of marker components was performed using a dry-ice-acetone cold trap and glass microcapillary tubules. The microcapillary tubules were immediately extracted with 1 ml hexane into scintillation vials to which scintillant was added and radioactivity was monitored.

The total hexane extractable products of 10 pooled control incubations and 10 pooled incubations subjected to brain-complex extracts (1.5 female equivalents, FE) were separated using preparative GC . Effluent collection of the marker compounds revealed a significant stimulation of incorporation into a peak co-eluting with Z-11-16:ald in the presence of brain-complex extracts (Fig. 1A, means$\pm$SEM of 4 replicate runs). Similar separations were also performed on pheromone gland incubations containing intact TAG in the presence and absence of brain-complex extracts (1.5FE). A significant increase in the incorporation of radioactivity occurred in the peak co-eluting with Z-11-16:ald as well as hexadecenal (Fig. 1B, means $\pm$SEM of 3 replicate runs) in the presence of TAG.

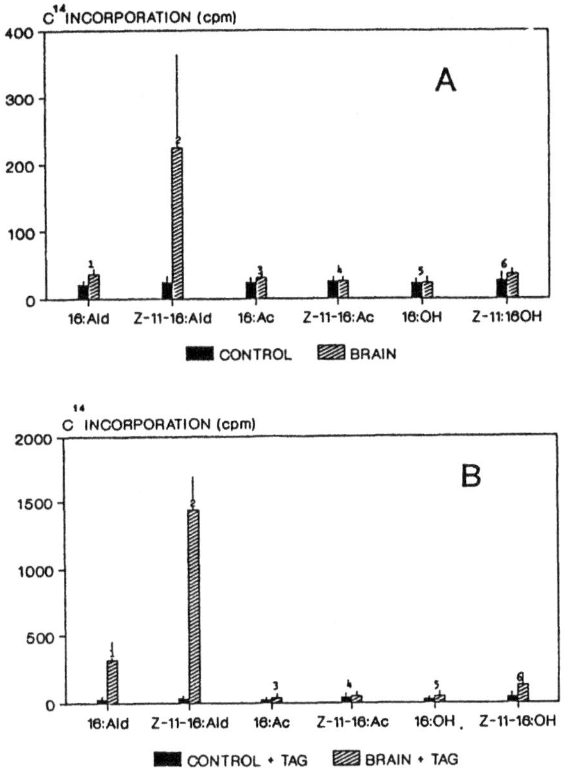

Fig. 1. Effect of brain extracts on the incorporation of radioactivity from ^{14}C sodium acetate into 1 Hexadecenal; 2 (Z)-11 hexadecenal; 3 Hexadecenyl acetate; 4 (Z)-11 hexadecenyl acetate; 5 Hexadecenol; 6 (Z)-11 hexadecenol by in vitro pheromone glands of *Heliothis armigera* and *H. zea*. Fig A shows the corresponding radioactivity present in the collected effluent by control and brain extract treated pheromone glands. Fig B shows the corresponding radioactivity present in the collected effluent by control and brain-extract treated pheromone glands with intact TAG.

The present study demonstrated *in vitro* stimulation of acetate incorporation by brain extracts in pheromone glands of *H. zea* thereby reconfirming our previous findings on *H. armigera*. In addition preparative radio-GLC separations of the total hexane extractable products, after in vitro incubations, clearly demonstrated that part of the radioactivity is incorporated specifically into the main pheromone component, Z-11-16:ald, of both species. The presence of TAG revealed an enhancement of the brain-extract stimulatory effect on in vitro incubations thus indicating its complementary role in pheromone production.

REFERENCES

Rafaeli A. and Soroker, V. (1989a) Effect of diel-rhythm and brain hormone on pheromone production in two Lepidopteran species. J. Chem. Ecol. 15, 447-455.

Rafaeli A. and Soroker, V. (1989b) cAMP mediation of the hormonal stimulation of ^{14}C-acetate incorporation by Heliothis armigera pheromone glands in vitro. Mol. & Cell. Endocr. (in press).

Soroker, V. and Rafaeli, A. (1989) In vitro hormonal stimulation of acetate incorporation by Heliothis armigera pheromone glands. Insect Biochem. 67, 1-6.

Tang J.D., Charlton, R.E., Carde, R.T. and Yin, C.-M. (1987) Effect of allatectomy and ventral nerve cord transection on calling, pheromone emission and pheromone production in Lymantria dispar. J. Insect Physiol. 33, 469-476.

ACKNOWLEDGMENTS

This research was funded in part by a BARD grant no. US 784-84 to J.A Klun and A. Rafaeli and by a fund for basic research administered by the Israeli Academy of Science & Humanities to A. Rafaeli.

From *Insect Neurochemistry and Neurophysiology • 1989* Edited by A. B. Borkovec and E. P. Masler

REGULATION OF SEX PHEROMONE TITER IN MATED GYPSY MOTH FEMALES

J.M. Giebultowicz[1,2], A.K. Raina[1], and E.C. Uebel[1]

[1]Insect Chemical Ecology, ARS, USDA, Beltsville MD 20705, [2]Dept. of Zoology, Univ. Maryland, College Park, MD 20742

Females of gypsy moth, Lymantria dispar, produce a sex pheromone, cis-2-decyl-3-(5-methylhexyl) oxirane, or disparlure, to attract their mates. After successful mating the females become nonreceptive to other males and commence oviposition. In Heliothis zea, it was reported that mating caused a rapid decline in the pheromone titer of the females but that the titer increased again on subsequent nights following the night of mating (Raina et al., 1986). The present study was initiated to investigate the fate of the sex pheromone in mated gypsy moth females and to determine the mechanism through which the switch from virgin to mated behavior operated.

The titer of sex pheromone (disparlure) was determined by extracting the abdominal tips of females in heptane containing a known quantity of an internal standard and analyzing the extract by capillary gas chromatography. Pheromone was measured in virgin females and in females mated to 3 types of males: (1) normal males kept in 16:8h light-dark cycles (LD), (2) males kept in constant light (LL), a condition causing significant reduction in the amount of sperm transferred to the female during mating (Giebultowicz, et al. 1989) and (3) castrated males. Pheromone titer decreased significantly in females mated to all 3 types of males within 2 h following mating (Fig 1). In females mated to LD males the titer remained low on subsequent days. In contrast, in females mated to LL or castrated males pheromone titer increased

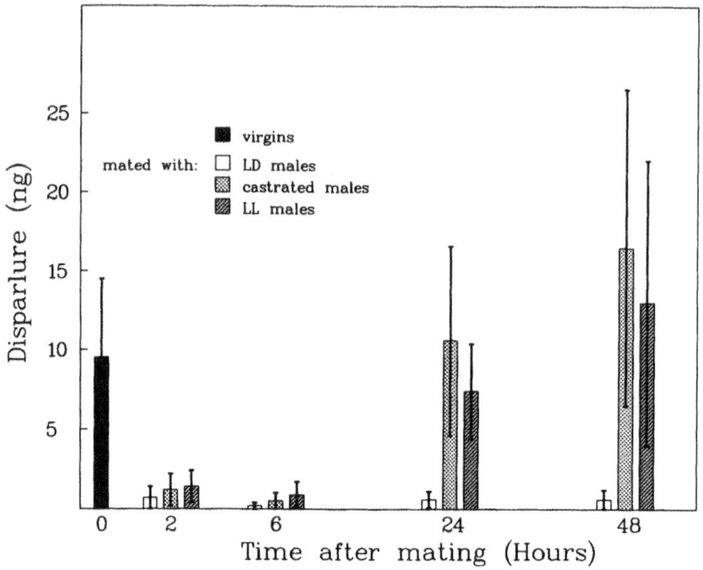

Fig. 1. Titer of sex pheromone, disparlure, in virgin and mated gypsy moth females. Each bar represents mean (±SE) from 8-12 females.

significantly on the day following mating.

Temporary decrease in pheromone titer seems to be caused by mechanical stimuli received during initial contact with male. Thus, in females mated for 2 minutes, that is time which does not allow transfer of spermatophore, pheromone titer decreased to 13% (n=8) of the virgin level. Moreover, in females that had no contact with male but instead a glass rod was inserted into their bursa copulatrix opening and held in place for 2 minutes pheromone titer decreased to 15% (n=10) of virgin level within 3 h after treatment.

For a more permanent decrease in pheromone production presence of sperm in spermatheca of mated females appears to be essential. Preliminary results indicated that removal of spermatheca within 1 h of mating with normal male caused subsequent increase in pheromone level, similar to increase observed after mating with castrated males (Fig 2). The gypsy

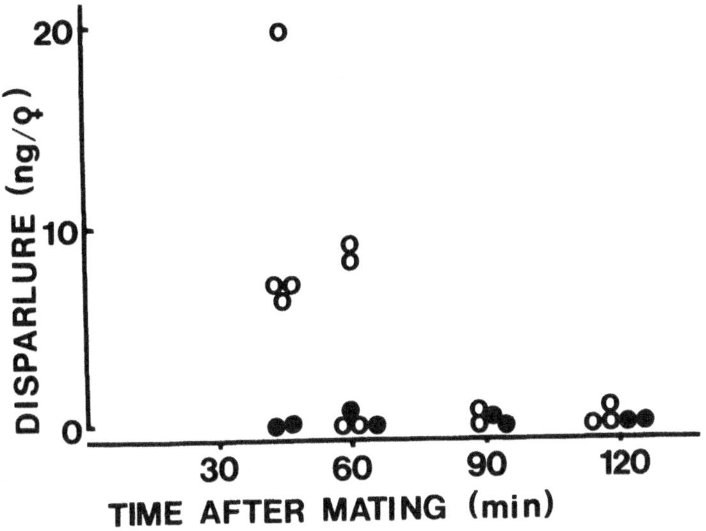

Fig. 2. Titer of disparlure in individual females that had spermatheca removed (open circles) or were sham-operated (closed circles) at different times after the beginning of mating. Disparlure was extracted 24 h after mating.

moth appears to differ from H. zea in which it is reported that the sperm play no essential role in termination of pheromone production following mating (Raina 1989).

In conclusion, the decline in pheromone titer in gypsy moth females after mating is regulated by a two step mechanism. An initial rapid decrease is mediated by mechanical stimuli, whereas long term suppression of pheromone production, appears to be dependent on the presence of sperm in spermatheca.

Supported in part by grant No. USDA88-37251-3970 to JMG.

REFERENCES

Giebultowicz J.M., Ridgway R.L. and Imberski R.B. (1989) Physiological basis for sterilizing effects of constant light in Lymantria dispar. Physiol. Entomol. (In press)

Raina A.K. (1989) Male-induced termination of sex pheromone production and receptivity in mated females Heliothis zea. J. Insect Physiol. (In press)

Raina A.K., Klun J.A., and Stadelbacher E.A. (1986) Diel periodicity and effect of age and mating on sex pheromone titer in Heliothis zea (Lepidoptera: Noctuidae). Ann. Entomol. Soc. Am. 79:128-131.

From *Insect Neurochemistry and Neurophysiology • 1989* Edited by A. B. Borkovec and E. P. Masler

CHANGES IN NUMBER AND SIZE OF CORPUS ALLATUM CELLS OF *BLATTELLA GERMANICA* DURING OOCYTE MATURATION

Ann-Shyn Chiang, Michal Gadot
and Coby Schal

Dept. of Entomology, Rutgers Univ.
New Brunswick, New Jersey 08903

The corpus allatum (CA) of the female German cockroach shows cyclic changes in both size and juvenile hormone III (JH III) biosynthesis during the gonotrophic cycle (Gadot et al., 1989). The increase in CA volume during oocyte maturation has been considered mainly due to increases in both cell number and cell size (Cassier, 1979; Tobe and Stay, 1985). To date, the number of nuclei in the CA has been estimated from sections of fixed thickness in fixed organs (Scharrer and von Harnack, 1958; Szibbo and Tobe, 1981; Tobe et al., 1984). Since nuclear size and shape vary with different developmental stages of the CA during the gonotrophic cycle in *Diploptera punctata* (Johnson et al., 1985) and other insects (Cassier, 1979), nuclear counts from sections may under- or over-estimate the total number of cells. The size of CA cells is usually presented as the nucleocytoplasmic ratio in fixed glands (Engelmann, 1957). We report on changes in number and size of CA cells of *Blattella germanica* during oocyte maturation by direct measurements of enzymatically dissociated cells.

Whole-mounts of partially dissociated CA monolayers (Fig. 1A) provided counts of the absolute number of cells in individual CA. Cell nuclei in the monolayer were intact and evenly distributed without overlap. Some nuclei were larger than others, especially in CA from females with mature oocytes. In contrast to reports of significant increases in CA cell number during oocyte maturation in *D. punctata* (Szibbo and Tobe, 1981; Tobe et al., 1984), cell number did

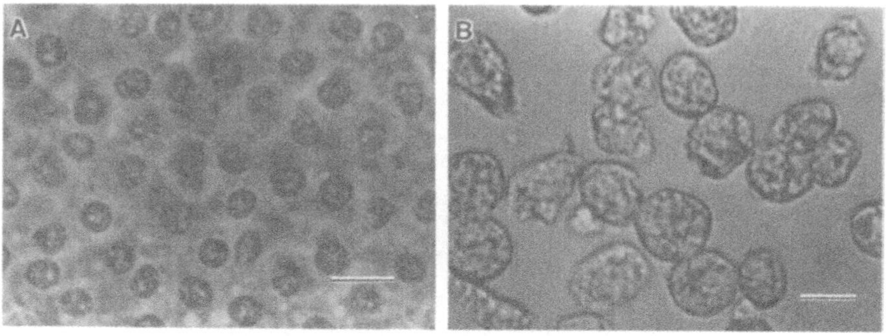

Fig. 1. (A) A whole-mount monolayer of CA cells. The CA from a day 0 female were desheathed by incubation with 0.1% collagenase, stained with 0.01% crystal violet in 0.1 M citric acid and then spread into a cell monolayer under a cover slip. (B) CA cells (from high-activity CA of day 6 grouped virgin females) in suspension. Individual CA were partially digested with 0.1% trypsin in cockroach saline and dissociated by vortexing. Bars = 10 µm.

not increase during this period in B. germanica. Cell number (mean ± SEM) per CA was 1927 ± 52 (n=12) on day 0, 2029 ± 63 (n=8) on day 3, and 2013 ± 68 (n=8) on day 6. Clearly, these changes in cell number do not account for the increase in CA volume and JH-III biosynthesis during oocyte maturation.

Changes in CA cell size during the period of oocyte maturation were directly measured in cell suspensions following enzymatic dissociation of freshly excised CA. Cells were randomly chosen by a hemocytometer grid and their maximum diameter was measured as an indicator of cell size. Dissociated CA cells were usually globular in suspension (Fig. 1B) and cell sizes from each of three age groups were normally distributed (Fig. 2). Cell diameter increased significantly during the period of oocyte maturation from a mean value of 8.9 ± 0.1 µm on day 0 (oocyte length < 0.5 mm) to 11.7 ± 0.2 µm on day 6 (oocyte length 1.6-1.9 mm) and to 12.7 ± 0.2 µm on day 7 (oocyte length 2.0-2.1 mm). Corpus allatum volume ($x10^6$ μm^3) increased from 0.87 ± 0.06 (n=8) to 1.22 ± 0.05 (n=12) and to 1.63 ± 0.09 (n=16) during the equivalent period. The pattern of increase in cell size in relation to oocyte length was similar to the increase in CA volume and JH-III biosynthesis.

Using the radiochemical assay for JH-III biosynthesis in vitro, during a 4 h incubation dissociated CA cells incorporated L-[methyl-^{3}H]methionine into JH-III at rates comparable to intact glands (2.16 ± 0.19 and 2.77 ± 0.22 pmol/hr/gland, respectively). HPLC analysis confirmed that JH-III was the only JH produced by these dissociated cells. Since methionine can not be utilized by homogenized CA for JH biosynthesis (Schooley and Baker, 1985), it is not plausible that released enzymes from disintegrated cells would account for this activity. These results indicate that the dissociation procedure did not impair the JH-III biosynthetic activity of CA cells and that the basal lamina and intercellular junctions between CA cells of adult female B. germanica are not required in the biosynthesis of JH-III, at least in short-term incubations.

In conclusion, we demonstrate methods to directly determine cell number and cell size in the CA. Total cell number in the corpora allata of B. germanica adult females does not increase as juvenile hormone synthesis and CA volume increase during oocyte maturation. Rather, cell size increases significantly during this period. Thus, CA activation in the first ovarian cycle of this species is associated mainly with an increase in cell size with minor changes in cell number. In addition, we demonstrate, for the first time, de novo synthesis of JH-III in a suspension of

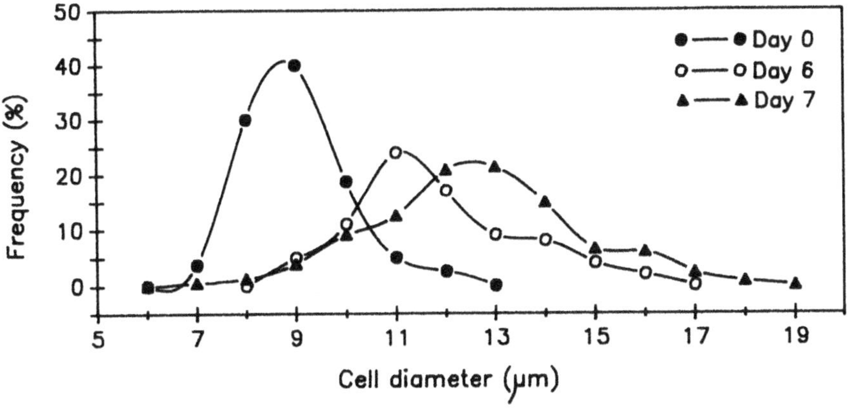

Fig. 2. Distributions of cell diameters during oocyte maturation. Each distribution was generated from a sample of 10 cells per gland from 8 different glands taken from 4 insects.

CA cells, which has important implications to future preparation of primary cell cultures.

Acknowledgments - We thank Edina Burns for technical assistance. Supported by PHS-NIH grant HD-21891 (to C. Schal) and the Thomas C. Headlee Fellowship (to A. Chiang). New Jersey Agricultural Experiment Station publication No. J-08170-01-89 supported by State funds and by the U.S. Hatch Act.

REFERENCES

Cassier P. (1979) The corpora allata of insects. Int. Rev. Cytol. **57**, 1-73.

Engelmann F. (1957) Die Steuerung der Ovarfunktion bei der ovoviviparen Schabe Leucophaea maderae (Fabr.). J. Insect Physiol. **1**, 257-278.

Gadot M., Chiang A. and Schal C. (1989) Farnesoic acid-stimulated rates of juvenile hormone biosynthesis during the gonotrophic cycle in Blattella germanica. J Insect Physiol. (in press).

Johnson G.D., Stay B. and Rankin S.M. (1985) Ultrastructure of corpora allata of known activity during the vitellogenic cycle in the cockroach Diploptera punctata. Cell Tissue Res. **239**, 317-327.

Scharrer B. and von Harnack M. (1958) Histophysiological studies on the corpus allatum of Leucophaea maderae. I. Normal life cycle in male and female adults. Biol. Bull. **115**, 508-520.

Schooley D.A. and Baker F.C. (1985) Juvenile hormone biosynthesis. In Comprehensive Insect Physiology, Biochemistry and Pharmacology (Edited by Kerkut G.A. and Gilbert L.I.), Vol. 7, 1st Edn, pp. 363-389, Pergamon press, Oxford.

Szibbo C.M. and Tobe S.S. (1981) Cellular and volumetric changes in relation to the activity cycle in corpora allata of Diploptera punctata. J. Insect Physiol. **27**, 609-613.

Tobe S.S. and Stay B. (1985) Structure and regulation of the corpus allatum. Adv. Insect Physiol. **18**, 305-432.

Tobe S.S., Clarke N., Stay B. and Ruegg P. (1984) Changes in cell number and activity of the corpora allata of the cockroach Diploptera punctata: a role for mating and the ovary. Can. J. Zool. **62**, 2178-2182.

From *Insect Neurochemistry and Neurophysiology • 1989* Edited by A. B. Borkovec and E. P. Masler

PATTERNS OF CORPORA ALLATA ACTIVITY IN OVARIECTOMIZED BLATTELLA GERMANICA

Michal Gadot, Ann-Shyn Chiang,
Edina Burns and Coby Schal

Dept. of Entomology, Rutgers Univ.
New Brunswick, New Jersey 08903

The corpora allata (CA) of adult female cockroaches, including Blattella germanica (Gadot et al., 1989a,b), synthesize juvenile hormone-III (JH-III) in a precisely regulated cycle which corresponds to the ovarian cycle. The brain, the ovary and the ootheca have been implicated as the main sources of regulatory factors which modulate the activity of the CA (Tobe and Stay, 1985). In Nauphoeta cinerea and Diploptera punctata, the maturing ovary activates the CA at the beginning of the first ovarian cycle, while the mature ovary (just before ovulation) has an inhibitory effect on JH-III synthesis (Lanzrein et al., 1981; Rankin, 1989). Herein we examine the role of the ovary in modulating CA activity in B. germanica.

Insect rearing, the radiochemical in vitro assay of JH-III synthesis and the chemicals used are described elsewhere (Gadot et al., 1989b). Ovariectomy was performed on early last instar nymphs. Newly-emerged ovariectomized females were placed daily with mature males until mating was observed; thereafter, each mated female was housed with two males.

The onset of receptivity in ovariectomized females, determined by their readiness to mate, was less synchronized than in intact grouped females: only 53.7% of 162 ovariectomized females mated by day 7, as compared with more than 95% of intact grouped females; most females in both groups mated between days 5-7. The rest of the ovariectomized females became receptive and mated at different ages. Because sexual receptivity in intact B. germanica females is related to CA activity, and mating has no effect on oocyte development and JH-III synthesis (Gadot et al., 1989a), these results suggest that the population of ovariectomized females may be more asynchronous with regard to CA activity.

The activity of the CA of ovariectomized females that refused to mate was compared to that of females that mated on day 6-7 (Fig. 1A). While the CA of unmated (non-receptive) females exhibited low activity at all ages, the CA of mated females exhibited a cyclic pattern of activity. The rates of JH-III synthesis of active CA from ovariectomized females are similar to those of active CA from intact females (compare Fig. 1 and data in Gadot et al., 1989a,b). However, the cyclic pattern is different in the two cases, since in intact grouped females peak activity of the CA is on days 5-8 (when basal oocyte length is 1.5-2.2 mm); by day 9 (ovulation) the CA are inactivated and they resume activity after the ootheca is dropped (Gadot et al., 1989a).

The cyclicity of JH-III synthesis by CA of ovariectomized females was confirmed by comparing the rates of JH-III synthesis between females which mated normally (on day 6-7) and females which mated later (on day 11 to 19, Fig. 1B). The onset of sexual receptivity proved to be a better predictor of CA activity than chronological age. The CA of individual ovariectomized females were activated at different ages, but once activated, the activity of CA in different females increased and decreased in a similar pattern.

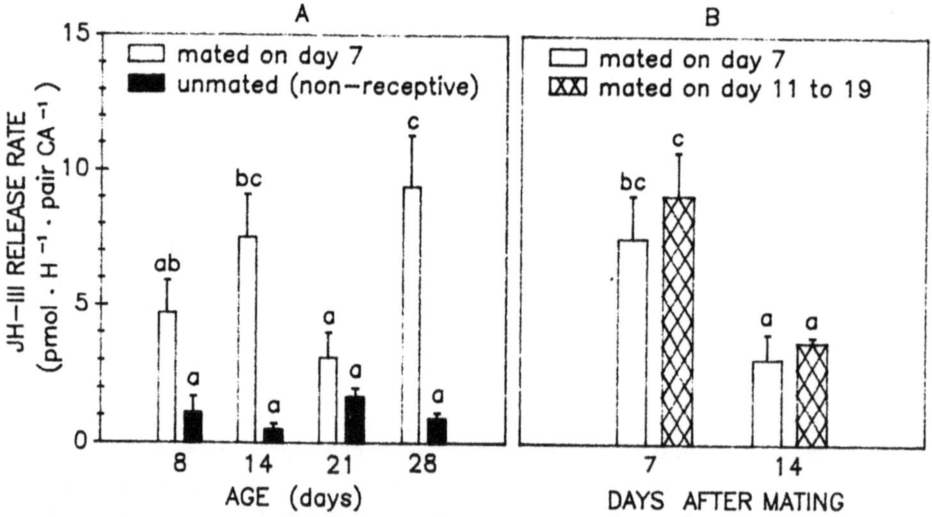

Fig. 1: Juvenile hormone-III synthesis rate, assayed by the in vitro release rate of radiolabeled JH-III (Gadot et al., 1989b) in CA of ovariectomized females: (A) in relation to chronological age in mated and non-receptive females that refused mating; and (B) in relation to onset of sexual receptivity, assayed by days after mating. Each bar represents the mean ± SEM of 4-9 females. Means with different letters differ significantly from each other (Duncan's Multiple Range Test, p<0.05).

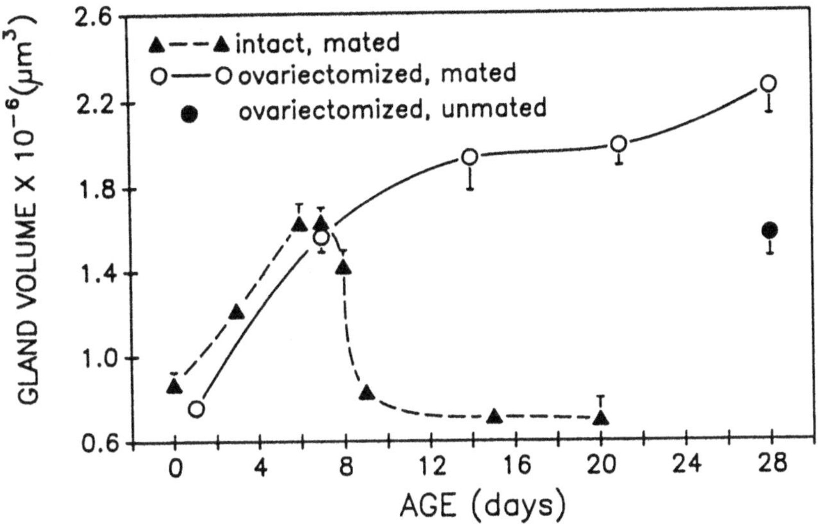

Fig. 2: Corpus allatum volume of intact and ovariectomized females in relation to age. Intact females were mated on day 6, and ovariectomized females as described in Fig. 1. Each point is the mean ± SEM of 8 individual glands.

The effects of ovariectomy in B. germanica are similar to those reported in two other cockroach species, Leucophaea maderae and Periplaneta americana, in which the CA of ovariectomized females exhibit varying levels of activity (Scharrer and von Harnack, 1961; Weaver, 1981). However, while in L. maderae histological and ultrastructural evidence suggests hypertrophy, accompanied by transient hyperactivity of the CA (Scharrer, 1978), in P. americana only low to moderate rates of JH-III synthesis in vitro are exhibited by CA of ovariectomized females (Weaver, 1981). In both cases, no cyclicity of CA activity in ovariectomized females was apparent. Our combined data from all B. germanica females would show a similar pattern, except that high-activity CA from ovariectomized females have the same rates of JH-III synthesis as active CA of intact females. However, by using mating as an indicator of physiological stage, we support the hypothesis of lack of population synchrony among ovariectomized females.

These data suggest that the ovaries are not required in B. germanica for full activation of the CA or for their subsequent inhibition, although they clearly accelerate CA activation and synchronize cyclicity in the population. This is in contrast to the effects of ovariectomy in N. cinerea and D. punctata, which prevents the activation of the CA (Lanzrein et al., 1981; Stay et al., 1983). In these two species, the inactive CA of ovariectomized females, nevertheless increase in volume with age (Lanzrein et al., 1981; Tobe et al., 1984). Similarly, the CA of ovariectomized B. germanica continued to increase in volume through day 28 and did not exhibit the normal cycle of volume changes (Fig. 2). Low-activity CA of ovariectomized unmated females also increased in size and on day 28 were as large as high-activity CA of intact females. Thus, the ovary may modulate developmental changes independently of its effects on overt CA activity, and may have different effects on each process even in closely related species.

Acknowledgments - Supported by PHS-NIH grant HD21891 (to CS) and the Headlee Fellowship (to AC). New Jersey Agricultural Experiment Station publication no. J-08170-02-89 supported by State Funds and by the U.S. Hatch Act.

REFERENCES

Gadot M., Burns E. and Schal C. (1989a) Juvenile hormone biosynthesis and oocyte development in adult female Blattella germanica: Effects of grouping and mating. Arch. Insect Biochem. Physiol. (in press).

Gadot M., Chiang A. and Schal C. (1989b) Farnesoic acid-stimulated rates of juvenile hormone biosynthesis during the gonotrophic cycle in Blattella germanica. J. Insect Physiol. (in press).

Lanzrein B., Wilhelm R. and Buschor J. (1981) On the regulation of the corpora allata activity in adult females of the ovoviviparous cockroach Nauphoeta cinerea. In Juvenile Hormone Biochemistry, Action, Agonism and Antagonism (Edited by Pratt G.E. and Brooks G.T.), 1st Edn, pp. 147-160, Elsevier, Amsterdam.

Rankin S. M. (1989) Regulation of corpus allatum activity in adult females of the viviparous cockroach, Diploptera punctata. In Cockroaches as Models for Neurobiology: Applications in Biomedical Research. (Edited by Huber I., Rao B.R. and Masler E.P.) 1st Edn, CRC Press, Boca Raton. (in press).

Scharrer B. (1978) Histophysiological studies on the corpus allatum of Leucophaea maderae. VI. Ultrastructural characteristics in gonadectomized females. Cell. Tiss. Res. **194**, 533-545.

Scharrer B. and Harnack von, M. (1961) Histophysiological studies on the corpus allatum of Leucophaea maderae. III. The effect of castration. Biol. Bull. **121**, 193-208.

Stay B., Tobe S.S., Mundall E.C. and Rankin S. (1983) Ovarian stimulation of juvenile hormone biosynthesis in the viviparous cockroach, Diploptera punctata. Gen. Comp. Endocrinol. **52**, 341-349.

Tobe S.S. and Stay B. (1985) Structure and regulation of the corpus allatum. Adv. Insect Physiol. **18**, 305-432.

Tobe S.S., Clarke N., Stay B. and Ruegg R.P. (1984) Changes in cell number and activity of the corpora allata of the cockroach Diploptera punctata: a role for mating and the ovary. Can. J. Zool. **62**, 2178-2182.

Weaver R.J. (1981) Radiochemical assays of corpus allatum activity in adult female cockroaches following ovariectomy in the last nymphal instar. Experientia **37**, 435-436.

From *Insect Neurochemistry and Neurophysiology • 1989* Edited by A. B. Borkovec and E. P. Masler

PROTEIN KINASE C AND CALCIUM ARE INVOLVED IN THE ACTION OF JUVENILE HORMONE ON FOLLICLE CELLS OF RHODNIUS PROLIXUS

V.L. Sevala and K.G. Davey

Department of Biology, York University

North York, Ontario, Canada, M3J 1P3

In Rhodnius prolixus, juvenile hormone (JH) is known to act at two sites. It acts on the fat body to initiate and sustain the synthesis and release of vitellogenin into the hemolymph. It also acts on the ovary, causing the appearance of large spaces (patency) between the follicular epithelial cells surrounding the developing oocytes; as a result, vitellogenin from the hemolymph gains access to the oocyte surface (Pratt and Davey, 1972). When JH is applied to follicles in vitro, patency is increased in a rapid, dose-dependent and reversible fashion. The development of patency by JH is unaffected by inhibitors of macromolecular synthesis, but is inhibited by ouabain, a specific inhibitor of Na^+,K^+-ATPase (Abu-Hakima and Davey, 1977; 1979). JH also exhibits saturable and specific binding to microsomal preparations of follicle cell membranes, and activates a JH-sensitive Na^+,K^+-ATPase in these preparations (Ilenchuk and Davey, 1982; 1985). These results suggest that JH binds to a receptor on the follicle cell membrane, but it is not yet clear what further processes in the membrane are interposed between the putative receptor and the activation of the ATPase. The present study explores the possible involvement of protein kinase C and calcium in the development of patency.

In order to determine the effect of various test agents on patency, ovarioles containing vitellogenic oocytes were incubated in Schneider's *Drosophila* medium in the presence and absence of a test compound. The appearance of spaces in the follicular epithelium was visualized by immersing the follicles briefly in a solution of 1 percent Evans' blue dye. Based on the size and distribution of spaces each follicle was assigned a value of 1 to 5 as indicated by Davey and Huebner(1974). In this method a patency index of 0 indicates an epithelium which is devoid of any visible spaces and 5 indicates a fully patent epithelium.

The data presented in the table confirm our earlier finding that JH increases the patency index of the follicles ($p < 0.001$) (Sevala and Davey, 1989). The addition of the calcium channel blocker $CoCl_2$ (Snyder et al., 1988) to the medium containing JH blocks the increase in patency index ($p < 0.001$). Furthermore, pretreating the ovarioles with $CoCl_2$, followed by thorough rinsing, does not block the effect of JH in increasing

The effect of various compounds on patency index of the follicle cells *in vitro*

Treatment	Dose (M)	Patency index (Mean ± SEM)
control	-	0.65 ± 0.25 (n = 15)
Juvenile hormone	10^{-7}	2.60 ± 0.32 (n = 12)
Cobalt chloride + JH	$5x10^{-3}$	0.90 ± 0.26 (n = 15)
Cobalt chloride	$5x10^{-3}$	0.54 ± 0.16 (n = 12)
JH ($CoCl_2$ rinsed off)	10^{-7}	1.80 ± 0.20 (n = 14)
H-7 + JH	10^{-4}	0.85 ± 0.19 (n = 12)
PDBU	10^{-7}	2.80 ± 0.29 (n = 15)
Ouabain + PDBU	10^{-3}	1.00 ± 0.28 (n = 15)

n = number of vitellogenic follicles examined

patency: this demonstrates that the effect of $CoCl_2$ is reversible.

Phorbol-12,13-dibutyrate, an activator of protein kinase C (Nishizuka, 1984), also elicited an increase in patency index ($p < 0.001$). This increase is inhibited by ouabain, an inhibitor of Na^+,K^+-ATPase ($p < 0.001$). Furthermore, 1-(5-isoquinolinylsulfonyl)-2 methyl piperazine (H-7), an inhibitor of protein kinase C (Hidaka et al., 1984), abolishes the JH induced patency of the follicular epithelium ($p < 0.01$).

The following conclusions can be drawn from this study. First, the binding of JH to the putative receptor leads to activation of the Na+,K+-ATPase via the activation of protein kinase C, since inhibition of protein kinase C blocks the activation by JH, and activation of protein kinase C induces patency via activation of Na^+,K^+-ATPase. Second, calcium appears to be essential for the increase in patency by JH, although the precise role is not clear.

REFERENCES

Abu-Hakima R. and Davey, K.G. (1977) Effects of hormones and inhibitors of macromolecular synthesis on the follicle cells of Rhodnius. J. Insect Physiol. 23, 913-917.

Abu-Hakima R. and Davey K.G. (1979) A possible relationship between ouabain sensitive (Na^+- K^+) dependent ATPase and the effect of juvenile hormone on the follicle cells of Rhodnius prolixus. Insect Biochem. 9, 195-198.

Davey K.G and Huebner E. (1974) The response of follicle cells of Rhodnius prolixus to juvenile hormone and antigonadotropin in vitro. Can. J. Zool. 52, 1407-1412.

Hidaka H., Inagaki M., Kawamoto S. and Sasaki Y. (1984) Isoquinolinesulfonamide, novel and potent inhibitors of cyclic nucleotide dependent protein kinase and protein kinase C. Biochemistry 23, 5036-5041.

Ilenchuk T.T. and Davey, K.G. (1982) Some properties of Na^+,K^+-ATPase in the follicle cells of Rhodnius prolixus. Insect Biochem. 12, 675-679.

Ilenchuk T.T. and Davey K.G. (1985) The binding of juvenile hormone to membranes of follicle cells in the insect Rhodnius prolixus. Can. J. Biochem. Cell Biol. 63, 102-106.

Nishizuka Y. (1984) The role of protein kinase C in cell surface signal transduction and tumor promotion. Nature 308, 693-698.

Pratt G.E. and Davey K.G. (1972) The corpus allatum and oogenesis in Rhodnius prolixus. I. The effects of allatectomy. J. Exp. Biol. 56, 201-204.

Sevala V.L. and Davey K.G (1989) Action of juvenile hormone on the follicle cells: Evidence for a novel regulatory mechanism involving protein kinase C. Experientia 45,355-356.

Snyder P.M., Krause K.H. and Welsh M.J. (1988) Inositol triphosphate isomers, but not 1,3,4,5-tetrakis phosphate, induce calcium influx in Xenopus laevis oocytes. J. Biol. Chem. 263, 11048-11051.

From *Insect Neurochemistry and Neurophysiology • 1989* Edited by A. B. Borkovec and E. P. Masler

VOLTAGE-DEPENDENT SODIUM AND CALCIUM CHANNELS OF THE CORPORA ALLATA OF THE COCKROACH *DIPLOPTERA PUNCTATA*.

A.R. McQuiston, C.S. Thompson and S.S. Tobe

University of Toronto, Department of Zoology, Toronto, Ontario M5S 1A1 Canada.

The corpora allata (CA) of *Diploptera punctata* are endocrine glands that synthesize and release juvenile hormone (JH). In mated females, cyclical changes in JH biosynthesis and titre have been correlated with the gonotrophic cycle (Tobe *et al.*, 1985). Calcium ions (Ca^{2+}) appear to contribute to control of JH biosynthesis in this species (Kikukawa *et al.*, 1987). For example, an optimal concentration (3-5mM) of extracellular Ca^{2+} is required for maximal rates of JH biosynthesis and in the absence of extracellular Ca^{2+}, JH biosynthesis is totally inhibited (Kikukawa *et al.*, 1987). Thus, it seems likely that Ca^{2+} modulates JH biosynthesis, in part by transport of Ca^{2+} across the plasma membrane of the CA cells. This paper describes types of ion channels that may be involved in Ca^{2+} transport into the cells of the CA. Previous work (Thompson and Tobe, 1986) has provided evidence for the existence of voltage-dependent Ca^{2+} channels in the CA. The following work has employed an experimental protocol modified from Thompson and Tobe (1986); the results confirm and extend previous observations.

The CA consists of several thousand small irregularly shaped parenchymal cells which are extensively coupled by low resistance gap junctions (Lococo *et al.*, 1986). Currents were passed through the gap junctions to other cells of the gland from which membrane voltage responses were recorded (as in Thompson and Tobe, 1986).

Depolarizing currents produced a nonlinear current-voltage relationship. This is attributable to an apparent decrease in input resistance (Fig. 1a). Large depolarizing currents (greater than 30nA) evoked small amplitude (2mV) and duration (30 msec) action potentials (APs) (Fig. 1a).

There was a decrease in apparent input resistance following the AP which is likely the result of a decrease in membrane resistance and/or gap junctional conductance. However, previous studies suggest that decrease in membrane resistance is the principal contributor to changes in input resistance (Thompson and Tobe, 1989). Depolarizing currents in the presence of voltage-dependent potassium channel blockers tetraethylammonium (TEA) and 4-aminopyridine (4-AP) lowered the threshold required to evoke APs (5-15nA) (Fig. 1b). The APs were of larger amplitude (6-8mV) and duration (110ms) and were followed by larger repolarizations (arrow, Fig. 1b), suggesting that TEA and 4-AP blocked currents involved in the rectification of depolarizing responses but did not affect others involved in the repolarization following the AP.

Sodium ions (Na^+) contribute to APs. In saline containing the voltage-dependent sodium channel blocker tetrodotoxin (TTX) or in sodium free saline, APs were replaced by a broad depolarizing bump followed by rectification (Fig. 1c). This suggests that Na^+ is required to evoke APs.

Calcium ions (Ca^{2+}) also contribute to APs. Duration of APs was increased following replacement of Ca^{2+} in saline by barium ions (Ba^{2+}) (Figs. 1d and 1e). This suggests that Ca^{2+} contributes to the AP, since Ba^{2+} is known to carry larger currents in some Ca^{2+} channels than Ca^{2+} (Hille, 1984) . Interestingly the Ca^{2+} channel blocker cadmium (Cd^{2+}) did not block APs but rather evoked multiple APs by a single current pulse (Fig. 1f). The larger the current pulse, the greater the number of APs that were evoked. Cd^{2+} also reduced the repolarization following individual APs (0.3mV in Cd^{2+} versus 11.2mV in TEA and 4-AP alone) (arrows, Figs. 1b and f). The effects of Cd^{2+} are likely due to blocking of Ca^{2+} channels because Cd^{2+} has been observed to inhibit the increased duration of APs in Ba^{2+} (data not shown). Thus, Cd^{2+} is likely evoking multiple APs and decreasing the repolarization by blocking the increase in intracellular Ca^{2+} accompanying depolarizations. This suggests that Ca^{2+}-activated currents may contribute to repolarization since they would not be activated in the presence of Cd^{2+}, thereby reducing repolarization as observed. Since a decrease in repolarizing currents can increase the frequency of APs in invertebrate neurons (Strong and Kaczmarek, 1985), it seems likely that the same mechanism is occurring in this preparation. This is supported by the observation that in the absence of Cd^{2+}, multiple APs can be evoked if the duration of the current pulse was increased to 6s from the normal 1.5s (data not

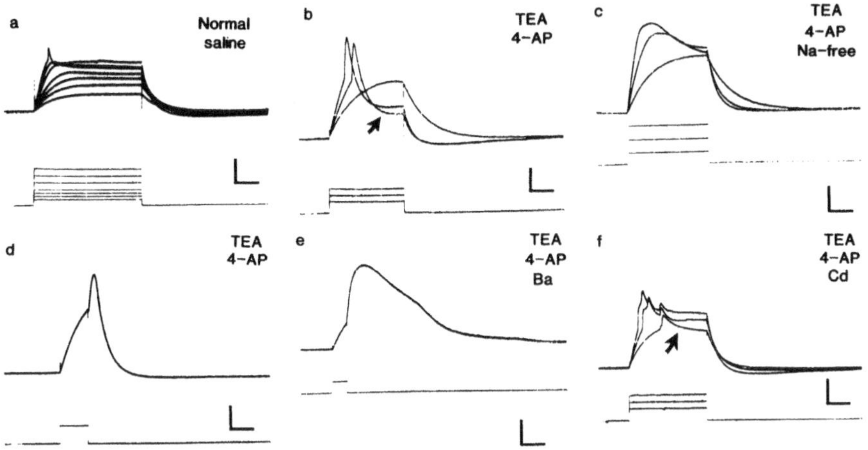

Figure 1. Electrotonic potentials (top traces) and depolarizing current pulses (bottom traces) of individual CA bathed in the indicated salines. (a) A nonlinear current-voltage relationship occurreds in CA exposed to normal saline as indicated by the increased rectification at larger current pulses. Large currents evoked an AP followed by an apparent decrease in input resistance. (b) Depolarizing pulses in TEA (40mM) and 4-AP (200uM) evoked APs at lower thresholds and were of larger amplitude, duration and magnitude of repolarization (see arrow) compared to (a). (c) Depolarizing pulses in TEA, 4-AP and TTX (50nM). Notice the absence of APs although a small depolarizing bump occurred, followed by a small rectification at larger current pulses. (d) Depolarizing pulses in TEA and 4-AP. The current pulse ended just as an AP was evoked. (e) Depolarizing pulse responses in TEA, 4-AP with Ba^{2+} (10mM) substituted for Ca^{2+}. The current pulse duration, as in (d), was just long enough to evoke an AP. Notice the increase in AP duration compared to (d). (f) Depolarizing pulses in TEA, 4-AP and Cd^{2+} (2mM). Notice that increased current pulses evoked multiple APs and reduced magnitudes of repolarization (arrow). Scale bar: vertical 10mV (voltage traces) and 10nA (current traces) (except in (a) 10mV (voltage) and 20nA (current)), horizontal 500msec.

shown). Latency to the second AP in the absence of Cd^{2+} was increased 8-12 fold. Thus it is likely that the multiple APs in Cd^{2+} are a result of inhibition of Ca^{2+}-activated repolarizing currents. Therefore, Ca^{2+} contributes to APs but is not essential for its activation. Ca^{2+} may also contribute to repolarization of APs which may influence the excitability of the cell membranes of the CA.

In conclusion, it appears that the plasma membrane of the CA contain both Na^+ and Ca^{2+} channels. Na^+ is essential to evoke APs but Ca^{2+} is not. There are several types of ion channels involved in rectifying depolarizing currents. Some are TEA- and 4-AP- sensitive whereas others are not. There may also be Ca^{2+}-activated currents involved in the repolarization of the AP.

REFERENCES

Hille, B. (1984) *Ion Channels of Excitable Membranes*. Sinauer Associates, Sunderland, MA.

Kikukawa, S., Tobe, S.S., Solowiej, S., Rankin, S.M. and Stay, B. (1987) Calcium as a regulator of juvenile hormone biosynthesis and release in the cockroach *Diploptera punctata*. *Insect Biochem*. **17**, 179-187.

Lococo, D.J., Thompson, C.S. and Tobe, S.S. (1986) Intracellular communication in an insect endocrine gland. *J. Exp. Biol*. **121**, 407-419.

Strong, J.A. and Kaczmarek, L.K. (1985) Multiple components of delayed potassium current in peptidergic neurons of *Aplysia*: Modulation by an activator of adenylate cyclase. *J. Neurosci*. **6**, 819-822.

Thompson, C.S. and Tobe, S.S. (1986) Electrical properties of membranes of cells of the corpora allata of the cockroach *Diploptera punctata*. In *Insect Neurochemistry and Neurophysiology*. (Ed. by Borkovec A.B. and Gelman D.B.), pp. 375-378. Humana Press, Clifton, NJ.

Thompson, C.S. and Tobe, S.S. (1989) Innervation and electrophysiology of the corpus allatum. In *Cockroaches as Model Systems for Neurobiology: Applications in Biomedical Research*. (Ed by Huber I., Masler E.P. and Rao B.P.) CRC Press, Boca Raton, FL.

Tobe, S.S., Ruegg, R.P., Stay, B.A., Baker, F.C. and Miller, C.A. and Schooley, D.A. (1985) Juvenile hormone titre and regulation in the cockroach *Diploptera punctata*. *Experientia*, 41, 1028-1034.

From *Insect Neurochemistry and Neurophysiology • 1989* Edited by A. B. Borkovec and E. P. Masler

Juvenile hormone production by wild type and *apterous*[4] mutant *Drosophila melanogaster* corpora allata in vitro

M. Altaratz[1,2], D. Segal[1], D.S. Richard[3], L.I. Gilbert[3] and S.W. Applebaum[2]

[1] Dept. of Microbiology, Tel Aviv University, Tel Aviv 69978, Israel. [2] Dept. of Entomology, The Hebrew University, Rehovot, 76100, Israel. [3] Dept. of Biology, University of North Carolina, Chapel Hill, NC 27599-3280, U.S.A.

The importance of the juvenile hormones (JHs) in the regulation of reproduction in insects is well documented (see Koeppe et al.,1985, for review), yet in the academically important species *Drosophila melanogaster*, very little is known. Recently, the major in vitro product of the larval ring gland, containing the corpus allatum (CA), was identified as JHB_3 (methyl 6,7;10,11-bisepoxyfarnesoate) (Richard et al.,1989a). Furthermore, production of this putative hormone is negatively regulated by the brain via two apparently discrete but interlinked mechanisms involving cAMP and calcium (Richard et al.,1989b; unpublished observations). Despite many morphological changes during the metamorphosis of larvae to adults, the corpus allatum is retained in the adult stage (King et al.,1966) while the ring gland degenerates. We now discuss aspects of CA maturation in adult females during early reproductive stages, and present data concerning JH production in the non-vitellogenic *apterous*[4] mutant. Vitellogenesis in this mutant can be rescued by the topical application of a JH analog (methoprene) or by the implantation of wild type CA (Postlethwait et al.,1976) and therefore is considered to be JH deficient. In this study, the *apterous*[4] mutation was maintained over an SM5 balancer chromosome which contains ap^+.

JH biosynthesis was determined by an in vitro radiochemical assay (Tobe and Pratt,1974) that involves the incorporation of the tritiated methyl group from methionine into JH precursors, as modified for the Diptera (Richard et al.,1989a). All experimental parameters were the same as for the larval ring gland assay. The identification of the radiolabelled products was accomplished using silica gel TLC although they had previously been identified by HPLC (methyl farnesoate, MF) and mass spectroscopy (JHB_3)(Richard et al.,1989a,1989b).

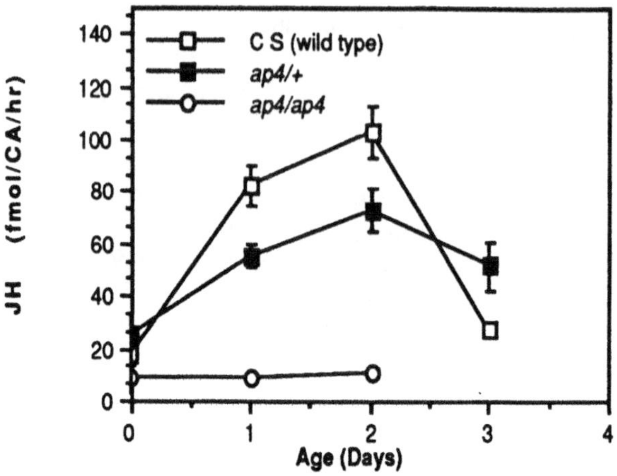

Figure 1. Developmental profile of total JH synthesis in vitro by CA from adult female wild type, *ap*^{4}/+ and *ap*^{4}/*ap*^{4} *Drosophila melanogaster.*

Total JH synthesis in vitro (JHB_3 and JH III) by Canton-S wild type (CS) corpora allata was found to be low at eclosion (20 fmol/h) yet increased within 2 days to 100 fmol/h (Figure 1). This rise is temporally associated with the onset of vitellogenesis, the histolysis of larval fat body and changes in sexual behavior, all processes considered to be dependent upon JH (Postlethwait and Giorgi,1985; Butterworth and King,1965; Jallon and Hotta,1979). However, the increase in JH production was slightly lower in *ap*^{4}/+ heterozygotes (which are not sterile), and absent in *ap*^{4}/*ap*^{4} homozygotes (which fail to survive more than 2-3 days after eclosion), strongly indicating that this mutation may result in the failure of the CA to synthesize normal amounts of JH.

The biosynthetic pathway of JHB_3 in *Drosophila* is believed to proceed via the epoxidation and methylation of the JH III precursor farnesoic acid (Richard et al.,1989a). It appears that these terminal steps are regulated by the adult brain following eclosion since brain-CA complexes produce less total JH than do CA alone (data not shown). This decrease is similar to that seen with larval brain-ring gland complexes in vitro where up to 24% of the JH produced was found to be MF (Richard et al.,1989b). This suggests that the larval brain may be affecting JH production by the regulation of the epoxidation steps. In the wild type adult, the intact brain-CA complexes produce up to 13%

MF (data not shown) whereas the CA alone only produce a maximum of 8% (Table 1). Furthermore, when isolated CA are treated with methanolic brain extracts, total JH synthesis is depressed and the proportion of MF is increased (data not shown). Control of the adult CA is also apparently axonally mediated by the brain via factors acting upon the epoxidation of the precursors.

Table 1. In vitro synthesis of methyl farnesoate by CA of adult *Drosophila melanogaster* at 25°C (expressed as % of total JH).

	Genotype		
Age (days)	Canton S (wild type)	*ap*^{4}/+	*ap*^{4}/*ap*^{4}
0	8	6	28
1	0	6	10
2	0	4	17

Levels of MF similar to wild type are produced by the phenotypically normal *ap*^{4}/+ heterozygotes. In the *ap*^{4}/*ap*^{4} homozygotes however, up to 28% of the already lowered levels of JH produced was found to be MF. This indicates that mutations of the *apterous* gene may also disrupt the terminal epoxidative steps in the production of JHB_3 resulting in an accumulation of this precursor. However, the question of whether or not the regulation of JH production by the brain and the *apterous* gene are functionally related remains a matter of conjecture.

Acknowledgements. This work was partly supported by grants to D. Segal from the Herz foundation and the Israel Cancer Research Fund.

References.

Butterworth, F.M., Bodenstein, D. and King, R.C. (1965) Adipose tissue of *Drosophila melanogaster.* An experimental study of larval fat body. *J. Exp. Zool.* **158**, 141-154.

Jallon, J.M. and Hotta, Y. (1979) Genetic and behavioral studies of female sex-appeal in *Drosophila. Behav. Genet.* **9**, 257-275.

King, R.C., Aggarwal, S.K. and Bodenstein, D. (1966) The comparative submicroscopic morphology of the ring gland of *Drosophila melanogaster* during the second and third larval instars. *Z. Zellforsch. Mikrosk. Anat.* **73**, 272-285.

Koeppe, J.K., Fuchs, M., Chen, T.T., Hunt, L.M., Kovalick, G.E. and Briers, T. (1985) The role of juvenile hormone in reproduction. In *Comparative Insect Physiology, Biochemistry and Pharmacology.* (Eds. Kerkut, G.A. and Gilbert, L.I.) Vol. **8**, pp.165-203. Pergamon Press, Oxford.

Postlethwait, J.H. and Giorgi, F. (1985) Vitellogenesis in insects. In *Developmental Biology: A Comprehensive Synthesis.* (Ed. Browder, L.W.) Vol **1**, pp. 85-126. Plenum Press, New York.

Postlethwait, J.H., Handler, A.M. and Gray, P.W. (1976) A genetic approach to the study of juvenile hormone control of vitellogenesis in *Drosophila melanogaster.* In *The Juvenile Hormones* (Ed. Gilbert, L.I.) pp. 449-469. Plenum Press, New York.

Richard, D.S., Applebaum, S.W., Sliter, T.J., Baker, F.C., Schooley, D.A., Reuter, C.C., Henrich, V.C. and Gilbert, L.I. (1989a) Juvenile hormone bisepoxide biosynthesis by the ring gland of *Drosophila melanogaster* : A putative juvenile hormone in the higher Diptera. *Proc. Natl. Acad. Sci. USA.* **86**, 1421-1425.

Richard, D.S., Applebaum, S.W. and Gilbert, L.I. (1989b) Devlopmental regulation of juvenile hormone biosynthesis by the ring gland of *Drosophila melanogaster. J. Comp. Physiol. B.* In Press.

Tobe, S.S. and Pratt, G.E. (1974) The influence of substrate concentrations on the rate of insect juvenile hormone biosynthesis by corpora allata of the desert locust in vitro. *Biochem. J.* **144**,107-113.

From *Insect Neurochemistry and Neurophysiology • 1989* Edited by A. B. Borkovec and E. P. Masler

DEVELOPMENTAL PROFILE OF PROTHORACICOTROPIC HORMONE (PTTH) ACTIVITY IN THE GYPSY MOTH, LYMANTRIA DISPAR.

+*Belgaum S. Thyagaraja, Thomas J. Kelly, Edward P. Masler, Robert A. Bell, *Richard B. Imberski and Alexej B. Borkovec

Insect Reproduction Laboratory, USDA, ARS, Beltsville, MD 20705, USA. +RSRS, Central Silk Board, C.R. Nagar 571 313, India, and *Department of Zoology, University of Maryland, College Park, MD 20742, USA

A profile of *L. dispar* brain PTTH activity during late embryonic to adult development has been generated using an *in vitro* prothoracic gland (PGs) stimulation bioassay. The existence and titer of PTTH in various lepidopteran insects has been discussed (Ishizaki, 1969; Nishiitsujii-Uwo and Nishimura, 1975; O'Brien *et al*, 1986; and Dorn *et al*, 1987). Our previous studies (Kelly *et al*., 1986) showed that in *L. dispar* the PGs respond to brain extract *in vitro* during day 3 to 5 of the fifth larval instar. Following a 180-min pre-incubation, day 4 PGs secrete 24 times more ecdysone/gland than their contralateral control glands when incubated for 90 min with 0.25 brain equivalents (Fig. 1). Since the day-4 PGs are equally responsive as those of day-5 glands (Fig. 1), day-4 PGs were used for *in vitro* assay with larval and pupal brains from day 1-11. Prothoracic gland activation was nearly constant using day 1-11 larval or pupal brain extracts, suggesting a relatively constant titer of PTTH. However, day- 7 and day-8 last-instar larval brain extracts showed maximum and minimum activity, respectively, at 1 brain equivalent/25μl (Fig. 2). Pupal brains showed maximum activity on day 3 at both 1 and 0.25 brain equivalents/25μl. Boiled extract from prehatch embryonic eggs showed mimimum activity at both concentrations.

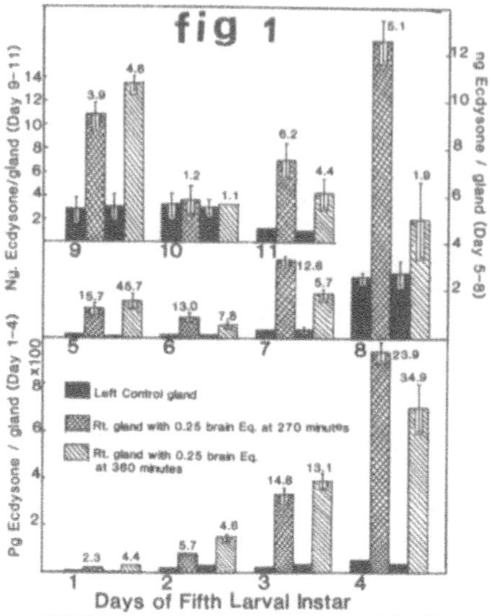

Fig. 1. Day-5 female larval PGs were incubated *in vitro* in Grace's medium. The medium was changed every 90 min, and 0.25 brain equivalents/25µl (day 5, boiled 2 min) was added to one gland of a pair at 180 and 270 min. Ecdysteroid was analyzed by RIA (Kelly *et al*, 1986). The number on each bar is the ratio of ecdysteroid (experimental vs. control) for the periods from 180-270 min and 270-360 min.

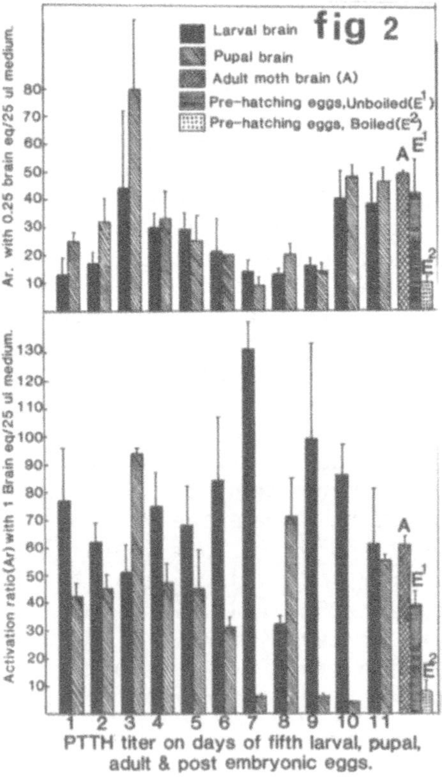

Fig. 2. Day-4 PGs were incubated *in vitro* for 2 h with 25µl Grace's medium. At 2 h one gland of a pair received Grace's and the other boiled neural tissue extract from egg to moth. Glands were further incubated for 2 h and RIA was done. 180 day gypsy eggs, New Jersey stock (NJ-180) extracted twice in HPLC grade water, centrifuged at 3000xg, supernatant dried, resuspended in Grace's, divided into two sets: one was unboiled (E^1); the other placed in boiling water for 2 min (E^2). Both were microcentrifuged at 4°C and the supernatant used.

Table 1. Comparision of PTTH titer in Gypsy moth Embryo to Adult.

Neural Tissue Extract	Tissue Eq. per 25 µl Grace's medium			
	Unboiled Extract		Boiled Extract	
	1	0.25	1	0.25
Pre-hatch Embryo Head	10.23± 3.08	1.81± 0.42	3.15± 0.95	1.80± 0.51
Pre-hatch Embryo Body*	3.90± 0.58	4.79± 1.34	4.59± 0.75	3.96± 0.48
Pre-hatch Eggs	38.92± 10.71	39.49± 4.55	7.58± 1.52	9.69± 1.44
Day-5 Larval Brain	75.11± 9.22	51.29± 5.87	37.69± 5.33	22.39± 3.79
Day-1 Pupal Brain	---	---	42.17± 5.10	24.96± 3.11
Adult Virgin Moth Brain	---	---	61.88± 5.82	49.33± 5.38
Day-5 Larval CC/CA complex	---	---	91.45± 13.82	97.43± 12.44

*Thorax & Abdomen only.

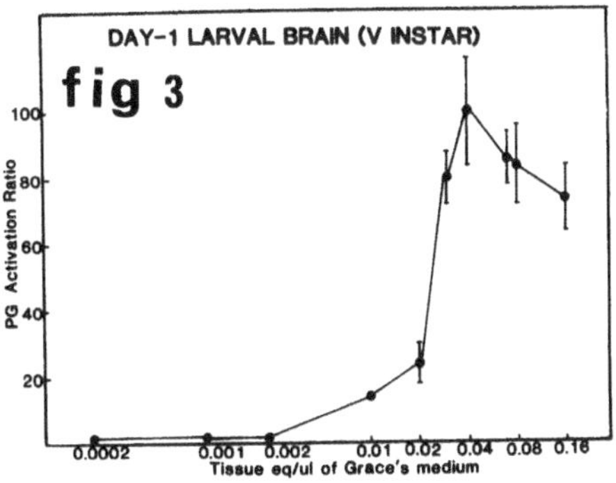

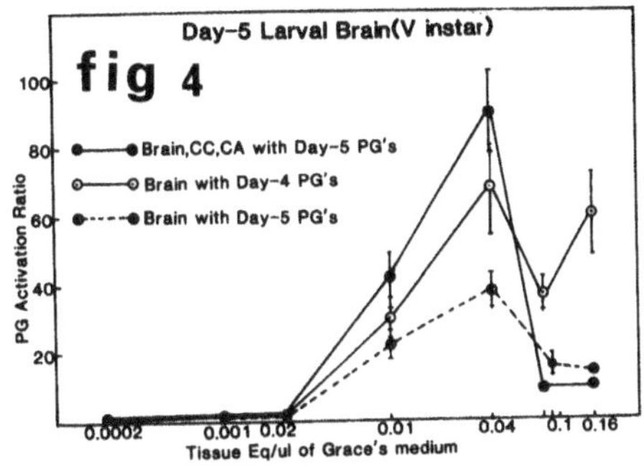

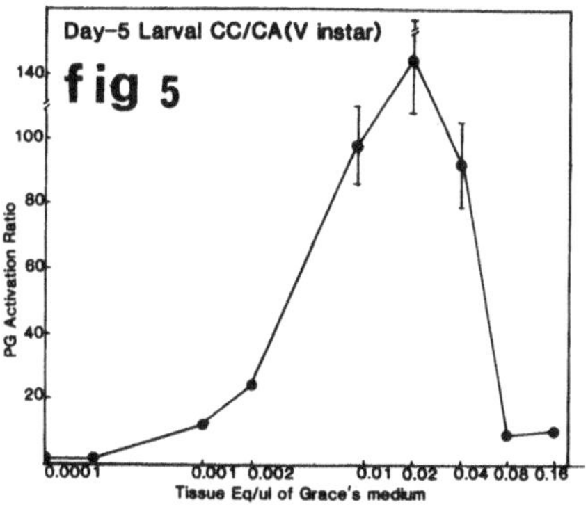

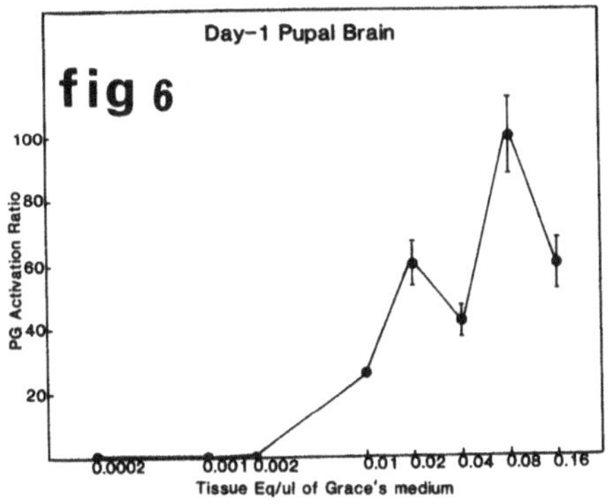

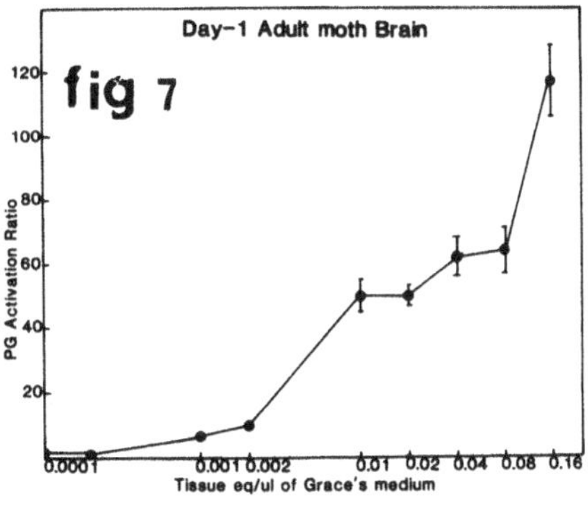

Table 1 and Figs. 3-7. Extracts and assays were done as described in Fig. 2. NJ-180 day embryos were dissected and washed in Bombyx saline, and the head and body separated. Day-5 larval PGs were used for assays.

Fig. 1-7 and Table 1. Each point represents the mean ± SE of at least 30 determinations. Activation ratio (Ar) is RIA activity of experimental vs. control.

The dose response curves generated with neural tissue extracts prepared at various stages of gypsy moth development revealed differences in PTTH activation with tissues from different sources (Figs. 3-7 and Table 1). Isolated heads and thorax/abdomens of pre-hatch embryos demonstrated head and possibly body-localized activity. Maximum PTTH activity was found with day-5 larval CC/CA complexes. Heat treatment of neural extracts reduced RIA activity, as revealed in Table 1, apparently by eliminating ecdysteroid ketoreductase activity (see Kelly et al. this volume). Overall, brain PTTH activity varied little from day-1 last instar larva to day-1 adults (Fig. 3-7) when based on approximate ED_{50} values. Thus, the present study reveals the existence of differential PTTH activity in neural tissues of L. dispar, which possibly exerted different stimulatory effects during development.

Supported in part by USDA Competitive Grant No. 86-CRCR-1-2111 to E.P.M. and T.J.K.

REFERENCES

Dorn A., Gilbert L.I. and Bollenbacher W.E. (1987) Prothoracicotropic hormone activity in the embryonic brain of the tobacco hornworn, Manduca sexta. J. Comp. Physiol. B. 157, 279-283.

Ishizaki H. (1969) Changes in titer of the brain hormone during development of the silkworm, Bombyx mori. Dev. Growth Diff. 25, 593-600.

Kelly T.J., Masler E.P., Thyagaraja B.S., Bell R.A. and Borkovec A.B. (1986) Prothoracicotropic hormone stimulation of ecdysone synthesis by the prothoracic glands of the gypsy moth, Lymantria dispar. In: Insect Neurochemistry and Neurophysiology 1986 (Ed. by Borkovec A.B. and Gelman D.B.), pp 327-330. Humana Press, Clifton, NJ.

Nishiitsutsuji-Uwo J. and Nishimura M.S. (1975) Brain hormone contained in the silkworm body of Bombyx mori. Experientia 31, 1105-1107.

O'Brien M.A., Granger N.A., Agui N., Gilbert L.I. and Bollenbacher W.E. (1986) Prothoracicotropic hormone in the developing brain of the tobacco hornworm, Manduca sexta: relative amounts of two molecular forms. J. Insect Physiol. 32, 719-725.

From *Insect Neurochemistry and Neurophysiology • 1989* Edited by A. B. Borkovec and E. P. Masler

ENDOCRINOLOGY OF EMBRYOGENESIS AND LATE EMBRYONIC DIAPAUSE IN THE GYPSY MOTH, LYMANTRIA DISPAR

Robert A. Bell, Thomas J. Kelly, Edward P. Masler, *Belgaum S. Thyagaraja, Albert B. DeMilo and Alexej B. Borkovec

Insect Reproduction Laboratory, USDA, ARS, Beltsville, MD 20705, USA *RSRS, Central Silk Board, C.R. Nagar 571 313, India

Our present knowledge regarding the hormonal control of insect development and diapause is based upon studies involving post-embryonic stages, i.e., late larval, pupal and adult forms. Except for a single species, the commercial silkworm, Bombyx mori, which enters diapause at a very early embryonic stage, little is known about the endocrinology of embryonic development in lepidopterous insects. The univoltine gypsy moth, Lymantria dispar, enters a prolonged obligatory diapause that occurs in late embryonic development after becoming a mature and fully differentiated pharate larva (Leonard, 1968). Little or no information is available on the endocrine regulation of diapause occurring at such a late stage of embryonic development. Thus the main purpose of the present study was to obtain fundamental information on the metabolic activity as well as the ecdysteroid and prothoracicotropic (PTTH) hormone titers during the course of embryogenesis and entry into diapause. The rationale for selecting this approach was to test the hypothesis that late embryonic diapause may be regulated by the brain-prothoracic gland system and therefore could be associated with a failure of synthesis or secretion of PTTH and development promoting ecdysteriod hormones in a manner similar to that found in larval and pupal diapause in post-embryonic stages (see Denlinger, 1984).

These studies were carried out with a laboratory strain of the gypsy moth reared according to procedures described previously (Bell et al., 1981; Bell, 1989). Egg masses were harvested within 24 hrs after oviposition, incubated at 25°C and sampled daily or at 5-day intervals during embryogenesis and entry into diapause. Oxygen consumption was measured with a Gilson respirometer (Bell, 1989) and ecdysteroid and PTTH titers were determined from whole egg extracts by radioimmunoassay (RIA) and by activation of prothoracic glands in vitro (Kelly et al., 1986; Masler et al., 1986).

The results of these studies, summarized in Fig. 1, provide some insight into hormonal mechanisms involved in embryogenesis and diapause in the gypsy moth. Early stages (day 1-5) were characterized by low hormonal levels and overall metabolic activity. Afterwards (day 6-10), there is a substantial increase in hormonal and metabolic activity that coincides with the time when the brain and associated endocrine systems are differentiated and presumably functional. Highest overall hormone titer and metabolic activity occurred later (day 11-15) coincident with synthesis and differentiation of the embryonic larval cuticle. These observations suggest that formation of the embryonic larval cuticle is mediated by de novo secretion of PTTH and ecdysteroid hormones from the newly differentiated brain neuroendocrine-prothoracic gland system and is in agreement with conclusions of Fugo et al. (1987) and Dorn et al. (1987). By day 10, high endogenous titers of 20-hydroxyecdysone were present and appears to be the principle ecdysteroid regulating embryonic development in the gypsy moth.

By day 15, the embryo had developed into a fully differentiated but overtly inactive pharate larva surrounded by scattered particles of remaining egg yolk. Afterwards, hormone and metabolic levels declined to a low, basal level indicative of entry into diapause by day 20. Almost all detectable ecdysteroids were conjugated by this time and 20-hydroxyecdysone was virtually absent, only 10-20 pg/mg or 1-2 percent of the amount found on day 10 was detected.

The foregoing observations lead to the conclusion that late embryonic diapause in the gypsy moth must be due, at least in part, to an impairment of the brain neuroendocrine-prothoracic gland system resulting in an ecdysteroid deficiency syndrome similar to that observed in post-embryonic larval and pupal diapause (Denlinger, 1984). Whether or not juvenile hormone and/or a diapause hormone or other as yet unidentified inhibitory mechanisms are also involved still remains to be determined.

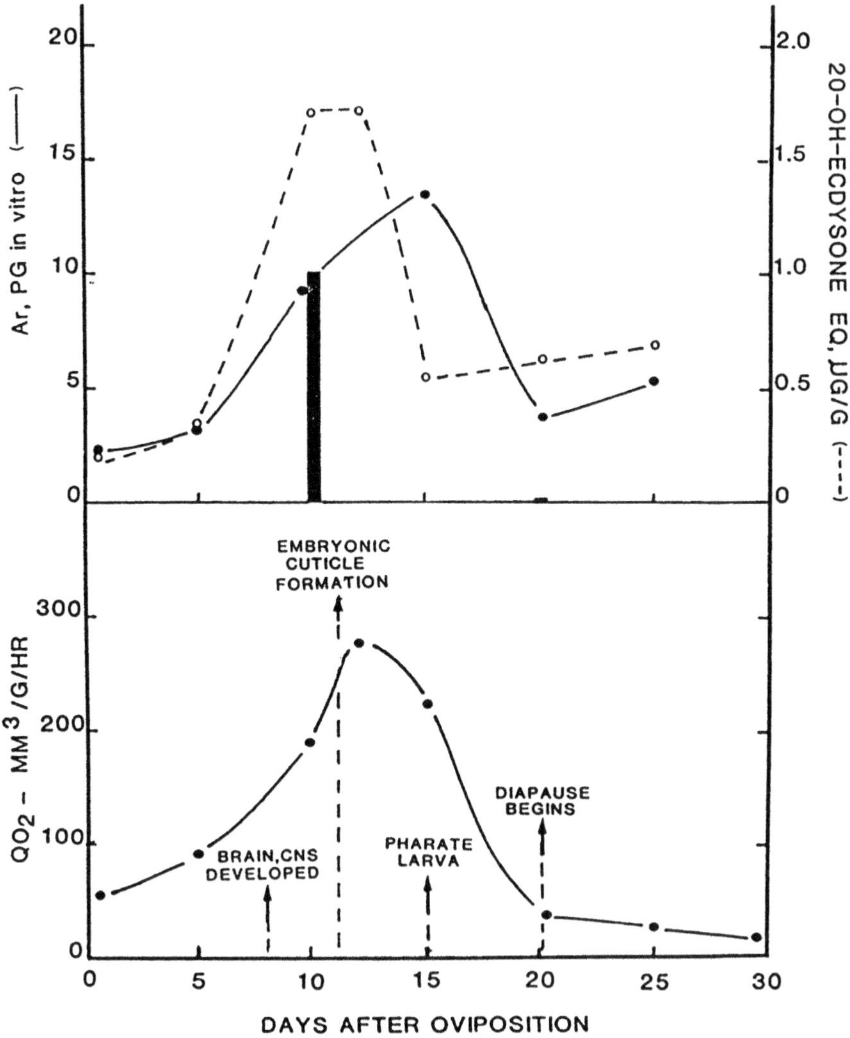

Fig. 1. Prothoracicotropic (PTTH) and ecdysteroid hormone titers (upper panel) and respiratory activity (lower panel) during embryogenesis and entry into diapause in the gypsy moth. PTTH titers shown were obtained by activation of ecdysone secretion of prothoracic glands <u>in vitro</u> with 1 mg of crude egg extract. The broken line (upper panel) shows the total immunoreactive ecdysteroid titer whereas the solid line is the PTTH levels and the solid bars (day 10 and 20) represent levels of 20-hydroxyecdysone. Respiratory metabolism is expressed as the quantity of oxygen consumed/gram live wt/hr at 25° C.

REFERENCES

Bell R.A., Owens C.D., Shapiro M. and Tardif J.D.R. (1981) Mass rearing and virus production. In: The Gypsy Moth: Research Toward Integrated Pest Management. USDA Tech. Bull. 1584, pp. 599-655.

Bell R.A. (1989) Respiratory activity during embryonic development in a diapausing and a selected non-diapausing strain of the gypsy moth, Lymantria dispar L. Comp. Biochem. Physiol.: In Press.

Denlinger D.L. (1989) Hormonal control of diapause. In: Comprehensive Physiol., Biochem. and Pharmacol., Vol. 8. Ed. Kerkut G.A. and Gilbert L.I. pp. 343-412, Pergamon Press, NY.

Dorn A., Gilbert L.I. and Bollenbacher W.E. (1987) Prothoracicotropic activity in the embryonic brain of the tobacco hornworm, Manduca sexta. J. Comp. Physiol. B 157: 279-283.

Fugo H., Chen J.H., Nakajima M., Nagasawa H. and Suzuki A. (1987) Neurohormones in the developing embryos of the silkworm Bombyx mori: the presence and characteristics of prothoracicotropic hormone-S. J. Insect Physiol. 33: 243-248.

Kelly T.J., Masler E.P., Thyagaraja B.S., Bell R.A. and Borkovec A.B. (1986) Prothoracicotropic hormone stimulation of ecdysone synthesis by the prothoracic glands of the gypsy moth, Lymantria dispar. In: Insect Neurochemistry and Neurophysiology 1986 (Ed. by Borkovec A.B. and Gelman D.B.), pp. 327-330, Humana Press, Clifton, New Jersey.

Leonard D.E. (1968) Diapause in the gypsy moth. J. Econ. Entomol. 61: 596-598.

Masler E.P., Kelly T.J., Thyagaraja B.S., Woods C.W., Bell R.A. and Borkovec A.B. (1986) Discovery and partial characterization of the prothoracicotropic hormones of the gypsy moth, Lymantria dispar. In: Insect Neurochemistry and Neurophysiology 1986 (Ed. by Borkovec A.B. and Gelman D.B.), pp. 331-334, Humana Press, Clifton, New Jersey.

From *Insect Neurochemistry and Neurophysiology • 1989* Edited by A. B. Borkovec and E. P. Masler

PROTHORACICOTROPIC HORMONE (PTTH) REGULATION OF THE LARVAL PUPAL MOLT IN THE GYPSY MOTH, *LYMANTRIA DISPAR*: OBSERVATIONS ON THE DEVELOPMENT OF A LARVAL *IN VIVO* BIOASSAY.

+*Belgaum S. Thyagaraja, Thomas J. Kelly, Edward P. Masler, Robert A. Bell, *Richard B. Imberski and Alexej B. Borkovec

Insect Reproduction Laboratory, USDA, ARS, Beltsville, MD 20705, USA. +RSRS, Central Silk Board, C.R. Nagar 571 313, India and *Department of Zoology, University of Maryland, College Park, MD 20742, USA

Since the initial discovery of Kopec (1922), the importance of a head factor to initiate molting/metamorphosis in lepidopterous insects has been studied by several workers either by neck ligation or brain removal (see references in Smith, 1985; also Kiguchi *et al*., 1985; Nagata *et al*., 1987). These findings lead to the development of bioassays for characterization of this factor, PTTH, from the brain (Kobayashi and Burdette, 1961; Kimura *et al*., 1962; Gibbs and Riddiford, 1977; Ishizaki *et al*., 1983; Suzuki and Ishizaki, 1986). To determine the feasibility of developing a similar assay, with *Lymantria*, the physiological and endocrinological promotion of larval-pupal development was examined by neck ligation of female fifth instar (last-stage) larvae on consecutive days. Newly molted larvae were collected between 2:00 pm to 8:00 pm and weighed daily. Weight and growth rate were maximum (4.7 times) on day 7 (Fig. 1). Larvae neck-ligated on day 7 did not pupate, even after 22 days, while larvae ligated on day 8 to 9 shrunk and pupated within 2 to 3 days (Fig. 2). A small pre-peak of hemolymph ecdysteroid (ca. 450 pg.) was recorded by radioimmunoassay (RIA) on day 7 and followed by another increase on day 8 which reached a maximum on day 9, two days before pupation (Fig. 2).

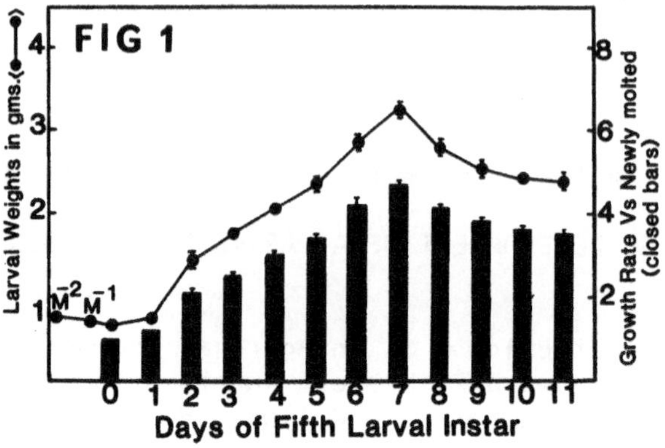

Fig. 1. Female larvae were separated from males after the third molt and kept 7 larvae per cup. Weights were taken from green-neck larvae 2 days prior to ecdysis (M^{-2}) which had started head slippage for the fourth molt, from brown-neck larvae (M^{-1}), from newly molted (0), and from each day of the fifth instar. Each point represents the mean ± SE for at least 40 larvae.

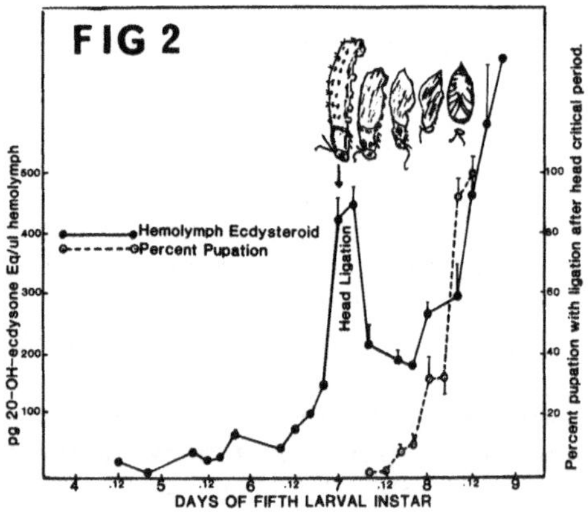

Fig. 2. The timing of PTTH release was examined by collecting 10μl of hemolymph in 490μl of 75% methanol at 4 h intervals from day 4 to 9 and analyzing by RIA. Neck ligations were done at 4 h intervals. Neck ligation after day 7 showed various percentages of body shrinkage and pupal transformation. Each point represents the mean ± SE for 90 larvae.

Fig. 3. Female fifth-instar larvae neck-ligated on day 7 were kept for five days. Larvae that did not show any body shrinkage were injected with various neural extracts from gypsy moth larval, pupal, and adult brains or with pre-hatch egg extract (see Kelly et al., this volume).

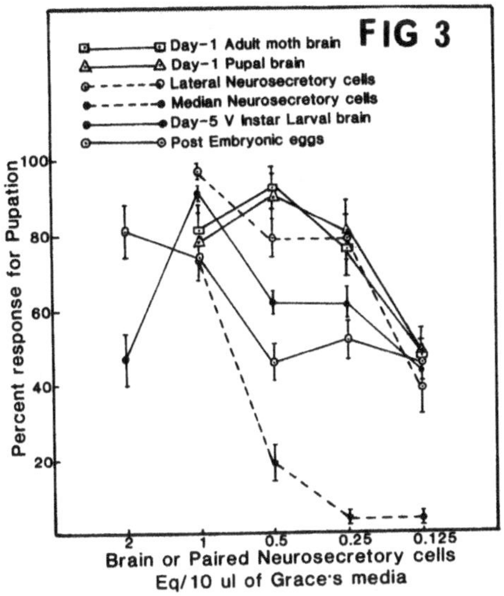

Brains from newly-ecdysed pupae were dissected in <u>Bombyx</u> saline, and lateral and median neurosecretary cells along with brain tissue carefully excised. All tissues were homogenized in Grace's medium, boiled 2 min, microcentrifuged 1 min, the supernatant taken, and appropriate dilution injected in a 10µl volume. Each point is the mean ± SE for 90 larvae.

The significance of a pre-peak of ecdysteroid during the fifth instar and its importance for pupal commitment has been described in several lepidopteran insects (see references in Smith, 1985; Nagata <u>et al</u>., 1987). Our studies show a similar pre-peak correlated with the developmental transformation from larvae to pupae. To confirm the necessity of a brain factor for commitment to pupal development, fifth-instar female larvae were head ligated on day 7 and injected 5 days later with various neural tissue extracts in 10µl of Grace's medium. Controls received 10µl of Grace's medium, alone. Injected larvae were scored 5 days later for pupal ecdysis. All insects injected with neural extracts responded positively and in a dose-dependent manner (Fig. 3). Controls showed no response. The present study demonstates a head critical period for molting/metamorphosis on day 7 after which head ligation results in pupation within 2 to 3 days. These experiments confirm the existence of PTTH activity in <u>Lymantria</u>, as previously demonstrated (Kelly <u>et al</u>., 1986; Masler <u>et al</u>., 1986). They also localize it to the lateral and median tissue of the gypsy moth brain, demonstrate activity in pre-hatch gypsy moth eggs and provide the basis for development of a <u>Lymantria</u> PTTH <u>in vivo</u> bioassay. The required period for feeding of fifth-instar larvae is also being examined and demonstrates a critical period on day 3 which may involve factors from the head (Thyagaraja <u>et al</u>., unpublished data). Supported in part by USDA Competitive Grant No. 86-CRCR-1-2111 to E.P.M. and T.J.K.

REFERENCES

Gibbs D. and Riddiford L.M. (1977) Prothoracicotropic hormone in Manduca sexta: Localization by a larval assay. J. Exp. Biol. 66, 255-266.

Ishizaki H., Suzuki A., Moriya I., Mizoguchi A., Fujishita M., O'oka H., Kataoka H., Isogai A., Nagasawa H. and Suzuki A. (1983) Prothoracicotropic hormone bioassay: pupal-adult Bombyx assay. Develop. Growth and Differ. 25, 585-592.

Kelly T.J., Masler E.P., Thyagaraja B.S., Bell R.A. and Borkovec A.B. (1986) Prothoracicotropic hormone stimulation of ecdysone synthesis by the prothoracic glands of the gypsy moth, Lymantria dispar. In Insect Neurochemistry and Neurophysiology 1986 (Ed. by Borkovec A.B. and Gelman D.B.), pp 327-330, Humana Press, Clifton, NJ.

Kiguchi K., Agui N., Kawasaki H. and Kobayashi M. (1985) Developmental time table for the last larval and pharate pupal stages in the silkworm, Bombyx mori, with special reference to the correlation between the developmental events and hemolymph ecdysteroid levels. Bull. Seri. Sci. Japan 30, 83-100.

Kimura J., Saito M. and Kobayashi M. (1962) Steroid hormone in an insect Bombyx mori. Nature 195, 729-730.

Kobayashi M. and Burdette W.J. (1961) Effect of brain hormone from B. mori on metamorphosis of C. erythrocephala. Proc. Soc. Exp. Biol. Med. 107, 240-242.

Kopec S. (1922) Studies on the necessity of the brain for the inception of insect metamorphosis. Biol. Bull. Wood's Hole 42, 322-342.

Masler E.P., Kelly T.J., Thyagaraja B.S., Woods C.W., Bell R.A. and Borkovec A.B. (1986) Discovery and partial characterization of the prothoracicotropic hormone of the gypsy moth, Lymantria dispar. In Insect Neurochemistry and Neurophysiology 1986. (Ed. by Borkovec A.B. and Gelman D.B.) pp 331-334, Humana Press, Clifton, NJ.

Nagata H., Ohtaki T. and Sakurai S. (1987) Role of low ecdysteroid titer in acquisition of competence for pupal transformation in Bombyx mori. J. Insect. Physiol. 33, 657-662.

Smith S.L. (1985) Regulation of ecdysteroid titer: synthesis: In: Comprehensive Insect Physiology, Biochemistry and Pharmacology. Vol. 7, (Ed. by Kerkut G.A. and Gilbert L.I.), pp 295-341. Pergamon Press, Oxford.

Suzuki C. and Ishizaki H. (1986) Prothoracicotropic hormone bioassay: Bombyx larva assay. Int. J. Invert. Rep. Develop. 10, 259-274.

From *Insect Neurochemistry and Neurophysiology • 1989* Edited by A. B. Borkovec and E. P. Masler

DEVELOPMENTAL PROFILES FOR THE REGULATION OF ECDYSTEROID PRODUCTION BY THE PROTHORACIC GLANDS OF FEMALE GYPSY MOTH, *Lymantria dispar*, PUPAE.

Howard W. Fescemyer, Edward P. Masler, and Thomas J. Kelly

Insect Reproduction Laboratory, USDA, ARS, Beltsville, MD 20705, USA

Activation of prothoracic glands (PGs) by their primary effector, prothoracicotropic hormone (PTTH), results in the production (synthesis and release) of ecdysteroids at precise times during larval and pupal life (Bollenbacher, 1988; Bollenbacher and Granger, 1985). An *in vitro* method has been developed by Bollenbacher *et al*. (1979, 1983) to quantify the PTTH-activation of PGs from larval or pupal *Manduca sexta*. This *in vitro* approach was used to quantify kinetic and developmental changes in the activity of PGs from female *L. dispar* pupae before and after PTTH stimulation. This information was used to establish an *in vitro* assay for PTTH using PGs from female *L. dispar* pupae.

Production of ecdysteroids by PGs from all ages of pupae tested increased exponentially during the first 2 h followed by a linear variation ($r^2 \geq 0.90$) in production (Fig. 1). This indicates that a 2 h pre-incubation with no stimulus would be a standard part of any *in vitro* assay for PTTH using PGs from female *L. dispar* pupae. The developmental pattern of *in vitro* ecdysteroid production, which correlates with the hemolymph ecdysteroid titer (Schnee *et al*., 1984), increased to a peak on day 3 and then

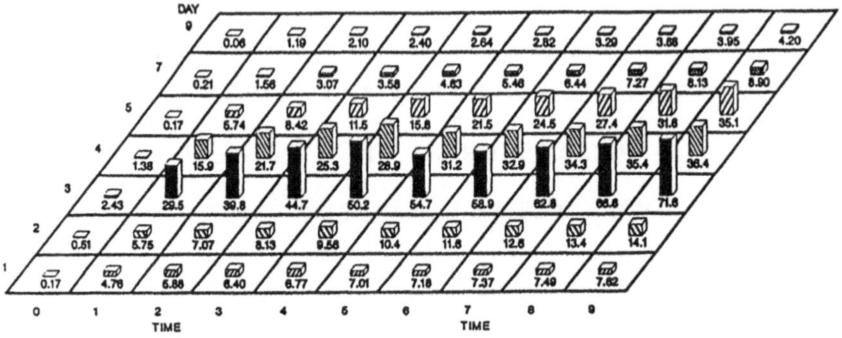

Figure 1. Time course for the in vitro production of ecdysteroid by unstimulated PGs from female L. dispar pupae. Glands were incubated in a humidified chamber at 26°C and constant darkness. The medium (25 µl of Grace's insect medium) was exchanged for fresh medium every hour and analyzed by radioimmune assay (Kelly et al., 1986) using ecdysone as the unlabelled ligand. Time zero represents the amount in freshly dissected glands. Bars represent the mean values (ng) printed in each square ($n \geq 5$; SE≤3% of the mean). Eclosion occurred on day 10.

declined through day 9 (Fig. 1).

A brain (Br) equivalent (eq) dose-response to extracts of one-day-old female pupal brains that were boiled (2 min; ED_{50}=0.0625 Breq/µl) or not boiled (ED_{50}=0.0575 Breq/µl) was determined using PGs from one-day-old female pupae that were pre-incubated for 2 h before the addition of extract followed by a 2 h incubation with extract. The dose-responses to boiled or unboiled brain extract did not differ (P=0.006), ranged from 3.3±1.5 (at 4×10^{-5} Breq/µl) to 21.4±2.0 ng ecdysone eq/gland (at 0.04 Breq/µl), had a maximum activation ratio (see Kelly et al., 1986) of 9.8±3.2 and were saturable.

Both PGs from an individual produce similar amounts of ecdysteroid in vitro before one member of the pair is stimulated at 2 h with brain extract (Fig. 2). Although the unstimulated gland continues to produce ecdysteroid at a linear rate, the stimulated

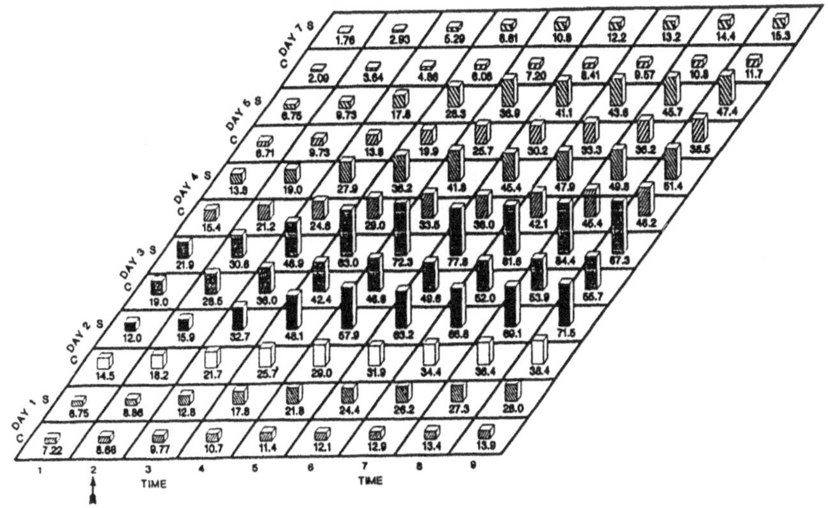

Figure 2. Kinetics of the <u>in vitro</u> production of ecdysteroid by PGs from female <u>L. dispar</u> pupae. Glands were incubated as in figure 1. The medium was exchanged for fresh medium or extract every hour. Beginning at hour 2 (arrow), one gland of each pair (C=control; S=stimulated) received boiled extract of one-day-old female pupal brains (0.0625 Breq/µl). Bars represent the mean values (ng) printed in each square ($n \geq 5$; SE≤13% of the mean).

gland increases its production of ecdysteroid in a non-linear manner for the next 2-3 h after which production continues at a higher linear rate that parallels the control gland. Only the time courses (hours 2-9) for ecdysteroid production by the stimulated member of PG pairs from 1-3-day-old pupae differed ($P \leq 0.03$) from those of the control PG. The activation ratios, determined at various times after stimulation, for PGs from 2-day-old pupae were greater ($P \leq 0.02$) than those observed for PGs from 1- and 3-day-old pupae which did not differ ($P \geq 0.17$) from each other. The developmental pattern of <u>in vitro</u> activation of pupal PGs correlates with the hemolymph ecdysteroid titer (Schnee <u>et al</u>., 1984)which undergoes its greatest increase/change during day 2.

An *in vitro* bioassay using PGs from *L. dispar* female pupae would use PGs from 2-day old pupae that were incubated for 2 hrs before the addition of stimulus followed by a 2-3 hr incubation with stimulus. A similar assay has also been developed for *L. dispar* larvae (Kelly *et al*., 1986). This information will be used in characterization and comparative physiolgical studies of PTTH. This research was supported by USDA Competitive Grant No. 86-CRCR-1-211 to E.P.M. and T.J.K.

REFERENCES

Bollenbacher W. E. (1988) The interendocrine regulation of larval-pupal development in the tobacco hornworm, *Manduca sexta*: A model. *J. Insect Physiol*. 34, 941-947.

Bollenbacher W. E. and Granger N. A. (1985) Endocrinology of the prothoracicotropic hormone. In *Comprehensive Insect Physiology, Biochemistry and Pharmacology* (Ed. by Kerkut G. A. and Gilbert L. I.), Vol. 8, pp. 109-151, Pergamon Press, New York.

Bollenbacher W. E., Agui N., Granger N. A. and Gilbert L. I. (1979) *In vitro* activation of insect prothoracic glands by the prothoracicotropic hormone. *Proc. Natl. Acad. Sci. USA* 76, 5148-5152.

Bollenbacher W. E., O'Brien M. A., Katahira E. J. and Gilbert L. I. (1983) A kinetic analysis of the action of the insect the prothoracicotropic hormone. *Mol. Cel. Endocr*. 32, 27-46.

Kelly T. J., Masler E. P., Thyagaraja B. S., Bell R. A. and Borkovec A. B. (1986) Prothoracicotropic hormone stimulation of ecdysone synthesis by the prothoracic glands of the gypsy moth, *Lymantria dispar*. In *Insect Neurochemistry and Neurophysiology* (Ed. Borkovec A. B. and Gelman D. B.), pp. 327-330, Humana Press, Clifton, New Jersey.

Schnee M. E., Ma M. and Kelly T. J. (1984) Hormonal basis of adult eclosion in the gypsy moth (Lepidoptera: Lymantriidae). *J. Insect Physiol*. 30, 351-356.

From *Insect Neurochemistry and Neurophysiology • 1989* Edited by A. B. Borkovec and E. P. Masler

DEVELOPMENTAL CHANGES IN LYMANTRIA DISPAR (L.) PROTHORACICOTROPIC HORMONE ACTIVITY FROM EMBRYO TO ADULT

Edward P. Masler, Thomas J. Kelly, *+Belgaum S. Thyagaraja, Robert A. Bell, Dale B. Gelman, *Richard B. Imberski and Alexej B. Borkovec

Insect Reproduction Laboratory, USDA, ARS, Beltsville, MD 20705, USA. +RSRS, Central Silk Board, C.R. Nagar 571 313, India and *Department of Zoology, University of Maryland, College Park, MD 20742, USA

Prothoracicotropic hormone (PTTH) in the brains of fifth instar *Lymantria dispar* larvae (Kelly *et al*., 1986; Masler *et al*., 1986) is presumably the factor(s) postulated by Kopec (1922) to control metamorphosis in *L. dispar*. We have extended our studies on *L. dispar* PTTH to other developmental stages (Masler *et al*., 1988) and now report PTTH activity in embryonic through adult tissue (see also Bell *et al*., Kelly *et al*, Thyagaraja *et al*., this volume). Embryonic PTTH has been reported for both *Bombyx mori* (Chen *et al*., 1986, 1987; Fugo *et al*., 1987) and *Manduca sexta* (Dorn *et al*, 1987) and developmental stage-related activation of *M. sexta* prothoracic glands (PGs) *in vitro* has been observed (Bollenbacher *et al*., 1984).

We examined the *in vitro* activation kinetics of *L. dispar* PTTH from five developmental stages (Table 1) using an extraction system developed for post-embryonic PTTH processing (Masler *et al*., 1986). Day 5, 5th larval instar (D_5L_5) PGs were stimulated most (highest Ar_{max}) by D_5L_5 brain extract followed, in descending order, by extracts of D_1L_5, D_1 pupal and D_1 adult brains, and embryonated eggs. Using other extraction regimens, further comparisons between embryonic and post-embryonic PTTH revealed that significant PTTH-like activity was present in embryonated eggs (Ar = 40; 58 percent of post-embryonic value) but that it was more susceptible to heating and organic solvent extraction than post-embryonic PTTH activity. Both embryonic and post-embryonic PTTH activities are destroyed by protease digestion.

Table 1. Values of Activation Kinetics Parameters Measured at Various Developmental Stages in L. Dispar: In Vitro Bioassay

Stage	Tissue[1]	Ar_{max}[2]	ED_{50}[3]
Embryonic	Whole Eggs	12(.02)	N.D.
D_1L_5	Brain	50(.10)	.030
D_5L_5	Brain	60(.05)	.018
D_1P	Brain	38(.10)	.040
D_1A	Brain	19(.10)	.035

[1] Egg masses were collected from the NJ-SS colony, chilled for 180 days at 5°C, then de-haired and homogenized in HPLC-H_2O (0.1 mg egg/ml) on ice. The homogenate was centrifuged at 1800xg, 4°C, 15 min, to remove debris, and the supernatant was dried under vacuum. The dried supernatant was dissolved in methanol: trifluroacetic acid: H_2O (90:0.1:9.9 by volume), centrifuged at 8000xg, 5 min and the supernatant dried under vacuum and stored at -15°C until use. Brains were dissected from ether anesthetized animals into cold Bombyx saline and stored at -15°C until extraction. Thawed brains were extracted in the acidic methanol system, centrifuged 8000xg, 5 min, dried under vacuum and stored at -15°C. [2] Ar_{max}: Maximum activation ratio obtained when methanolic extracts were dissolved in Grace's medium and tested in the in vitro prothoracic gland bioassay (Kelly et al., 1986; Masler et al., 1986). Number in parentheses indicates tissue dose (equivalent/µl) at which Ar_{max} was observed. [3] ED_{50}: Dose at which one-half maximal activation was observed.

Physiological variations in PTTH activity exist among developmental stages in L. dispar and we suspect that biochemical differences may exist as well. The different labilities of embryonic vs. post-embryonic PTTH activities under denaturing conditions (Table 2) suggests, for example, that a greater portion of crude embryonic (vs. post-embryonic) "PTTH" activity detected in vitro may be due to non-hormonal, labile factors such as ketoecdysteroid reductase. However, the increased recovery of embryonic PTTH activity with lowered methanol concentrations (Table 2) suggests that extraction and chromatographic methods developed for post-embryonic ecdysiotropic preparation (Masler et al., 1983, 1986) can be successfully adapted for extraction and analysis of embryonic PTTH.

Table 2. Comparison of the Response of Embryonic and Post-Embryonic PTTH Activities to Various Extraction Regimens

Treatment	Embryonic[1]	%Max	Post-Embryonic[2]	%Max
Grace's	40.0±11.0	100	69±9	100
Grace's/ Heat	9.3 ±4.7	23	29±7	42
CH_3OH(90%)[3]	3.6 ±0.2	9	60±19	87
CH_3OH(40%)[3]	15.1 ±4.3	38	----	----

[1] Eggs were water-extracted as described in Table 1. Dried extracts dissolved in Grace's were tested in the bioassay or heated (5 min in a boiling water bath) then tested in the bioassay. Methanol treatments were as described in Table 1. [2] Brains were dissected from D_5L_5 females as described in Table 1. Grace's extracts were tested directly in the bioassay or heated then tested. Methanolic extracts were as described in Table 1. [3] In 0.1% TFA.

All doses are 0.04 tissue equivalent/µL.

This work was supported in part by USDA competitive Grant No. 86-CRCR-1-2111 to E.P.M. and T.J.K.

REFERENCES

Bollenbacher W.E., Katahira E.J., O'Brien M.A., Gilbert L.I., Agui N. and Baumhover A.H. (1984) Insect prothoracicotropic hormone: evidence for two molecular forms. Science 224: 1243-1245.

Chen J.H., Fugo H., Nakajima M., Nagasawa H. and Suzuki A. (1986) The presence of neurohormonal activities in embryos of the silkworm, Bombyx mori. J. Seric. Sci. Jpn. 55: 54-57.

Chen J.H., Fugo H., Nakajima M., Nagasawa H. and Suzuki A. (1987) Neurohormones in developing embryos of the silkworm, Bombyx mori: the presence and characteristics of prothoracicotropic hormone-B. J. Insect Physiol. 33: 407-411.

Dorn A., Gilbert L.I. and Bollenbacher W.E. (1987) Prothoracicotropic hormone activity in the embryonic brain of the tobacco hornworm, Manduca sexta. J. Comp. Physiol.-B. 157: 279-283.

Fugo H., Chen J.H., Nakajima M., Nagasawa H. and Suzuki A. (1987) Neurohormones in developing embryos of the silkworm, Bombyx mori: the presence and characteristics of prothoracicotropic hormone-S. J. Insect Physiol. 33: 243-248.

Kelly T.J., Masler E.P., Thyagaraja B.S., Bell R.A. and Borkovec A.B. (1986) Prothoracicotropic hormone stimulation of ecdysone synthesis by the prothoracic glands of the gypsy moth, Lymantria dispar. In Insect Neurochemistry and Neurophysiology-1986. (Ed. by Borkovec A.B. and Gelman D.B.) pp. 327-330, Humana Press, Clifton, New Jersey.

Kopec S. (1922) Studies on the necessity of the brain for inception of insect metamorphosis. Biol. Bull. Woods Hole 42: 322-342.

Masler E.P., Hagedorn H.H., Petzel D.H. and Borkovec A.B. (1983) Partial purification of egg development neurosecretory hormone with reverse-phase liquid chromatographic techniques. Life Sciences 33: 1925-1931.

Masler E.P., Kelly T.J., Thyagaraja B.S., Woods C.W., Bell, R.A. and Borkovec A.B. (1986) Discovery and partial characterization of the prothoracicotropic hormones of the gypsy moth, Lymantria dispar. In Insect Neurochemistry and Neurophysiology - 1986 (Ed. by Borkovec A.B. and Gelman, D.B.), pp. 331-334, Humana Press, Clifton, New Jersey.

Masler E.P., Kelly T.J., Thyagaraja B.S., Davis R.E., Bell, R.A. and Borkovec A.B. (1988) Ecdysteroid activity in the gypsy moth, Lymantria dispar: identification of prothoracicotropic hormones from embryo through adult. Proc. Int'l. Congr. Entomology. 18: 120.

From *Insect Neurochemistry and Neurophysiology • 1989* Edited by A. B. Borkovec and E. P. Masler

PROTHORACICOTROPIC HORMONE AND ECDYSTEROID KETOREDUCTASE FROM PRE-HATCH EGGS OF THE GYPSY MOTH, *LYMANTRIA DISPAR*

Thomas J. Kelly, Edward P. Masler, [+]*Belgaum S. Thyagaraja, Robert A. Bell, Dale B. Gelman, *Richard B. Imberski and Alexej B. Borkovec

Insect Reproduction Laboratory, USDA, ARS, Beltsville, MD 20705, USA. [+]RSRS, Central Silk Board, C. R. Nagar 571 313, India and *Department of Zoology, University of Maryland, College Park, MD 20742, USA

The identification of prothoracicotropic hormone (PTTH) activity in *Lymantria* (Kelly *et al*., 1986; Masler *et al*., 1986) verified the existence of a molt/metamorphosis-regulating brain secretory factor, first postulated by Kopec (1922). Since Kopec's discovery, this factor has received considerable attention in post-embryonic stages of *Bombyx mori* (Ishizaki *et al*., 1987) and *Manduca sexta* (Flanagan *et al*., 1988). Recently, PTTH activity was demonstrated in embryonic stages of *Bombyx* and *Manduca* (Chen *et al*., 1987; Fugo *et al*., 1987; Dorn *et al*., 1987). We have demonstrated similar activity in embryonic eggs of *Lymantria* (see also Bell *et al*., Masler *et al*. and Thyagaraja *et al*., this volume).

Initial dose-response analyses of extracts prepared from pre-hatch whole eggs of *Lymantria* demonstrated a high degree of ecdysteroid activity in the incubation media of cultured prothoracic glands (PGs) as assayed by radioimmunoassay (RIA) (see Masler *et al*., this volume). This activity was reduced by ca. 75% when extracts were heated or solvent-extracted (Table 1), in contrast to the effect of heat or solvent treatment on post-embryonic brains of *Lymantria* (see Masler *et al*., this volume). This reduction in the production of RIA-detectable ecdysteroid after incubation with treated embryonic egg extract was due, at least in part, to the elimination of ecdysteroid ketoreductase

Table 1. Prothoracicotropic and Ketoecdysteroid Reductase Activities of Lymantria dispar Egg Extracts Subjected to Heat or Organic Solvent

EXTRACT[1]	PROTHORACIC GLAND[2] ACTIVATION	REDUCTASE[3] ACTIVITY
Unboiled	100%	100%
Boiled	25%	2%
Solvent Extracted	27%	3%

[1]All extracts were tested at 0.01mg egg equivalents/µl. Unboiled: extracts were prepared in Grace's medium from H_2O-extracted, vacuum-dried, pre-hatch eggs (Harper's Ferry wild strain) frozen at -20°C 1-2 days prior to hatch. Boiled: unboiled extract, prepared as above, was immersed in boiling water for 2 min, followed by centrifugation for 1 min, 12,000g. Solvent Extracted: dried H_2O extract was dissolved in methanol:TFA:H_2O (90:0.1:9.9), stored overnight at -20°C, centrifuged for 1 min, 12,000g, vacuum-dried and re-solubilized in Grace's medium.

[2]Values represent percentage of activation relative to unboiled extract. Activation ratios (A_r; see Kelly et al., 1986) were calculated for each extract and compared with the A_r for the unboiled (control) extract. Incubation conditions were as described in Kelly et al., 1986 (Figure 2) using female 5th instar, day 5 PGs.

[3]Values represent percentage of activity relative to unboiled extract. Ketoecdysteroid reductase activity was determined by drying 1ng of 3-dehydroecdysone (a gift from M. Thompson, Insect Hormone Lab, Beltsville, MD, and further purified by HPLC) in a 6x50 mm glass tube and adding 10µl of the appropriate test extract. Incubations were for 2 hours at 25°C and were stopped by freezing at -20°C. Samples were thawed and analyzed for ecdysteroid by RIA. Conversion was complete by 2 hours. PGs from 5-day old, 5th instar Lymantria secrete in vitro an ecdysteroid which co-elutes on reverse-phase HPLC with 3-dehydroecdysone (Kelly et al., unpublished results).

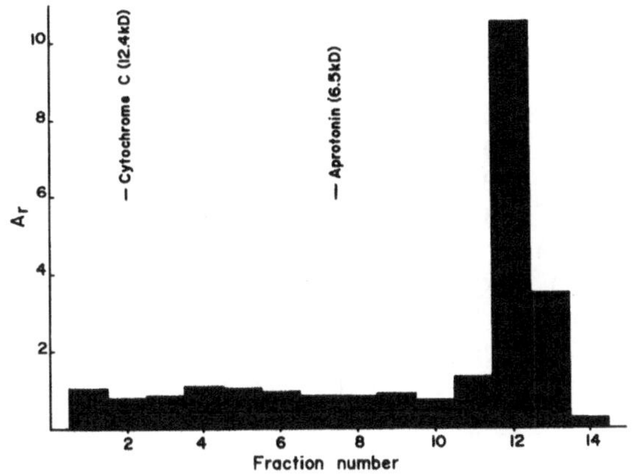

Fig 1. HP-SEC fractionation of Lymantria dispar egg extract. An extract was prepared in H_2O (1.0 gm egg equivalents in 3.0ml H_2O) from H_2O-extracted, vacuum-dried, pre-hatch eggs (Harper's Ferry wild strain). The extract was passed through a C_{18} SEP-PAK (Water's Assoc.), and the 60% CH_3CN eluate collected according to Masler et al. (1983) and vacuum dried. Two hundred mg egg equivalents were redissolved in 200µl 0.1% TFA in 40% CH_3CN, fractionated in the same solvent on a tandem Water's I-125 HP-SEC column system at 1ml/min (room temperature), 0.5ml fractions collected, split in half, and vacuum-dried. One set (i.e. 100mg egg equivalents) was redissolved in Grace's medium to give ca. 1.0mg egg equivalents/µl and assayed according to Table 1.

activity (Table 1). As with Manduca hemolymph, (Warren et al., 1988), ecdysone, derived from reductase conversion of 3-dehydroecdysone, may constitute a considerable portion of the total RIA activity obtained with unheated, ketoecdysteroid reductase-containing extracts. Ketoreductase activity is reduced by 97% or more in Lymantria embryonic egg extracts treated with heat or solvent (Table 1). Analysis of the activity remaining in embryonic extracts following organic solvent treatment revealed a single peak of activity on HP-SEC with a molecular weight of <5kD (Fig. 1). This <5kD fraction presumably represents a low molecular weight form of embryonic PTTH in Lymantria.

Supported in part by USDA Competitive Grant No. 86-CRCR-1-2111 to E.P.M. and T.J.K.

REFERENCES

Chen J.H., Fugo H., Nakajama M., Nagasawa H. and Suzuki A. (1987) Neurohormones in developing embryos of the silkworm, Bombyx mori: the presence and characteristics of prothoracicotropic hormone B. J. Insect Physiol. 33: 407-411.

Dorn A., Gilbert L.I. and Bollenbacher W.E. (1987) Prothoracicotropic activity in the embryonic brain of the tobacco hornworm, Manduca sexta. J. Comp. Physiol. B 157: 279-283.

Flanagan T.R., Tomioka K., O'Brien M.A., Westbrook A.L., Agui N. and Bollenbacher W.E. (1988) Neuroendocrine regulation of insect development. In Biomechanisms Regulating Growth and Development (Ed. by Steffens G.L. and Ramsey T.S.), pp. 119-135, Kluwer Academic Press, Boston.

Fugo H., Chen J.H., Nakajima M., Nagasawa H. and Suzuki A. (1987) Neurohormones in developing embryos of the silkworm Bombyx mori: the presence and characteristics of prothoracicotropic hormone-S. J. Insect Physiol. 33: 243-248.

Ishizaki H., Mizoguchi A., Halta M., Suzuki A., Nagasawa H., Kataoka H., Isogai A., Tamura S., Fujino M. and Kitada C. (1987) Prothoracicotropic hormone (PTTH) of the silkworm, Bombyx mori: 4K-PTTH. In Molecular Entomology (Ed. by Law J.H.), pp. 119-128, Alan R. Liss, New York.

Kelly T.J., Masler E.P., Thyagaraja B.S., Bell R.A. and Borkovec A.B. (1986) Prothoracicotropic hormone stimulation of ecdysone synthesis by the prothoracic glands of the gypsy moth, Lymantria dispar. In Insect Neurochemistry and Neurophysiology 1986 (Ed. by Borkovec A.B. and Gelman D.B.), pp. 327-330, Humana Press, Clifton, New Jersey.

Kopec S. (1922) Studies on the necessity of the brain for inception of insect metamorphosis. Biol. Bull., Wood's Hole 42: 322-342.

Masler E.P., Hagedorn H.H., Petzel D.H. and Borkovec A.B. (1983) Partial purification of egg development neurosecretory hormone with reverse-phase liquid chromatographic techniques. Life Sci. 33: 1925-1932.

Masler E.P., Kelly T.J., Thyagaraja B.S., Woods C.W., Bell R.A. and Borkovec A.B. (1986) Discovery and partial characterization of the prothoracicotropic hormones of the gypsy moth, Lymantria dispar. In Insect Neurochemistry and Neurophysiology 1986 (Ed. by Borkovec A.B. and Gelman D.B.), pp. 331-334, Humana Press, Clifton, New Jersey.

Warren J.T., Sakurai S., Rountree D.B., Gilbert L.I., Lee S.-S. and Nakanishi K. (1988) Regulation of the ecdysteroid titer of Manduca sexta: reappraisal of the role of the prothoracic glands. Proc. Natl. Acad. Sci. 85: 958-962.

From *Insect Neurochemistry and Neurophysiology • 1989* Edited by A. B. Borkovec and E. P. Masler

ECDYSIOTROPINS AND KETOREDUCTASE IN THE BRAIN OF THE EUROPEAN CORN BORER, OSTRINIA NUBILALIS PROMOTE THE PRODUCTION OF ECDYSONE AND 3-DEHYDROECDYSONE BY PROTHORACIC GLANDS OF THE GYPSY MOTH, LYMANTRIA DISPAR.

Dale B. Gelman, *Belgaum S. Thyagaraja, Thomas J. Kelly, Edward P. Masler, Robert A. Bell and Alexej B. Borkovec

Insect Reproduction Laboratory, USDA, ARS, Beltsville, MD 20705. USA. *RSRS, Central Silk Board, C.R. Nagar 571 313, India

Prothoracicotropic hormone (PTTH) is an important insect ecdysiotropin that is produced in the brain, and upon release into the hemolymph, stimulates the prothoracic glands (PGs) to produce ecdysteroid (Gilbert et al., 1980). Until recently, ecdysone was believed to be the only ecdysteroid secreted by the PGs. However, it is now known for both the tobacco hornworm, Manduca sexta (Warren et al., 1988) and the gypsy moth, Lymantria dispar (Kelly et al., unpublished results) that the PGs produce 3-dehydroecdysone which is converted by a ketoreductase to ecdysone and finally by peripheral tissues to the molting hormone, 20-hydroxyecdysone.

Using a newly developed triple incubation in vitro assay, we have found that the brain of Ostrinia contains peptidergic factors that stimulate the prothoracic glands to secrete both ecdysone and 3-dehydroecdysone in a dose-dependent manner. Brain tissue also contains a ketoreductase which converts 3-dehydroecdysone to ecdysone. Lymantria PGs do not (Table 1).

Prothoracic glands from day-5 Lymantria 5th instars were preincubated in 25µl of Grace's medium for 2-h after which time the production of RIA-detectable ecdysteroid dropped to low levels (Kelly et al., 1986). PGs were then transferred

Table 1. KETOREDUCTASE ACTIVITY IN SELECTED INSECT ORGANS

Insect	Organ	pg E/µg protein/2h
European Corn Borer Prepupa	Brain	217
	Proctodaeum	156
Gypsy Moth (Day 5, 5th Instar)	Prothoracic Gland	0

Extracts were incubated for 2 h with 1 ng of 3-dehydroecdysone (a gift from M. Thompson, Insect Hormone Lab., USDA, ARS) and the reaction was stopped with 100% methanol. HPLC-RIA confirmed that 92% of the 3-dehydroecdysone had been converted to an RIA-detectable ecdysteroid that had the same retention time as ecdysone. E = ecdysone; Prepupa = wandering 5th instar that had completed gut purge.

to fresh Grace's medium (controls), or to Grace's medium which contained an Ostrinia prepupal brain extract that had been boiled for 2.5 min to destroy ketoreductase. After an additional 2 h of incubation the PGs were removed from both the experimentals and the controls, and the incubation fluid from each was divided into two 11-µl drops. One drop, "G", received Grace's medium; the second drop, "P", received unheated Ostrinia prepupal brain or proctodaeal extract. These previously had been shown to contain ketoreductase activity (Table 1). After an additional 2 h of incubation, the reaction was stopped with 100 percent methanol. HPLC-RIA (Gelman and Woods, 1986) of a pooled sample of incubation medium revealed that an RIA-detectable ecdysteroid with the same retention time as ecdysone was produced by the experimental glands after both treatments "G" and "P". However, 3-5x more ecdysone was generated after treatment "P". In a separate experiment, pooled samples of "G" and "P" again were subjected to HPLC. Fractions that had the same retention time as 3-dehydroecdysone (an ecdysteroid that is not detected by the ecdysone antiserum used in the RIA) were pooled, dried, exposed to ketoreductase (unheated proctodaeal extract), subjected to HPLC and finally to RIA. For "G" an RIA detectable-ecdysteroid that had the same retention time as ecdysone was observed. Thus, the 3-dehydroecdysone that had been produced during the 2nd 2-hour incubation with the heated brain extract now had been converted to ecdysone. For "P" there was very little detectable ecdysone produced because almost all of the 3-dehydroecdysone had already been converted to ecdysone during the final 2-hour incubation with the unheated proctodaeal or brain extract.

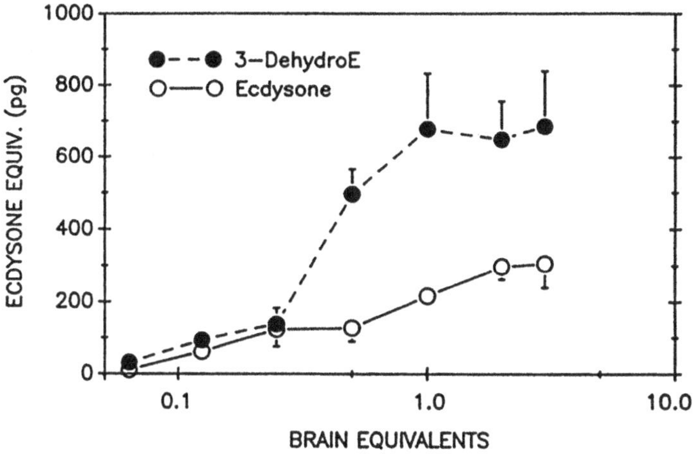

Fig. 1. Production of ecdysone (Treatment "G"-control[1]) and 3-dehydroecdysone (Treatment "P"-"G"- control[1]) by Lymantria PGs that had been exposed to heated brain extracts from Ostrinia prepupae (wandering 5th instars that had completed gut purge). For treatment "G", final 2-h incubation of medium was with added Grace's medium. For treatment "P", final 2-h incubation of medium was with added unheated protodaeal or brain extract (contained ketoreductase). pg = picograms. Each point represents the mean of at least 6 separate determinations. Error bars = standard errors of the mean.

[1] Control = basal levels of RIA-detectable ecdysteroid produced by unstimulated PGs (incubated in Grace's medium instead of heated brain extract during the 2nd 2-hour incubation; incubation medium then subjected to either treatment "G" or "P" during the 3rd 2-hour incubation.)

Dose-response curves for the production of ecdysone and 3-dehydroecdysone (produced during the 2nd 2-hour incubation with heated brain extract) are illustrated in Fig. 1. The amount of RIA-undetectable 3-dehydroecdysone could only be determined after a 3rd 2-hour incubation with ketoreductase (unheated brain or proctodaeal extract, treatment "P") in which the 3-dehydroecdysone was converted to RIA-detectable ecdysone. The value for the ecdysone produced during the 2nd 2-hour incubation (as measured after treatment "G") was subtracted so that results represent net 3-dehydroecdysone synthesized.

The brain's ecdysiotropins that stimulate both the production of ecdysone and 3-dehydroecdysone were stable to boiling and to treatment with acetonitrile-TFA and methanol-TFA. They were destroyed upon exposure to pronase. Thus, the factors are probably peptides or small proteins. After centrifugation with Amicon micro-concentrator filters (Amicon, Danvers MA)[1], activity was found in the 10-20 and ≥30 KD fractions.

[1]Mention of a commercial product does not imply endorsement by the U.S. Department of Agriculture.

REFERENCES

Gelman D.B. and Woods, C.W. (1986) Ecdysteroid conjugates in pupal and pharate adult haemolymph of the European corn borer, Ostrinia nubilalis (Hubner). Insect Biochem. 16: 99-108.

Gilbert L.I., Bollenbacher W.E. and Granger N.A. (1980) Insect Endocrinology: Regulation of endocrine glands, hormone titre, and hormone metabolism. Ann. Rev. Physiol. 42: 493-510.

Kelly T.A., Masler E.P., Thyagaraja B.S., Bell R.A. and Borkovec A.B. (1986) Prothoracicotropic hormone stimulation of ecdysone snythesis by the prothoracic glands of the gypsy moth, Lymantria dispar. In Insect Neurochemistry and Neurophysiology 1986 (Ed. by Borkovec A.B. and Gelman D.B.) pp 327-330, Humana Press, Clifton, New Jersey.

Warren, J.T., Sakurai S., Rountree D.B., Gilbert L.I., Lee, S.S. and Nakanishi K. (1988) Regulation of the ecdysteroid titre of Manduca sexta: Reappraisal of the role of the prothoracic glands. Proc. Natl. Acad. Sci., USA. 85: 958-962.

From *Insect Neurochemistry and Neurophysiology* • *1989* Edited by A. B. Borkovec and E. P. Masler

MOLT INHIBITION BY AZADIRACHTIN; EFFECTS ON PROTHORACICOTROPIC HORMONE, PROTHORACIC GLANDS AND ECDYSTEROID TITERS IN MANDUCA SEXTA PUPAE

M. P. Pener*, D. B. Rountree**,
S. T. Bishoff** and L. I. Gilbert**

*Department of Zoology, The Hebrew University, Jerusalem, 91904, Israel

**Department of Biology, The University of North Carolina, Chapel Hill, NC 27514, USA

Azadirachtin (AZA) acts as an antifeedant and growth regulator that inhibits molt in many insects (cf. Jacobson, 1986). We report here on experiments related to the mode of molt-inhibitory action of AZA in Manduca sexta pupae.

Eight-day-old, gate 2 last-instar larvae were injected with 4 μg of AZA per g fresh weight in 4 μl of 90% aqueous ethanol, or with ethanol alone (controls). Those molting to normal or almost normal pupae within 2 days after injection were used for further studies (80% of AZA-injected and 91% of ethanol-injected insects). All experiments were made on pupae obtained from the treated larvae. As the larvae had ceased feeding 4 days before injection, starvation or semi-starvation inducing effect of AZA was eliminated. Pupae obtained from AZA-injected larvae exhibited retarded adult development; they survived for considerable periods, but eventually died without emergence of adults. Normal adults, at normal time, emerged from the control pupae. Ecdysteroid titers in hemolymph samples and in the incubation medium of isolated prothoracic glands (PG) were assessed by RIA using H-22 antibody (cf. Warren and Gilbert, 1988.)

Brain extracts were prepared as described (O'Brien et al., 1986) from day 1 and day 6-7 AZA-treated and control pupae. Extracts were diluted serially; prothoracicotropic hormone (PTTH) activity of the extracts was assessed by in vitro ecdysteroid production of a PG, freshly dissected from day 0 normal (untreated) pupae. One PG was tested with a known amount of brain extract in the medium and the contralateral PG without brain extract. Incubation time was 2 h at 25°C. No difference in PTTH activity was found between AZA-injected and control insects in day 1 pupae (Fig. 1A), nor in day 6-7 pupae (Fig. 1B). Thus, AZA does not interfere with PTTH biosynthesis which occurs from day 8 of the last larval instar to day 1 of the pupa (O'Brien et al., 1986) and it does not seem to inactivate or deplete PTTH content of the brain.

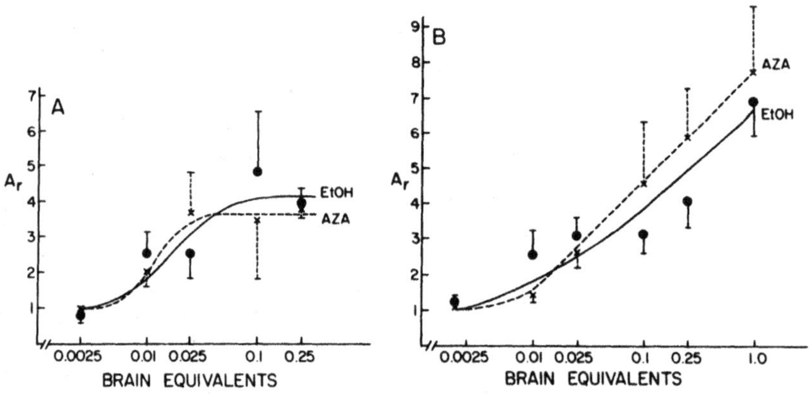

Fig. 1. Stimulation of PG from day 0 normal (untreated) pupae by brain extracts (abscissa) from day 1 (A) and day 6-7 (B) pupae which had been injected on day 8 of the last larval instar with azadirachtin (AZA, crosses and broken line), or with ethanol (EtOH, dots and continuous line). Stimulation is expressed as an activation ratio (Ar, ordinate) which is the amount of ecdysteroid produced by a gland with brain extract in the medium divided by the amount produced by the contralateral gland without brain extract in the medium. Each point presents the average ± SEM (marked only in one direction for the sake of clarity) obtained from four independent tests.

In vitro ecdysteroid production by isolated PG from AZA-treated and control insects was assessed. One gland was unstimulated, whereas the contralateral gland was stimulated by adding 0.25 units of big PTTH to the medium. Incubation time was 4 h at 25°C. The PG appeared normal in both AZA-treated and control pupae of day 1 and 3. In day 6-7 controls the PG were degenerating (not tested). The PG appeared normal in day 6-7 AZA-treated insects. In day 14-15 no traces of PG were found in the controls, while in

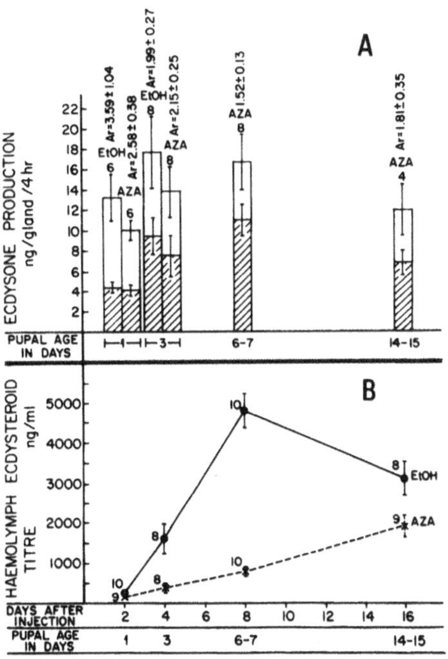

Fig. 2A (top). In vitro ecdysteroid production by PG of pupae injected on day 8 of the last larval instar with AZA or with ethanol (EtOH). One gland was unstimulated (shaded area), whereas the contralateral gland was stimulated by 0.25 units of big PTTH in the medium (shaded plus white areas). Division of the latter by the former yields the activation ratio (Ar) noted on the top of columns with SD. Fig. 2B (bottom). Hemolymph ecdysteroid titers (same age scale as in A) in similarly treated insects (AZA, crosses and broken line; EtOH, dots and solid line). Height of columns (A) and points (B) are means, vertical bars are ± SD; n is noted on top of bars (A) or at points (B).

50% of the AZA-treated insects the PG appeared normal. Comparing AZA-treated and control insects on day 1 and 3, Fig. 2A reveals that previous injection of AZA did not reduce ecdysteroid production by the PG, nor affected rate of stimulation of the PG by PTTH. The long-surviving PG of AZA-treated insects were active in day 6-7 and even in day 14-15, and were responsive to PTTH stimulation (Fig. 2A).

Despite the activity of the long-surviving PG in the AZA-treated insects, hemolymph ecdysteroid titers were much lower in these pupae than in the controls (Fig. 2B).

We conclude that AZA does not inhibit directly the functioning of the PG, does not interfere with PTTH biosynthesis, neither inactivates brain PTTH activity. Nevertheless, hemolymph ecdysteroid titers are much lower in AZA-treated than in control insects. Our results do not exclude possible effects of AZA on PTTH release, or on post-PG ecdysteroid metabolism. The exact mode of molt-inhibitory action of AZA remains to be elucidated.

We thank Dr. M. Jacobson for a sample of AZA. The purified PTTH was prepared with Dr. W. E. Bollenbacher. The research was financed by NIH grant DK-30118 to L.I.G.

REFERENCES

Jacobson M. (1986) The neem tree: Natural resistance par excellence. In Natural Resistance of Plants to Pests, Roles of Allelochemicals (Ed. by Green M. B. and Hedin P. A.), ACS Symposium Series, No. 296, pp. 220-232.

O'Brien M. A., Granger, N. A., Agui, N., Gilbert L. I. and Bollenbacher W. E. (1986) Prothoracicotropic hormone in the developing brain of the tobacco hornworm, Manduca sexta: Relative amounts of two molecular forms. J. Insect Physiol. **32**, 719-725.

Warren J. T. and Gilbert, L. I. (1986) Ecdysone metabolism and distribution during pupal-adult development of Manduca sexta. Insect Biochem. **16**, 65-82.

From *Insect Neurochemistry and Neurophysiology* • *1989* Edited by A. B. Borkovec and E. P. Masler

METABOLIC CHANGES IN THE NERVE TISSUE OF THE SILK WORM, BOMBYX MORI FOLLOWING REPEATED TOPICAL APPLICATION OF ENDOSULFAN

Geeta K. Jadhav & Vasant L. Kallapur

Department of Zoology
Karnatak University
Dharwad-580003

The fifth instar silkworm, Bombyx mori (pure Mysore breed) served as experimental animal. Sublethal doses (7μg/larval) of technical grade endosulfan (Yadwad and Kallapur, 1988) dissolved in acetone were applied topically to the dorsal side of the larva at intervals of 24h from the 2nd day to 6th day of larval development, whereas five-day old larvae were used for all single dose experiments. Larvae treated topically with acetone alone served as controls. The entire nerve cord along with brain was dissected and homogenized in appropriate medium and the clear supernatant obtained after centrifugation was used as a source of enzyme. The glycogen content was determined.

Results and Discussion

Bombyx mori shows a phenomenal growth rate during the 5th instar stage and is the most active feeding stage in development. High metabolic activities are anticipated in the nervous system to coordinate various activities of the larva. The present study revealed that the glycogen content of the nerve tissue increased during its development (Table-1). The accumulation of glycogen reached a maximum (40%) on the 7th day of development. This phenomenal increase in glycogen level is probably needed to meet the energy demand of the tissue, since the glycogen serves as chief source of energy in the nerve tissue of insects (Lund, 1985). The depletion of glycogen stored in the nerve tissue of the insecticide treated insects (Table-1) further supports this contention. It is interesting to note that a single topical application of insecticide caused a reduction in glycogen content during the initial stage (up to 3h) but in the subsequent periods of 6h and 12h the level is substantially increased. This increase may be the result of its synthesis from the precursors available in the hemolymph. The hemolymph shows a high level of trehalose during those hours. The hyperglycemic condition of the larva may be due to the breakdown of glycogen store of the fat body in response to insecticide action on the cc.

Hour after Treatment	% Change in Glycogen Value	Day after Treatment	% Reduction in Glycogen
1/2	14 ± 3.0*	3	33 ± 5.8
1-1/2	40 ± 6.6	4	42 ± 5.8
3	30 ± 5.0	5	36 ± 4.6
6	55 ± 8.6+	6	36 ± 6.5
12	13 ± 2.5+	7	34 ± 5.5

Table 1. Glycogen content of the nerve tissue following endosulfan treatment. (*) ug/mg tissue (+) increase.

	Active Phosphorylase (% of total)			
Hour after Treatment	Insecticide Treated	cc extract	Day after Treatment	Insecticide Treatment
1/2	14.8 ± 1.6	37.0 ± 6.6	3	24.0 ± 4.6
1-1/2	36.0 ± 8.0	57.0 ± 6.0	4	29.0 ± 3.8
3	48.0 ± 9.0	12.5 ± 3.0	5	21.0 ± 5.0
6	29.0 ± 7.0	---	6	23.0 ± 6.0
12	---	---	7	12.0 ± 3.4

Table 2. Stimulation of phosphorylase activity of the nerve tissue by insecticide treatment and cc extract.

The present study has shown that phosphorylase activity in the nerve tissue is stimulated by a factor from the cc (Table-2). Insecticides have been shown to stimulate cc which are known to release hormone(s) that regulate glycogen metabolism in insects (Ziegler et al. 1979; Maddrell, 1980; Sing and Orchard, 1982). It may be argued from the present study that the action of endosulfan on the activation of phosphorylase appears to be mediated through the stimulation of the cc.

Hour after Treatment	% Increase in specific activity of enzyme*	Day after Treatment	% Increase in specific activity of enzyme
1/2	5 ± 1.2	3	27 ± 4.0
1-1/2	9.5 ± 3.0	4	35 ± 5.0
3	30 ± 4.0	5	23 ± 4.6
6	20 ± 4.0	6	39 ± 6.6
12	27 ± 5.0	7	20 ± 4.8

Table 3. Induction of GSH S- transferase activity following endosulfan treatment. (*)DCNB conjugated/min/mg protein.

Both single and repeated topical applications of a sublethal dose of endosulfan caused activation of GSH S-transferase (Table-3). A single dose produced a 22% increase in the enzyme level 3h after the treatment and was further increased to 39% at 12h of post-treatment. Interestingly, the response of the enzyme as well as GSH content to repeated doses was cyclic throughout the experimental period. The level of GSH seems to be a rate-limiting factor in the induction of GSH S-transferase. In the present study, a single dose of endosulfan has no inhibitory effect on the activity of the enzyme, whereas repeated doses produced a moderate increase. However, the increase is found to be insignificant. A similar increase in the enzyme activity following DDT treatment has been reported in cockroach nerve cord homogenate (Cutkomp, et al. 1976). It appears that the sublethal dose of endosulfan has no inhibitory effect on the nerve cord Na^+-K^+ATPase of the silkworm, B. mori.

The interesting observation of the present study is the finding that that cocoon shell weight of the insecticide treated larva (450±15 mg) is significantly higher when compared to untreated (375±12 mg). Both size and weight of the cocoon depend upon the nutritional state of 5th instar larva. Low concentration of organophosphate compounds when administered along with food to 5th instar larva of B. mori resulted in decreased metabolic rate while conversion efficiency increased resulting in more production of body tissues and also silk proteins (Radhakrishna and Delvi, 1988). Whether cyclodiene compounds have the similar effect on the conversion efficiency of the ingested food and augment the silk protein production in the multivoltine B. mori requires further investigation.

REFERENCES

Cutkomp, L.K., Desaiah, D., Chang, E., Vea, V. and Koch, R.B. (1976) In vivo sensitivity of ATPases tro DDT, DDE and physical immobilization in American cockroaches. Pestic Biochem. Physiol. 6, 203-208.

Lund, A. (1985) Insecticides: Effects on the nervous system. In Comprehensive Insect Physiol. Biochem. and Pharmacol. (Ed. Kerkut, G.A. and Gilbert, L.I. 12, pp. 9-56. Pergamon Press, New York, Tononto, Sydney, Paris, Frankfurt.

Maddrell, S.H.P. (1980) The insect neuroendocrine system as a target for insecticides. In Insect Neurobiology and Pesticide Action. SCI Monoor. 329-334.

Singh, C.J.P. and Orchard, I. (1982) Is insecticide induced release of neurohormone a secondary effect of hyperactivity of central nervous system? Pestic. Biochem.Physiol. 17, 232-236.

Radhakrishna, P.G. and Delvi, M.R. (1988) Effect of organophosphorus insecticides on food utilization in different races of Bombyx mori (L) (Lepidoptera: Bombycidae). Abs. In National Symp. Nutritional Ecology of Insects and Environment. pp 24.

Yadwad, V.B. and Kallapur, V.L. (1988) Induction of glutathione S-transferase in the castor semilooper, Achaea janata (Lepidoptera, Noctuidae) following fenetrothione treatment. J. Biosci. 13(2), 139-146.

Ziegler, R., Ashida, M., Fellon, A., Wimer, L.T., Wyatt, S. and Wyatt, G.R. (1979) Regulation of glycogen phosphorylase in fat body of cecropia silk moth pupae. J. Comp. Physiol. 131, 321-332.

From *Insect Neurochemistry and Neurophysiology • 1989* Edited by A. B. Borkovec and E. P. Masler

PRECOCENE II SUPPRESSES THE BIOSYNTHESIS OF JUVENILE HORMONE, ECDYSTEROIDS AND VITELLOGENIN IN THE BLACK BLOWFLY, *PHORMIA REGINA* MEIGEN

C.-M. Yin, B.-X. Zou and J. G. Stoffolano, Jr.

Dept. of Entomology, Univ. of Massachusetts

Amherst, MA 01003, USA

Vitellogenesis in most insects is contingent on vitellogenin (Vg) production by the fat body and/or follicle cells which are stimulated by ecdysteroids, juvenile hormones (JH), or both in addition to other hormones. Earlier literatures on *Phormia regina* emphasize the role of the corpus allatum or JH on Vg production.

We report here our published and unpublished results about the effect of precocene II (P-II) on the biosynthesis of JH, ecdysteroids and Vg by *P. regina*. This blowfly is an anautogenous species which requires a proteinaceous meal to supply raw materials and to trigger the neuroendocrine cascade leading to the Vg production by the fat body. During vitellogenesis all terminal oocytes accumulate vitellin (Vt) synchronously. Because of this gonotrophic concordance, *P. regina*, has been used in many studies on the dietary induced hormonal regulation of oogenesis (Yin *et al.*, 1989a). We found that a single topical treatment of P-II shortly after adult emergence could suppress oogenesis in flies subsequently fed on liver (Yin *et al.*, 1989b). A radiochemical assay (RCA) for JH biosynthesis *in vitro* (Zou *et al.*, 1989) and a radioimmunoassay (RIA) for ecdysteroid content (Yin *et al.*, 1989a) were used to show that P-II suppressed the production of both hormones. An enzyme-linked immunosorbent assay (ELISA, Zou *et al.*, 1988) for Vg and Vt detected that the biosynthesis of Vg and accumulation of Vt were also greatly affected in these flies.

P-II was dissolved in acetone and given topically to each fly tested while control flies were given acetone only. All flies were continuously provided with water and sugar, except during a brief 3 to 4 h period of liver feeding which commenced at 72 h post-emergence. In some cases, methoprene (a JH analog) was administered to P-II-treated, liver-fed flies 3 days after the liver meal to see if the effect of P-II could be reversed.

Oocyte development (length) was inhibited in flies treated with P-II (20 or 40 ug/fly) at 4 or 12 h post-emergence. The length of terminal oocytes of treated, liver-fed flies was, in average, less than 0.25 mm, while that of control, liver-fed flies reached 1.12 ± 0.05 mm (Yin *et al.*, 1989b). A sensitive period about 30 h post-emergence was observed for the responsiveness to P- II (20 ug/fly). Older flies were much less responsive (Yin *et al.*, 1989b).

Using the RCA mentioned above, JH biosynthesis by the corpus allatum *in vitro* was estimated for P-II-treated, liver-fed flies and control, liver-fed flies. P-II at 10, 20 and 40 ug/fly suppressed JH synthesis and release from the corpus allatum by 35, 81 and 99%, respectively (Zou *et al.*, 1989). Further, fully developed eggs were observed in 77, 96 or 100% of the P-II-treated liver-fed flies if they also were topically treated with methoprene at 2.5. 5.0 or 10 ug/fly on day 3 post liver-feeding. Thus far, our data appeared to be consistent with earlier results emphasizing the role of JH in the oogenesis of *P. regina*. However, our following results suggested otherwise.

RIA results showed that the hemolymph ecdysteroids profile in P-II-treated (20 ug/fly), liver-fed flies was quite different from that of normal liver-fed flies (Fig. 1). Since ecdysteroids regulate vitellogenin production in many dipterans (Hagedorn, 1985), and P-II suppressed these steroids in *P. regina*, the role of ecdysteroids in oogenesis of this species must be studied further.

Profiles of JH biosynthesis, hemolymph ecdysteroids, Vg, and Vt in control, liver-fed flies revealed some interesting points (Fig. 2). First, ecdysteroid titer peaked 24 h earlier than JH. Second, the temporal profiles of Vg and Vt resembled closely to those of ecdysteroids and JH, respectively. When Vg and Vt patterns were determined for P-II-treated, liver-fed flies, they differed greatly from those of the controls (Fig. 3) showing a slower production yet steady accumulation of Vg and a nearly non-existence of Vt. Methoprene application (Fig. 3, ZR-515) caused an immediate decrease of Vg and increase in Vt.

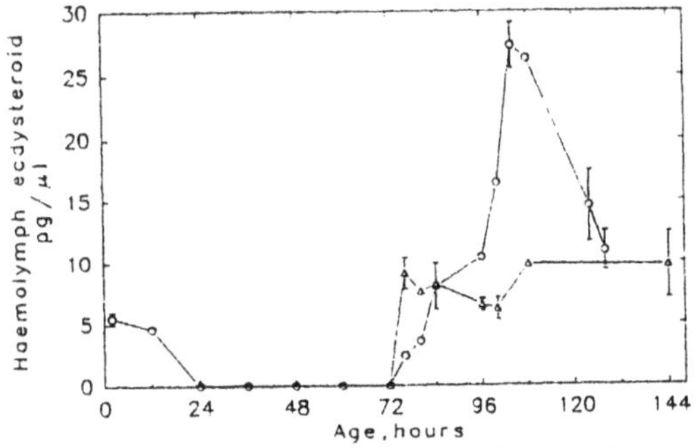

Fig. 1. Profiles of hemolymph ecdysteroids in control, liver-fed (circles) and P-II-treated, liver-fed (triangles) *P. regina* females. Liver was provided at 72 h of age.

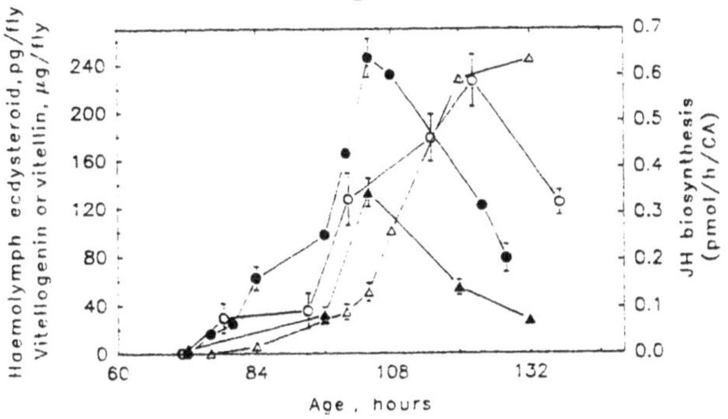

Fig. 2. Profiles of JH biosynthesis (open circles), hemolymph ecdysteroid (filled circles), Vg (filled triangles), and Vt (open triangles) in control, liver-fed *P. regina* females.

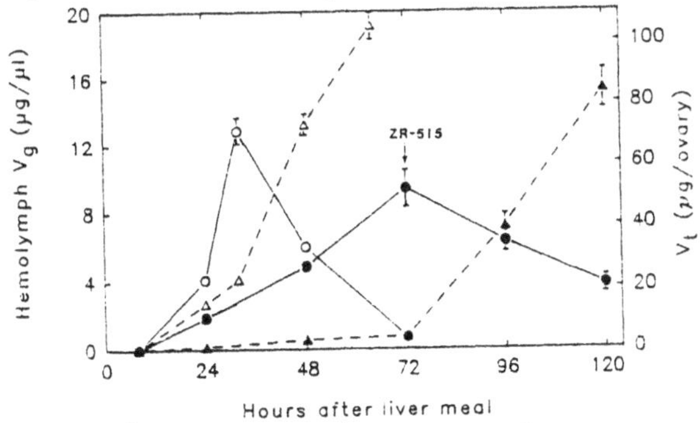

Fig. 3. Vg (solid lines) and Vt (dash lines) in control, liver-fed (open symbols), and P-II treated, liver-fed (filled symbols) flies. ZR--515 applied at 72 h after liver meal.

It appeared that the essential roles of ecdysteroids and JH in *P. regina* are to regulate the Vg production and Vg uptake (to form Vt), respectively. We do not know presently if JH might also play some role in the ecdysteroids regulated Vg production, nor do we know if JH may stimulate the follicle cells to produce Vg in the later phases of oogenesis like that reported for *Drosophila melanogaster* (Postlethwait and Kunert, 1986). More investigations, including some at molecular level, are needed to clarify these uncertainties.

Acknowledgements--We thank NSF (DCB-8409378) and Mass. Agric. Exp. Station (Hatch 570 and Hatch 571) for supports. Published as journal No. 2933 from the Station.

REFERENCES

Hagedorn H.H. (1985) The Role of Ecdysteroids in Reproduction. In "*Comprehensive Insect Physiology, Biochemistry and Pharmacology*", Vol. 8 (Edited by G.A. Kerkut and L.I. Gilbert), pp. 205-262. Pergamon Press, Oxford.

Postlethwait J.H. and Kunert C.J. (1986) Endocrine and genetic regulation of vitellogenesis in *Drosophila*. In "Comparative Endocrinology: Developments and Directions" (Edited by C.L. Ralph), pp. 33-52. Alan R. Liss, N.Y.

Yin C.-M., Zou B.-X. and Stoffolano J.G., Jr. (1989a) Dietary induced hormonal control of terminal oocyte development in *Phormia regina*. In "*Host Regulated Developmental Mechanisms in Vector Arthropods*" (Ed. by D. Borovsky and A. Spielman), pp. 81-88. Univ. Florida.

Yin C.-M., Zou B.-X. and Stoffolano J.G., Jr. (1989b) Precocene II treatment inhibits terminal oocyte development but not vitellogenin synthesis and release in the black blowfly *Phormia regina*. *J. Insect Physiol.*, in press.

Zou B.-X., Stoffolano J.G., Nordin J.H. and Yin C.-M. (1988) Subunit composition of vitellin and concentration profiles of vitellogenin, and vitellin in *Phormia regina* (Meig) following a protein meal. *Comp. Biochem. Physiol.* 90B, 861-867.

Zou B.-X., Yin C.-M., Stoffolano J.G., Jr. and Tobe S.S. (1989) Juvenile hormone biosynthesis and release during oocyte development in, *Phormia regina* Meigen. *Physiol. Ent.*, in press.

From *Insect Neurochemistry and Neurophysiology • 1989* Edited by A. B. Borkovec and E. P. Masler

REGULATION OF VITELLOGENESIS IN GYPSY MOTH, LYMANTRIA DISPAR (L.), FIFTH-INSTAR LARVAE.

Robin E. Davis, Howard W. Fescemyer, Thomas J. Kelly, Edward P. Masler, Robert A. Bell, *Belgaum S. Thyagaraja, and Alexej B. Borkovec

Insect Reproduction Laboratory, USDA, ARS, Beltsville, MD 20705, USA. *RSRS, Central Silk Board, C. R. Nagar 571 313, India, and Department of Zoology, University of Maryland, College Park, MD 20742, USA

Gypsy moth larvae initiate vitellogenesis at an atypical time in development, during the last larval instar (Davis et al., submitted). Between days 2 and 4 of the instar, female larvae begin to synthesize vitellogenin (Vg) in the fat body. Vg accumulates in the hemolymph of female larvae throughout the remainder of the instar where it can be detected by 1) co-migration of the apo-vitellogenin polypeptides (apoVg180 and apoVg165) on Coomassie-stained SDS-PAGE with the apo-vitellin (Vt) polypeptide components (apoVt180 and apoVt165) and/or 2) recognition with a rabbit polyclonal anti-Vt antisera on Western blots of Native- or SDS-PAGE analyzed hemolymph samples (Figure 1).

In most insects, vitellogenesis is initiated in pupal or adult stages and juvenile hormones (JH) and/or ecdysteroids are active in the induction and/or maintenance of Vg synthesis (Engelmann, 1984; Hagedorn, 1985; Bownes, 1986). During larval stages, these hormones are involved in controlling metamorphosis. Since the gyspy

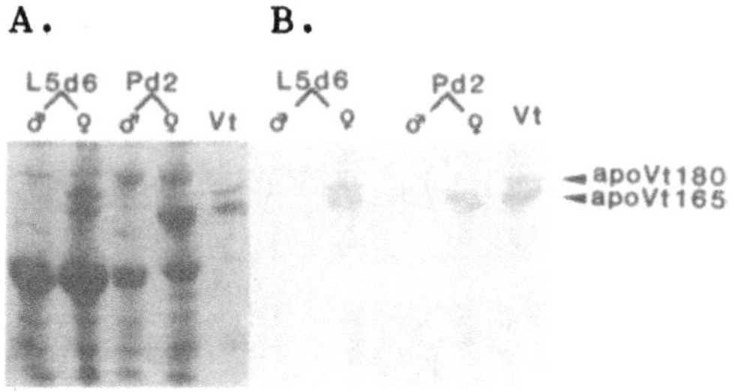

Figure 1. Gypsy Moth Vitellogenin and Vitellin. A. Coomassie-stained SDS-PAGE (6% acrylamide) of hemolymph from male and female day 6 fifth instar larvae (L5d6) and day 2 pupae (Pd2) gypsy moths. Gypsy moth vitellin (Vt) isolated from pupal follicles by high salt extraction and H_2O precipitaion was loaded at 4 ug per lane. Hemolymph samples were loaded at 0.2 ul hemolymph equivalents per lane. Arrows indicate the positions of 180,000 (apoVg180) and 165,000 dalton (apoVg165) apo-Vg components. B. Western blot of SDS-PAGE hemolymph and Vt probed with a rabbit anti-vitellin rabbit polyclonal antisera. The antisera was blocked with 20% male hemolymph in TBST (10 mM Tris-HCl, pH 8.0, 150 mM NaCl, 0.05% Tween 20) for 30 min at 0° C prior to incubation with the nitrocellulose filter bound gypsy moth proteins.

moth larva is inducing reproductive and metamorphic processes simultaneously, we set out to determine what role, if any, JH and ecdysteroids play in the initiation of Vg synthesis in the last larval instar in contrast to their role in metamorphosis.

Vg hemolymph accumulation was found to be blocked in a stage-dependent manner by head-ligation, starvation and/or treatment with a JH analogue (JHA), methoprene. Larvae treated before day 3 did not synthesize Vg, but larvae treated after day 3 did. Treatment on day 3 resulted in different and variable degrees of reduction in Vg synthesis as measured by hemolymph accumulation of apo-Vg detected by Coomassie-stained SDS-PAGE. The starved and ligated larvae did not grow and showed reductions in the accumulation of non-Vg hemolymph proteins as compared to controls. The response to JHA was dose-dependent, and unlike starvation and ligation treatments, JHA-treatment selectively blocked Vg synthesis and accumulation without any apparent affect on non-Vg hemolymph proteins. The JHA-treated larvae attained normal larval weights. Injection of starved, ligated or JHA-treated larvae with up to 2.5 ug 20-hydroxyecdysone failed to restore Vg hemolymph accumulation in larvae treated on day 2. However, this does not rule out a role for ecdysteroids if other hormonal and nutritive conditions are permissive. Although the ecdysteroid titers are very low early in the fifth instar gypsy moth, a small peak in ecdysteroid titer occurs between days 2 and 3 (B. S. Thyagaraja, personal communication).

From these experiments, we conclude that a low (or declining) JH titer during a critical time early in the last larval instar (between days 2 and 4) is necessary, but not sufficient, for Vg synthesis and accumulation to proceed normally in the gypsy moth. In addition to JH titer, Vg synthesis is influenced by one or more of the following factors including factors from the CNS or brain, nutritive factors, factors induced by the process of feeding, and/or not eliminated by the physiological response to starvation.

It is difficult to separate out the requirements for these factors since we cannot remove and replace them selectively _in vivo_ without disrupting the growth and development of the animal. Therefore, we have developed an _in vitro_ culture system in which Vg synthesis can be assayed by SDS-PAGE and fluorography of ^{35}S-methionine labelled fat body protein products. We have also established cDNA clones from L. dispar fat body, four of which are female-specific and may be specific for L. dispar Vg. With the availability of an in vitro fat body assay and the tools of molecular biology, it should now be possible to determine which factors regulate L.dispar vitellogenin synthesis and to determine how they function at the molecular level.

REFERENCES

Bownes M. (1986) Expression of the genes coding for vitellogenin (yolk protein). Ann. Rev. Entomol., 31: 507-531.

Davis R. E., Kelly T. J., Masler E. P., Thyagaraja B. S., Paramasivan V., Fescemyer H. W., Bell R. A., and Borkovec A. B. Vitellogenesis in the gypsy moth, Lymantria dispar (L.): characterization of hemolymph vitellogenin, ovarian weight, follicle growth and vitellin content. Int. J. Invert. Reprod. & Dev. (submitted).

Engelmann F. (1984) Regulation of vitellogenesis in insects: the pleiotropic role of juvenile hormones. In Biosynthesis, Metabolism and Mode of Action of Invertebrate Hormones. (Ed. by Hoffmann J. and Porchet M.), pp. 444-453, Springer, Berlin.

Hagedorn H. H. (1985) The role of ecdysteroids in reproduction. In Comprehensive Insect Physiology, Biochemistry and Pharmacology. Vol. 8 (Ed. by Kerkut G. A. and Gilbert L. I.), pp.205-262, Pergamon Press, New York.

From *Insect Neurochemistry and Neurophysiology* • *1989* Edited by A. B. Borkovec and E. P. Masler

6-HYDROXYDOPAMINE RETARDATION OF OVARIAN DEVELOPMENT IN THE HOUSEFLY, *MUSCA DOMESTICA* L.

Jeffery W. Gerst and T. S. Adams

Department of Zoology, North Dakota State University and USDA Biosciences Research Laboratory, Fargo, ND 58105

INTRODUCTION

Drug treatments affecting biogenic amine systems interfere with the processes of vitellogenesis and egg maturation in dipterans. 6-hydroxydopamine (6-OHDA) acts as a "chemical castrator" in female *Sarcophaga bullata* (De Loof et al., 1981) presumably by inhibiting the release of a gonadotrophic factor (Huybrechts and De Loof, 1981). Musca domestica were treated with 6-OHDA to assess the drug's overall effects on ovarian development.

Insects. Orlando regular strain houseflies were reared and provided with a diet containing nonfat dry milk, sugar and whole powered egg.

6-OHDA treatment. 6-OHDA was dissolved in fly saline at a concentration of 25μg/μL. Flies were injected (1μL) at 8 and 24 h post emergence with 6-OHDA or fly saline. Hemolymph samples were collected at 48 and 72 h post emergence. Ecdysteroid and vitellogenin titres were determined, and the stage of ovarian development was scored as described by Adams (1974).

Ecdysteroid Radioimmunoassay (RIA). RIA's were performed using an antiserum provided by James Buckner, USDA-ARS Biosciences Laboratory, Fargo, ND. Results are expressed as pg ecdysteroid/μL hemolymph.

Vitellogenin Immunoelectrophoresis. Rocket immunoelectrophoresis was performed as described by Adams and

Filipi (1983). Results are expressed as μg vitellogenin/μL hemolymph.

6-OHDA treatment resulted in a significant retardation of ovarian development (Table 1). At 72 h post emergence, oocytes of 6-OHDA injected flies were just beginning to accumulate yolk granules in the ooplasm. Their ovaries were at a developmental stage less than that observed in 48 h old saline injected flies.

Table 1. Ovarian developmental stage of houseflies injected with 25μg 6-hydroxydopamine (6-OHDA). Stage of the ultimate follicles were estimated at 48 and 72 h post emergence. Values represent mean +/- standard error (number). Treatment means followed by an asterisk are significantly different from control means ($p < 0.05$).

	Ovarian Stage	
	48 hours	72 hours
Saline injected	4.9 +/- 0.07 (208)	7.6 +/- 0.16 (165)
6-OHDA injected	3.8 +/- 0.05 (116)*	4.4 +/- 0.10 (110)*

No treatment differences in hemolymph ecdysteroid concentrations were observed (Table 2). Total immunologically detectable ecdysteroid concentrations were within the range observed in houseflies having vitellogenic ovaries (Adams, Hagedorn and Wheelock, 1985) at both 48 and 72 h post emergence.

Fluctuations in hemolymph vitellogenin concentrations generally correlate with changes in total hemolymph ecdysteroid concentration (Huybrechts and De Loof, 1981; Adams and Filipi, 1983; Adams, Hagedorn and Wheelock,1985). Yet, in spite of having similar levels of ecdysteroids, 6-OHDA injected houseflies had significantly lower vitellogenin concentrations than saline injected controls (Table 3). Huybrechts and De Loof (1981) observed a similar depression of vitellogenin concentrations in 6-OHDA treated *S. bullata*. They pointed out that if 250ng 20-OH were simultaneous injected with 6-OHDA, the depressing effects 6-OHDA could be partially overcome.

Given that the hemolymph of newly emerged female houseflies have "vitellogenic" 20-OH concentrations (Adams, Kelly and Woods, 1988), but no immunologically detectable vitellogenin (Adams and Filipi, 1983), it is reasonable to conclude that vitellogenic hemolymph 20-OH titres alone are not sufficient to support vitellogenin synthesis. Juvenile hormone and 20-OH both appear to be

Table 2. Hemolymph ecdysteroid concentrations of female houseflies injected with 25μg 6-hydroxydopamine (6-OHDA). Ecdysteroid titres were estimated at 48 and 72 h post emergence.

	Hemolymph Ecdysteroid, pg/μL	
	48 hours	72 hours
Saline injected	29.6 +/- 2.83 (11)	25.1 +/- 3.00 (12)
6-OHDA injected	25.2 +/- 1.24 (28)	30.4 +/- 2.91 (18)

necessary prerequisites for normal vitellogenin synthesis in the housefly (Adams and Filipi, 1988). It is possible that the 6-OHDA effects are mediated by an inhibition of the activity of the corpus allatum resulting in a suppression of juvenile hormone titres. Experiments are currently underway which are designed to test this hypothesis.

Table 3. Hemolymph vitellogenin concentrations of female houseflies were injected with 25μg 6-hydroxydopamine (6-OHDA). Hemolymph vitellogenin titres were estimated at 48 and 72 h post emergence. Treatment means followed by an asterisk are significantly different from control means ($p < 0.05$).

	Hemolymph Vitellogenin, μg/μL	
	48 hours	72 hours
Saline injected	9.9 +/- 1.66 (7)	4.9 +/- 0.86 (7)
6-OHDA injected	0.2 +/- 0.15 (7)*	1.7 +/- 0.88 (7)*

REFERENCES

Adams T. S. (1974) The role of juvenile hormone in housefly ovarian follicle morphogenesis. *J. Insect Physiol.* **20**, 263-276.

Adams T. S., Hagedorn, H. H. and Wheelock, G. D. (1985) Hemolymph ecdysteroid in the housefly, *Musca domestica*, during oogenesis and its relationship with vitellogenin levels. *J. Insect Physiol.* **31**, 91-97.

Adams T. S. and Filipi P. A. (1983) Vitellin and vitellogenin concentrations during oogenesis in the first gonotrophic cycle of the housefly, *Musca domestica*. *J. Insect Physiol.* **29**, 723-733.

Adams T. S. and Filipi P. A. (1988) Interaction between juvenile hormone, 20-hydroxyecdysone, the corpus cardiacum-allatum complex, and the ovaries in regulating vitellogenin levels in the housefly, *Musca domestica*. *J. Insect Physiol.* **34**, 11-19.

Adams T. S., Kelly, T. J. and Woods, C. W. (1988) Relationship between the corpus cardiacum-allatum complex and ovaries with the hemolymph ecdysteroid profile in the housefly, *Musca domestica*. *J. Insect Physiol.* **34**, 1105-1109.

De Loof A., Huybrechts R., Briers T. and Stoppie P. (1981) The effect of 6-OH dopaminehyrochloride and ecdysterone on vitellogenesis in *Sarcophaga bullata* and *Leptinotarsa decemlineate* (Insecta). In *Advances in Invertebrate Reproduction* (Ed. by Clark W. H., Jr. and Adams T. S.), p. 386. Elsevier/North-Holland, New York.

Huybrechts R. and De Loof A. (1981) Effect of ecdysterone on vitellogenin concentration in haemolymph of male and female *Sarcophaga bullata*. *Inter. J. Invert. Reprod.* **3**, 157-168.

From *Insect Neurochemistry and Neurophysiology • 1989* Edited by A. B. Borkovec and E. P. Masler

THE HYPOLIPAEMIC HORMONE OF THE LOCUST

Barry G. Loughton

Dept. of Biology, York University

North York, Ontario, Canada M3J 1P3

Locust hypolipaemic hormone is a small peptide released from the storage lobe of the corpus cardiacum (CC) upon stimulation of nervous corporis cardiacum II, (NCC II). The source of the hormone is the lateral cells of the protocerebrum (Orchard and Loughton 1980, Loughton 1987). Hypolipaemic hormone can be detected in the haemolymph after haemolymph lipid levels are raised by excitation, flight, or injection of adipokinetic hormone (Loughton 1987).

The action of hypolipaemic hormone in causing the uptake of lipid by the fatbody suggested a similarity to vertebrate insulin. Experiments showed that porcine, bovine or human insulin mimicked the action of hypolipaemic hormone. This activity was restricted to the insulin ß chain. Digestion of insulin ß chain with Staphylococcus aureus protease V8 indicated that the activity resided at the N-terminal end of the molecule. This deduction was confirmed by the lack of activity displayed by synthetic carboxyterminal fragments (Amino acids 21-26 and Amino acids 22-30). The similarity between locust hypolipaemic hormone and vertebrate insulin was further exemplified by the ability of labelled insulin to bind to locust fatbody membranes in a specific manner.

In order to pursue the apparent similarities between hypolipaemic hormone and vertebrate insulin its

TABLE 1

Locust haemolymph lipid and cyclic AMP in locust fatbody 10 minutes after injection of test substance.

Treatment	Haemolymph lipid/ μg/μl	cAMP pmol/mg protein
Saline	6.5	50
AKH (2μg)	21	40
AKH (3μg)	21	60
Insulin (1μg)	2	190
Forskolin 100 μM	2.7	120
Hypolipaemic hormone (1/10 brain equiv.)	3	140

mode of action was investigated. Partially purified hypolipaemic hormone injected in vivo raised fatbody cyclic AMP levels (Table 1). This was a surprising result since earlier workers (eg. Gade and Beenakkers 1977, Asher et al 1984) had reported AKH acted on fatbody to cause lipid mobilization via cAMP, though Goldsworthy et al 1986 showed only minor increases in fatbody cAMP after injection of low doses of AKH. Fat body membranes incubated in vitro with AKH in an ATP regenerating medium failed to increase cAMP over control levels. By contrast, forskolin, AKH II, insulin, insulin ß chain and hypolipaemic hormone all increased cyclic AMP production by crude fatbody membrane fractions after incubation (Table 2). It would seem that hypolipaemic hormone in locusts, which has some similarities to vertebrate insulin acts at the fatbody membrane to activate adenylate cyclase. This is in direct contrast to the circumstance in vertebrates, where insulin usually depresses intracellular cAMP levels. A second insect molecule with similarities to vertebrate insulin is PTTH (Nagasawa et al 1984). Smith et al (1985 have shown that PTTH acts on Manduca prothoracic gland membranes to cause an influx of Ca^{++} and subsequently to activate calmodulin sensitive adenylate cyclase.

We set out to localise insulin-like molecules in

TABLE 2

cAMP produced by locust fatbody membrane preparations in the presence of test reagents.

Treatment	cAMP pmol/mg protein
Control	7.0 ± 6.29
AKHI (3μg)	2.0 ± 1.6
Insulin 2 μg	25 ± 15
Insulin ß chain 2μg	46 ± 26
Forskolin (100 μM)	166 ± 60
Octopamine	216 ± 100
Hypolipaemic hormone	262 ± 130

Fatbody membranes were incubated in an ATP regenerating medium in the presence of GTP and IBMX (50 mM) for 30 minutes. cAMP was estimated by RIA.

the locust brain with polyclonal anti-bovine insulin serum. The antiserum stained median and lateral neuro-secretory cells in the brain. Other neurosecretory cells were observed in the tritocerebrum and suboesophageal ganglion. Preabsorption of the antiserum with bovine insulin eliminated staining in the median and lateral cells and in the suboesophageal ganglion. Separate pre-absorption of the antiserum with either A and B chains of insulin failed to alter the observable staining of either median or lateral neurosecretory cells. Only the lateral cells contain hypolipaemic (insulin-like) activity and on physio-logical grounds appear to be the source of hypolipaemic hormone. The insulin-like immunoreactivity in the median cells and in the suboesophageal ganglion is apparently not due to the presence of hypolipaemic hormone. The failure of separate pre-absorption of the antiserum with A or ß chain to eliminate staining of insulin positive cells suggests the presence of an insulin analog in all the stained cells. At this time we propose the existence of an insulin-like precursor molecule which may give rise to hypolipaemic hormone, a PTTH-like molecule (in the median cells) and yet another molecule in the suboesophageal ganglion whose function is as yet unknown.

REFERENCES

Asher, C., Moshitzky, P., Ramachandran, J. and Applebaum, S.W. (1984). The effects of Synthetic locust adipokinetic hormone on dispersed fatbody cell preparations: cAMP induction, lipid mobilization and inhibition of protein synthesis. Gen. Comp. Endocr. **55**, 167-173.

Gade, G. and Beenakkers, A.M. Th. (1977). Adipokinetic hormone-induced lipid mobilization and cyclic AMP accumulation in the fatbody of Locusta migratoria during development. Gen. Comp. Endocr. **32**, 481-487.

Goldsworthy, G.J., Mallison, K. and Wheeler, C.H. (1986). The relative potencies of two known locust adipokinetic hormones. J. Insect Physiol. **32**, 95-101.

Loughton, B.G. (1987). Studies on locust hypolipaemic hormone. J. Insect Physiol. **33**, 569-573.

Nagasawa, H., Kataoka, H., Isogai, A., Tamura, S., Suzuki, A., Mizoguchi, A. and Fujiwara, Y.A. (1986). Amino-acid sequence of a prothoracotropic hormome of the silkworm, Bombyx mori. Proc. Natl. Acad. Sci. USA. **83**, 5840-5843.

Orchard, I. and Loughton, B.G. (1980). A hypolipaemic factor from the corpus cardiacum of locusts. Nature, Lond. **286**, 494-496.

Saltiel, A., Fox, J.A., Sherline, P. and Cuatrecasas, P. (1986). Insulin-stimulated hydrolysis of a novel glycolipid generates modulator of cAMP phosphodiesterase. Science, **233**, 967-972.

Smith, W.A., Gilbert, L.I. and Bollenbacher, W.E. (1985). Calcium-cAMP interactions in prothoracicotropic stimulation of ecdysone synthesis. Molec. Cell. Endocr. **39**, 71-78.

From *Insect Neurochemistry and Neurophysiology • 1989* Edited by A. B. Borkovec and E. P. Masler

DEMONSTRATION OF A HUMAN ATRIAL NATRIURETIC PEPTIDE-LIKE MATERIAL IN THE STABLE FLY, *STOMOXYS CALCITRANS*.

Andrew C. Chen

USDA, ARS, VTERL

Route 5, Box 810, College Station, TX 77840

Atrial natriuretic peptide (ANP) is a peptide first reported in humans (hANP) and later in other mammals (Flynn, *et al.*, 1987). More recently, similar peptides have been isolated from lower vertebrates (Miyata, *et al.*, 1988; Sakata, *et al.*, 1988). Although the precise mechanism of action of this peptide is not well understood, the gross effects when administered in these animals are vasodilation and increases in the rates of excretion of water and electrolytes, particularly sodium. The mammalian ANP's are extremely well conserved 28 amino acid peptides with only one amino acid substitution at the 12th position. Due to my interest in stable fly diuresis, I decided to examine the possible existence of similar peptides in this insect. In this communication I shall report evidence for the presence of hANP-immunoreactivity in the stable fly.

A commercial radioimmunoassay (RIA) kit (Peninsula Laboratories, Belmont, CA) using antiserum developed in the rabbit against the α-hANP was employed in this study. Only adult stable flies emerging within a 2 h period were used. Twenty-four hours after emergence flies were anaesthetized by chilling at 4°C. Enough flies were removed from the cage for pre-feeding samples, and the rest were fed citrated cow blood for 1 h. Five hours after the blood meal, flies were chilled, and all subsequent steps were carried out at 4°C. Body parts (see Table 1) from 5 to 20 flies were homogenized in the RIA buffer (19 mM NaH_2PO_4; 81 mM Na_2HPO_4; 50 mM NaCl; 0.1% BSA; 0.1% Triton X-100; 100 μl per fly equivalent) in a Potter-Elvehjem type micro-homogenizer, centrifuged at 8,000 xg, and the supernatant removed for assay. Phenylmethylsulfonylfluoride (PMSF, final concentration, 20 μM) was added to the RIA buffer at the onset of homogenization to prevent proteolytic degradation. Hemolymph (30 μl each) was collected from the dorsal sinus of ca. 30 flies by puncturing the cervical membrane, aspirating with a calibrated capillary tube and mixing with 570 μl of RIA buffer containing 20 μM PMSF. RIA was performed in duplicate tubes according to the vendor's instruction and the radioactivity was measured and concentration calculated with a gamma counter (Model 28019, Micromedic Systems, Atlanta, GA).

Table 1 shows the amount of hANP-immunoreactivity in the stable fly. There is a dramatic increase in hANP-immunoreactivity after a blood meal. The mean increase of hANP-immunoreactivity of 4 replicates is ca. 8-fold. The increase in different batches of flies varied from 4-fold to as much as 30-fold (not shown). There was no difference between male and female flies. When various body parts were examined for hANP-immunoreactivity after a blood meal, most of the activity was found in the gut. When the gut was further divided into midgut and hindgut, most of the activity was found in the midgut. The circulating level of the hANP-immunoreactivity in the stable fly is two orders of magnitude higher than hANP in human plasma which has been reported to be 23-67 pg/ml (Yamaji, *et al.*, 1985; Shenker, *et al.*, 1985).

Bovine ANP has an identical sequence to hANP (Ong, *et al.*, 1986), and therefore is expected to have a 100% cross immunoreactivity with hANP. At 5 h after a blood meal the stable fly midgut still contains much of the ingested blood. As a control, a sample of the bovine blood that each batch of flies was fed was collected and hANP-immunoreactivity measured. Immunoreactive hANP in the bovine whole blood or plasma was less than 1 pg/μl (Table 1). Assuming that a maximum of 15 μl of blood was ingested, less than 6 pg of the hANP-immunoreactive substance in the fly after the blood meal is attributable to the bovine blood. This accounts for only a small fraction of the hANP-immunoreactivity found in the stable fly after a blood meal.

Results of this study suggest that the stable fly has a substance(s) that immunologically resembles the hANP. Although the physiological significance of the hANP-immuno-

Table 1. α-hANP-immunoreactivity in the stable fly

	Immunoreactivity*	
Body part	Pre-feeding	Post-feeding
Whole body (M)	54.3 ± 20.4 (4)	470.2 ± 176.8 (4)
Whole body (F)	58.2 ± 23.9 (4)	456.5 ± 190.6 (4)
Head (M)	27.4 ± 7.1 (3)	20.5 ± 0.9 (4)
Head (F)	29.5 ± 8.1 (3)	22.9 ± 1.6 (4)
Thorax-abdomen (M)	47.5 ± 19.9 (3)	362.0 ± 38.7 (3)
Thorax-abdomen (F)	48.5 ± 20.5 (3)	474.7 ± 54.0 (3)
Hemolymph (M)		3.4 ± 1.2 (4)
Hemolymph (F)		3.5 ± 1.2 (4)
Thoracic ganglion (M)		3.1 ± 0.1 (2)
Thoracic ganglion (F)		2.6 ± 0.1 (2)
Heart (M)		7.5 ± 2.4 (2)
Heart (F)		9.8 ± 4.5 (2)
Gut (M)		648.3 ± 19.5 (2)
Gut (F)		711.7 ± 132.9 (2)
Malpighian tubule (M)		4.1 ± 0.1 (2)
Malpighian tubule (F)		7.1 ± 1.2 (2)
CC-CA (M)		1.4 (1)
CC-CA (F)		2.2 (1)
Salivary gland (F)		2.2 (1)
Midgut (F)		238.3 (1)
Hindgut (F)		17.0 (1)
Whole bovine blood	0.33 ± 0.05 (3)	
Bovine plasma	0.20 ± 0.06 (4)	

*Immunoreactivity is expressed in pg/fly equivalent except the hemolymph and bovine samples which are in pg/μl. Values given are means ± SE of replicates, the number of which is indicated in parentheses. M and F denote male and female, respectively. Only the whole body, head and thorax-abdomen were analyzed before the blood meal.

reactive substance in the stable fly is unknown it is tempting to speculate that this substance may play a role similar to that in mammals, namely, as a diuretic hormone, especially in view of the fact that hANP-immunoreactivity increases dramatically after a blood meal. Recent reports show that some mammalian hormone-immunoreactive peptides not only are found in insects but also have physiological functions (Proux, *et al.*, 1987). Research is underway to isolate the hANP-immunoreactive peptide(s) from the stable fly for physiological studies.

REFERENCES

Flynn T. G., Laing L. P., Vuolteenaho O. and Yip C. C. (1987) Atrial natriuretic factor: physiology and biochemistry. In *Neuropeptides and Their Peptidases* (Edited by Turner A. J.) pp. 134-154. Ellis Horwood, Chichester and VCH, Weinheim.

Miyata A., Minamino N., Kangawa K. and Matsuo, H. (1988) Identification of a 29-amino acid natriuretic peptide in chicken heart. *Biochem. Biophys. Res. Comm.* **155**, 1330-1337.

Ong H., McNicoll N., Lazure C., Seidah N., Chretien M., Cantin M. and De Lean A. (1986) Purification and sequence determination of bovine atrial natriuretic factor. *Life Sciences* **38**, 1309-1315.

Proux J. P., Miller C. A., Li J. P., Carney R. L., Girardie A., Delaage M. and Schooley D. A. (1987) Identification of an arginine vasopressin-like diuretic hormone from *Locusta migratoria*. *Biochem. Biophys. Res. Comm.* **149**, 180-186.

Sakata J., Kangawa K. and Matsuo H. (1988) Identification of new atrial natriuretic peptides in frog heart. *Biochem. Biophys. Res. Comm.* **155**, 1338-1345.

Shenker Y., Sider R. S., Ostafin E. A. and Grekin R. J. (1985) Plasma levels of immunoreactive atrial natriuretic factor in healthy subjects and in patients with edema. *J. clin. Invest.* **76**, 1684-1687.

Yamaji T., Ishibashi M., and Takaku F. (1985) Atrial natriuretic factor in human blood. *J. Clin. Invest.* **76**, 1705-1709.

From *Insect Neurochemistry and Neurophysiology • 1989* Edited by A. B. Borkovec and E. P. Masler

DIURESIS IN ADULT *HELIOTHIS ZEA* : NEURAL AND HUMORAL REGULATION.

David W. Bushman and Judd O. Nelson . Department of Entomology and Center for Agricultural Biotechnology, University of Maryland College Park, MD 20742.

Hormonal control of diuresis has been reported in approximately 20 insect species from 6 different orders (Maddrell, 1986). Studies of diuresis in Lepidoptera have been predominantly concerned with a post-eclosion diuretic response following emergence from the pupa to the adult (Nicolson, 1976; Dores et al., 1979; Nicolson, 1980; Bushman et al., 1986). Recently, a diuretic hormone has been identified in pharate adult Manduca sexta which stimulates post-eclosion diuresis in the butterfly Pieris rapae (Kataoka, et al., 1989). In these newly emerged insects, this diuretic response is responsible for a marked reduction in total body weight that may facilitate flight.

Diuresis following feeding has typically been observed following blood meals in hematophagous species, although a post-feeding diuretic response has been reported in the nectar-feeding adult Monarch butterfly Danaus plexippus (Dores et al., 1979). We report here a post-feeding diuretic response in the adult corn earworm moth Heliothis zea. Relatively large meals of blood or nectar represent potentially damaging challenges to the composition of the insect's haemolymph. In addition to weight reduction, post-feeding diuresis may also serve to maintain the ionic balance of the haemolymph within acceptable tolerances.

Diuresis in adult H. zea has been examined at the whole insect level using several *in vivo* bioassays. Diuresis was examined following adult eclosion from the pupa and later in adult life following feeding or intrahaemocoelic injection of saline. Regulation of these behaviorally distinct diuretic responses apparently involves neural and humoral mechanisms. Diuresis following eclosion or intrahaemocoelic injection of saline is stimulated by homogenates of the brain (BR) or the abdominal ganglia (ABG). Molecular weight estimates of the active soluble factor based on size exclusion HPLC indicate an approximate size of 3500-4000D. Post-feeding diuresis can be markedly reduced by an antidiuretic factor present in the corpora cardiaca/corpora allata complex (CC/CA). This factor appears to act in conjunction with the frontal ganglion in regulating the rate at which fluid leaves the crop and is subsequently excreted.

We have previously reported a post-eclosion diuresis in H. zea (Bushman, et al., 1986). Intact females typically lost approximately 24% of their body weight during the first 6 hours after eclosion. That this weight loss was due primarily to diuresis was confirmed by sealing the anus of newly emerged females with wax. In this way, only evaporative losses could occur. At the end of the 6 hour assay, these insects had lost only 6% of their body weight. Subsequent removal of the wax plug resulted in the immediate loss of a large volume of fluid, making the total weight loss indistinguishable from that in intact insects. Ligation immediately following emergence resulted in a significant reduction in normal weight loss. Injection of homogenates of BR or ABG into ligated insects restored weight loss to a level similar to that of intact insects.

Two day old starved adult females when presented with a 10% (w/v) sucrose solution would typically feed for 3-5 minutes during which time they ingested a meal (about 100ul) weighing nearly half of their initial body weight. These insects then lost approximately 50% of the weight of the ingested meal (WIM) during the first hour following feeding. Ligation or frontal ganglionectomy immediately following feeding each resulted in a significant and permanent reduction in normal weight loss. While no tissue injection could restore weight loss in ligated insects, injection of homogenates of CC/CA into unligated insects immediately after feeding did significantly reduce post-feeding weight loss to levels comparable to those seen in ligated or frontal ganglionectomized insects. However, following injection of CC/CA, the inhibition of weight loss was transient and disappeared after one hour (Fig. 1). These data are consistent with the presence of a soluble antidiuretic factor in the CC/CA that is slowly degraded or otherwise removed from the haemolymph. Dissection and weighing of the crop from either ligated, frontal ganglionectomized, or CC/CA injected insects confirmed the crop as the predominant site of fluid retention in each case. We suggest that a soluble antidiuretic factor from the CC/CA acts in conjunction with the frontal ganglion to control the rate of crop emptying and subsequent diuresis by regulating the volume of ingested fluid that is passed into the haemolymph from the crop/midgut.

In order to examine the role of putative diuretic factors without the complicating influences of crop emptying, an assay was developed to investigate diuresis following intrahaemocoelic injection of saline. In this

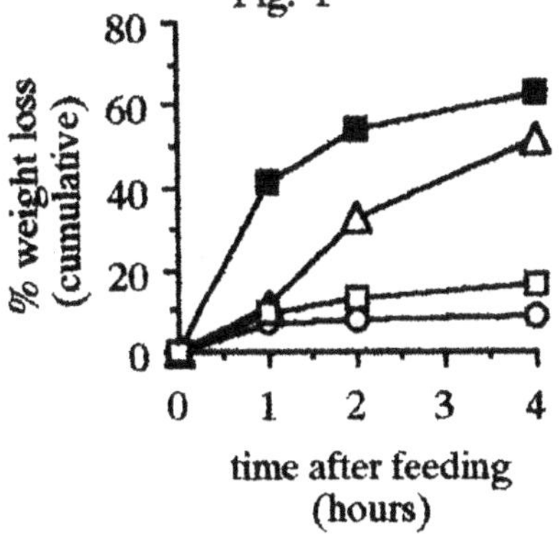

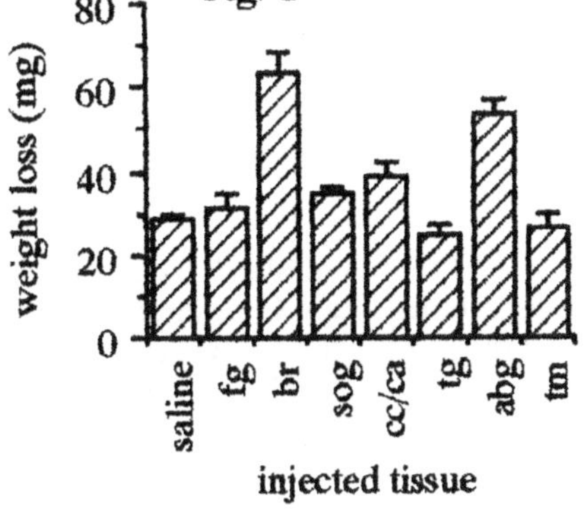

Figure 1. The effect of ligation (○), frontal ganglionectomy (□), or injection of CC/CA homogenate (△) on post-feeding weight loss. Weight loss for controls (■) and treatments expressed as a % of the weight of the ingested meal.

Figure 2. Post-injection diuresis monitored by fluid weight loss in ligated adult H. zea. Injected tissues include: frontal ganglion, fg; brain, br; suboesophageal ganglion, sog; corpora cardiaca/corpora allata complex, cc/ca; thoracic ganglia, tg; abdominal ganglia, abg; thoracic muscle, tm.

assay, the fluid to be excreted was provided by the injection, rather than by feeding. Intact two day old adults injected with 75ul (75 mg) of saline lost 55-60 mg of fluid in the 4 hours following injection. When two day old adults which had been ligated the previous day were injected with 75ul of saline, they lost only 27 mg of fluid during the same 4 hour assay. As was the case in the post-eclosion assay, injection of homogenates of BR or ABG into ligated insects resulted in significant increases in diuretic weight loss (Fig. 2). Indeed, the weight loss in these insects was comparable to that in intact insects in magnitude and exhibited a similar time course. Preliminary studies have indicated that homogenates of BR and ABG from M. sexta also stimulate diuresis in ligated H. zea recipients.

Adult H. zea exhibit periods of diuresis following emergence from the pupa and later in adult life following feeding. The post-eclosion and post-feeding diuretic responses appear to be regulated by both neural and humoral mechanisms. A diuretic factor present in the brain and abdominal ganglia stimulates diuresis following eclosion and intrahaemocoelic injection of saline. An antidiuretic factor present in the corpora cardiaca/corpora allata complex strongly inhibits weight loss following feeding. This antidiuretic factor may act in conjunction with the frontal ganglion in regulating the rate at which fluid leaves the crop and is subsequently excreted. Further studies aimed at the isolation and identification of both the diuretic and the antidiuretic factors are now underway.

ACKNOWLEDGEMENTS

This material is based upon work supported under a National Science Foundation Graduate Fellowship to DWB. We acknowledge the assistance of Ashok K. Raina of the Insect Chemical Ecology Laboratory, ARS, USDA. Scientific Article No. A-4981, Contribution No. 8026 of the Maryland Agricultural Experiment Station.

REFERENCES

Bushman, D. W., Raina, A. K., and Nelson, J. O. (1986). The biology and regulation of post-eclosion diuresis in adult Heliothis zea. In: Insect Neurochemistry and Neurophysiology, 1986. A. B. Borkovec and D. B. Gelman, eds., The Humana Press, New Jersey, 359-362.

Dores, R. M., Dallmann, S. H., and Herman, W. S. (1979). The regulation of post-eclosion and post-feeding diuresis in the monarch butterfly, Danaus plexippus. J. Insect Physiol. 25, 895-901.

Kataoka, H., Troetschler, R. G., Li, J. P., Kramer, S. J., Carney, R. L., and Schooley, D. A. (1989). Isolation and identification of a diuretic hormone from the tobacco hornworm, Manduca sexta. Proc. Natl. Acad. Sci. 86, 2976-2980.

Maddrell, S. H. P (1986). Hormonal control of diuresis in insects. In: Insect Neurochemistry and Neurophysiology, 1986. A. B. Borkovec and D. B. Gelman, eds., The Humana Press, New Jersey, 79-90.

Nicolson, S. W. (1976). The hormonal control of diuresis in the cabbage white butterfly Pieris brassicae. J. Exp. Biol. 65, 565-575.

Nicolson, S. W. (1980). Diuresis and its hormonal control in butterflies. J. Insect Physiol. 26, 841-846.

From *Insect Neurochemistry and Neurophysiology • 1989* Edited by A. B. Borkovec and E. P. Masler

CHARACTERIZATION OF FMRF-AMIDE-LIKE PEPTIDES IN THE LOCUST, SCHISTOCERCA GREGARIA

Peter D. Evans[+], Sandra Robb[+] and Isabelle Cournil[*]

[+]AFRC Unit of Insect Neurophysiology and Pharmacology, Dept. of Zoology, University of Cambridge, Downing Street, Cambridge, CB2 3EJ, U.K.
[*]Laboratoire de Neurobiologie et Physiologie Comparées, Université de Bordeaux 1, C.N.R.S., Arcachon, France.

FMRFamide was the first peptide isolated and characterized in a growing interphyletic family of neuropeptides (see Greenberg et al., 1988). In the locust, Schistocerca gregaria, the distribution of FMRFamide-like immunoreactivity has been described using immunocytochemistry and radioimmunoassay, and suggests both neurohormonal and neuromodulatory roles for peptides from this family (see Evans and Myers, 1986; Robb and Evans, submitted).

HPLC analysis of extracts (Fig 1) reveals that a different spectrum of FMRFamide-like peptides are expressed in different parts of the locust nervous system. Material co-eluting with the peaks designated C,D and E potentiate neuromuscular transmission in the locust extensor tibiae muscle and are also bioactive on the locust heart producing concentration dependent biphasic effects.

In thoracic nerve cord extracts from older sexually mature locusts, the major immunoreactive peak elutes in the region of peak E shown in Fig 1. Material from this peak in mature locusts has been isolated and subjected to a second phase separation on HPLC using a 10 - 40% acetonitrile gradient in the presence of 10mM triethylamine titrated to pH 6.5 with trifluoroacetic acid. Under these conditions it ran as a single immunoreactive peak, 10 pmoles of which were sequenced using a pulsed liquid phase sequencer and the sequence PDVDHVFLRF-NH_2 was deduced (Robb et al., 1989). We named this peptide SchistoFLRFamide and it has a strong structural similarity with leucomyosuppressin,

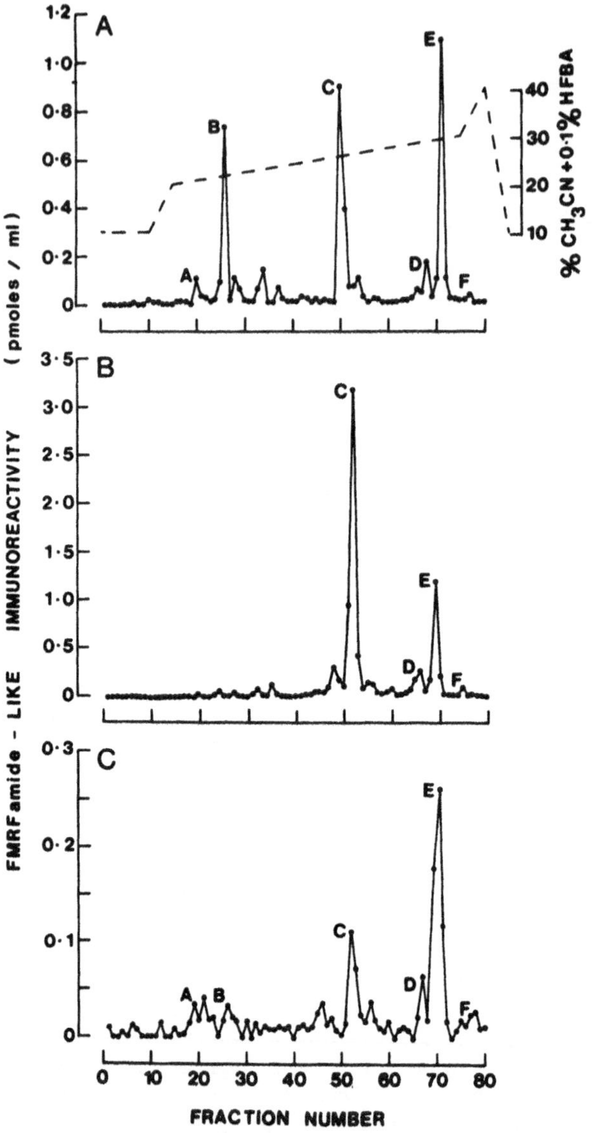

Fig 1 A comparison of the FMRFamide-like immunoreactive profiles of extracts from different parts of the locust nervous system (animals 2-7 days old) subjected to HPLC analysis on a 10-40% acetonitrile gradient plus 0.1% hydrofluorobutyric acid gradient indicates a differential expression of immunoreactive peaks. A, brain extract; B, thoracic nervous system extract; and C, abdominal nerve cord extract.

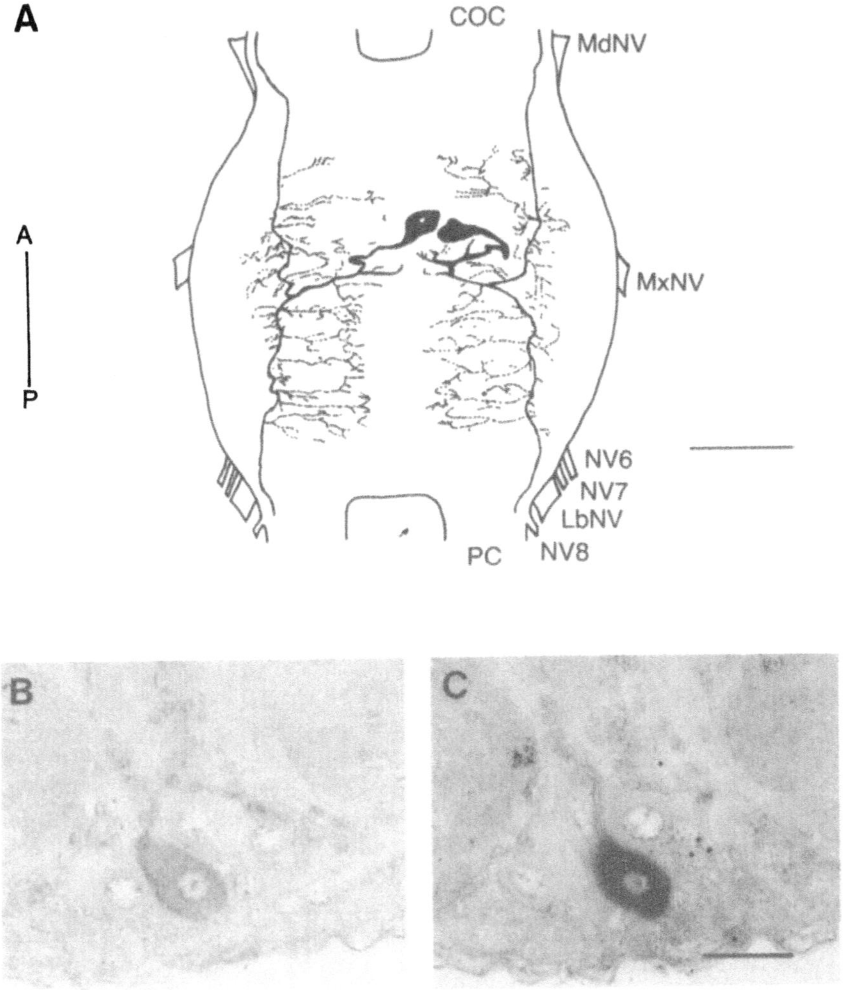

Fig 2 A, Innervation of suboesophageal ganglion by the pair of ventral median neurones showing FLRFamide-like immunoreactivity reconstructed from 10µm serial sections. Scale bar: 200µm. B and C consecutive transverse 10 µm sections of ventral portion of suboesophageal ganglion just anterior to level where maxillary nerve (MxNV) exits the ganglion. B was stained with the anti-FLRF antibody at a dilution of 1:500 and C with a vasopressin antibody at a dilution of 1:200. One of the pair of neurones drawn in A is stained with both antibodies. Scale bar 50µm.

(pQDVDHVFLRF-NH_2) isolated from the cockroach, Leucophaea maderae (Holman et al., 1986). SchistoFLRFamide potentiates neuromuscular transmission in the locust extensor tibiae muscle and has biphasic concentration dependent actions on locust heart.

An antibody raised against the amino acid sequence, FLRF was found to stain only two cells in the whole locust nervous system, with cell bodies occurring ventro-medially in the sub-oesophageal ganglion (Evans and Cournil, submitted). These cells appear similar to the vasopressin-immunoreactive cells in the suboesophageal ganglion of Locusta migratoria (Rémy and Girardie, 1980). Staining of alternate serial sections of Schistocerca suboesophageal ganglia (Fig 2B and C) with our anti-FLRF antiserum and with an anti-serum raised against vasopressin shows that both antibodies stain the same pair of cells. However, there are marked differences in the detailed anatomy of the pair of cells in these two species of locust. Thus a set of two homologous neurones in the suboesophageal ganglia of two species of locust may contain both a vasopressin-like molecule and a peptide related to the FMRFamide family, but the neuropeptides released from these cells could have different functional roles in the two species.

REFERENCES

Evans,P.D. and Myers,C.M. (1986) Peptidergic and aminergic modulation of insect skeletal muscle. J.exp.Biol., 124, 143-176.

Greenberg,M.J., Payza,K., Nachman,R.J., Holman,G.M.and Price,D.A. (1988) Relationships between the FMRFamide-related peptides and other peptide families. Peptides, 9, Suppl. 1, 125-135.

Holman,G.M., Cook,B.J. and Nachman,R.J. (1986) Isolation, primary structure and synthesis of leucomyosuppressin, an insect neuropeptide that inhibits spontaneous contractions of the cockroach hindgut. Comp.Biochem.Physiol., 85C, 329-333.

Rémy,C. and Girardie,J. (1980) Anatomical organization of two vasopressin-neurophysin-like neurosecretory cells throughout the central nervous system of the migratory locust. Gen.Comp.Endocrinol., 40, 27-35.

Robb,S., Packman,L.C. and Evans,P.D. (1989) Isolation, primary structure and bioactivity of SchistoFLRFamide, a FMRFamide-like neuropeptide from the locust, Schistocerca gregaria. Biochem.Biophys.Res.Comm., 160, 850-856.

From *Insect Neurochemistry and Neurophysiology • 1989* Edited by A. B. Borkovec and E. P. Masler

FMRF-AMIDE IN THE CORN EARWORM (*H. ZEA*): IMMUNOREACTIVITY IN THE MIDGUT AND CEREBRAL NERVOUS SYSTEM

Joe W. Crim[1], Alan C. Jenkins[1], Mark R. Brown[2], Gary A. Herzog[2], and Arden O. Lea[2]

Departments of Zoology[1] and Entomology[2], University of Georgia, Athens, GA

The midgut is the functional interface between insects and their food. Although the digestive and excretory functions of the midgut likely are highly regulated, surprisingly little is known about the chemical mediators - neural, endocrine, or paracrine - which coordinate these vital processes.

The midgut has been recently recognized (Brown *et al.*, 1986) as perhaps the largest endocrine organ in the insect. The midgut probably secretes a variety of peptide hormones that may themselves be regulators of other endocrine organs, a variety of physiological functions, and even behavior, in addition to controlling digestive processes.

We have focused our studies on peptides related to Phe-Met-Arg-Phe-amide (FMRFa), first described by Price and Greenberg (1977) as a molluscan cardioexcitory peptide. A group of related peptides from vertebrates and invertebrates share with FMRFa the common feature of the carboxyl Arg-Phe-amide sequence, and thus constitute a recognized family of peptides (Greenberg *et al.*, 1988).

Because of its world-wide economic importance as a foliage feeding pest, we have chosen the corn earworm (*Heliothis zea*) as our model. First, we used anti-FMRFa sera in immunocytochemistry and radioimmunoassay (RIA) to examine midgut and cerebral nervous tissues of fifth instar larvae (Jenkins *et al.* 1989a,b). Second, larvae were subjected to short term starvation, and FMRFa immunoreactivities in tissues and hemolymph then were determined (Jenkins et al. 1989a,b). Third, tests of bioactivity of FMRFa peptides on midgut/hindgut contractions in vitro have been initiated (Crim et al., unpublished).

In the midgut of larval H. zea endocrine cells visualized with antiserum were situated adjacent to the basal lamina and had apical extensions (Jenkins et al., 1989a). Thus, these cells are well-positioned to monitor the luminal contents of the midgut and to transmit hormonal signals via the hemolymph. This characteristic morphology, resembling that of endocrine cells of the vertebrate gut, is the same as that described for endocrine cells in the midgut of other insects (cf. Brown et al., 1986). We detected FMRFa immunoreactivity in many endocrine cells scattered throughout the midgut of H. zea (Jenkins et al., 1989a). Midgut endocrine cells exhibiting immunoreactive FMRFa also have been described for another larval lepidopteran, Manduca sexta (Copenhaver and Taghert, 1989), as well as an adult dipteran, Aedes aegypti (Brown et al., 1986). Thus, FMRFa-related peptides may be widely occurring hormones of the insect midgut.

We also localized axons immunoreactive for FMRFa innervating the musculature of the posterior regions of the midgut of H. zea (Jenkins et al. 1989a). This finding suggested that FMRFa-related material may affect the motility of the posterior midgut (see below). Immunoreactive FMRFa in the developing enteric nervous system of M. sexta has been detailed by Copenhaver and Taghert (1989). Embryonic expression of immunoreactivity proceeds stereotypically, resulting in a broad occurrence in the midgut.

Immunoreactive FMRFa is also widely distributed in the cerebral nervous system of H. zea (Jenkins et al., 1989b). Perikarya and axons immunostained for FMRFa occurred in the brain, subesophageal ganglion, and first thoracic ganglion. Immunoreactive axons also were located in the frontal ganglion and the corpora cardiaca. A variety of other insects also exhibit immunoreactive FMRFa in the brain and other neural structures (reviewed in Jenkins et al., 1989b), including neurosecretory regions from which neuropeptides may be released into the hemolymph.

FMRFa immunoreactivity in H. zea larvae subjected to short term starvation was evaluated by immunocytochemistry and RIA. FMRFa immunoreactivity exhibited changes in mid-

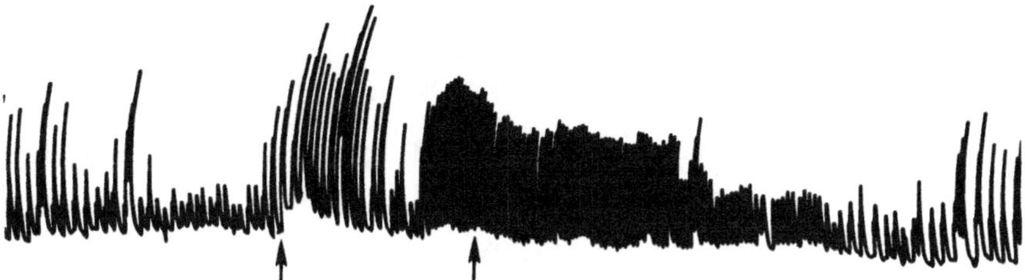

Figure 1. A preparation of isolated midgut/hindgut from a fifth instar Heliothis zea larva exhibits increased contractile activity in response to FMRFa perfusion. The arrows indicate the times at which the FMRFa or wash perfusates entered the bath.

gut and in hemolymph (Jenkins et al. 1989a), but not in brain tissues (Jenkins et al. 1989b). Starved larvae were isolated from food 12 hours prior to ecdysis and were dissected 48 hours after food removal. Fed larvae were given diet ad lib. Midgut immunoreactivity was approximately doubled for fed vs. starved larvae, but brain tissues were not different. RIA of hemolymph showed coincident changes: ca. 60 vs. 250 fmol immunoreactive FMRFa/µl, fed vs. starved larvae. Thus, midgut is a likely source of hemolymph immunoreactivity, which may be released coincident with starvation. This pattern suggests an endocrine response of the midgut

Localization of axons immunoreactive for FMRFa in the musculature of the posterior midgut of H. zea identified a possible target of peptide regulation of digestion in vivo. An assay of gallbladder muscle contraction (Rajjo et al., 1988) has been adapted for continuous perfusion of the isolated midgut/hindgut with Heliothis saline (Bindokas and Adams, 1988). Such preparations exhibit a pattern of spontaneous contractility, which is stimulated by FMRFa (Fig. 1; Crim et al. unpublished).

Interestingly, the myoactive peptides leucomyosuppressin and leucosulfakinins of cockroaches are also members of the RF-amide family (Greenberg et al., 1988). Study of the distribution and activities of FMRFa-related peptides thus may provide insights into regulation by peptides of digestive processes in H. zea. In turn, bioassays of processes such as motility may allow determination of physiologically active endogenous peptides of H. zea.

Acknowledgements. Supported by U.S.D.A. Grant No. 88-37252-3972 to J.W.C., G.A.H. and A.O.L.

REFERENCES

Bindokas V.P. and Adams M.E. (1988) Hemolymph composition of the tobacco budworm, Heliothis virescens F. (Lepidoptera: Noctuidae). Comp. Biochem. Physiol. **90A**, 151-155.

Brown M.R., Crim J.W. and Lea A.O. (1986) FMRFamide- and pancreatic polypeptide-like immunoreactivity in midgut endocrine cells of a mosquito. Tissue & Cell **18**, 419-428.

Copenhaver P.F. and Taghert P.H. (1989) Development of the enteric nervous system in the moth. I. Diversity of cell types and the embryonic expression of FMRFamide-related neuropeptides. Devel. Biol. **131**, 70-84.

Greenburg M.J., Payza K., Nachman R.J., Holman G.M. and Price D.A. (1988) Relationships between the FMRFamide-related peptides and other peptide families. Peptides **9**, 125-135.

Jenkins A.C., Brown M.R. and Crim J.W. (1989a) FMRF-amide immunoreactivity and the midgut of the corn earworm (Heliothis zea). J. Exp. Zool. In press.

Jenkins A.C., Brown M.R. and Crim J.W. (1989b) FMRF-amide immunoreactivity in a moth larva (Heliothis zea): the cerebral nervous system. Tissue & Cell. In press.

Price D.A. and Greenburg M.J. (1977) Structure of a molluscan cardioexcitatory neuropeptide. Science **197**, 670-671.

Rajjo I.M., Vigna S.R. and Crim, J.W. (1988) Actions of cholecystokinin-related peptides on the gallbladder of bony fishes in vitro. Comp. Biochem. Physiol. **90C**, 267-273.

From *Insect Neurochemistry and Neurophysiology • 1989* Edited by A. B. Borkovec and E. P. Masler

UPTAKE AND BINDING OF PONASTERONE A IN THE BRAIN OF FIFTH STADIUM *MANDUCA SEXTA* LARVAE

Hans-Jürgen Bidmon, Walter E. Stumpf and
Noelle A. Granger
Department of Cell Biology and Anatomy
University of North Carolina at Chapel Hill
Chapel Hill, North Carolina 27599

The neuroendocrine-endocrine systems that regulate insect post-embryonic development and reproduction are controlled by complex interendocrine regulatory mechanisms. For the juvenile hormone (JH) and ecdysteroid endocrine systems of the tobacco hornworm, *Manduca sexta*, these mechanisms function at various levels to modulate the synthesis/secretion of the JHs and ecdysteroids (Bollenbacher, 1988). One example is the brain-mediated control by 20-hydroxyecdysone (20-HE) of JH synthesis during larval-pupal development: the pupal commitment peak in the hemolymph ecdysteroid titer stimulates the post-commitment increase in JH acid synthesis by the CA (Watson et al., 1986), while the large pre-molt peak in the titer a few days later apparently inhibits synthesis (Granger et al., 1987).

Little is known about the cellular and molecular aspects of the processes by which these interendocrine mechanisms are expressed. To probe the processes involved in 20-HE regulation of CA activity during larval-pupal development, a thaw mount autoradiographic approach with [^{3}H]-ponasterone A (PNA), a 20-HE agonist, has been used to identify stage-specific cellular and subcellular binding sites for 20-HE in brains of fifth stadium *Manduca* larvae. The results from this approach should ultimately provide insight into the neural pathway(s) by which 20-HE controls synthesis of JH/JH acid by the CA.

For this study, gate II larvae on day 3 plus 12 hr and day 4 plus 18 hr were used because these represent times just before and after the commitment peak in the hemolymph ecdysteroid titer. To confirm the developmental status of these larvae, hemolymph ecdysteroid titers were determined by radioimmunoassay. For the day 3 plus 12 hr and the day 4 plus 18 hr larvae, the titers were 25.6 ng ml^{-1} and 174.4 ng ml^{-1}, respectively, which are approximately the same as those reported previously (Watson et al., 1986; Granger et al., 1987). Brains from these larvae were preincubated in Grace's medium (27°C, 2 hr) to remove endogenous ecdysteroids, and then incubated (1 hr) in [^{3}H]-PNA (4 nM, sp. act. 178 Ci $mmol^{-1}$, gift of Drs. P. Maroy and L.I. Gilbert). A time course of specific binding revealed that saturation occurred within 40-50 min. The tissues were then rinsed (20 min) in Grace's medium to remove unbound [^{3}H]-PNA. The day 3 brain bound 0.2 fmol PNA (162 $\pm$ 6 dpm), and the day 4 brain bound 0.4 fmol (322 $\pm$ 53 dpm), suggesting that the number of PNA-specific binding sites doubled during the pupal commitment period. The specificity of [^{3}H]-PNA binding was demonstrated by its reduction to background levels in the presence of a hundred-fold excess of unlabelled PNA (gift of Drs. D. Horn and M. Galbraith) or 20-HE. The increase in binding sites during the commitment period was not growth-dependent, since the size and protein content of the brain were essentially unchanged during this time. A similar apparent increase in ecdysteroid binding sites during an increase in the ecdysteroid titer has been reported for Drosophila melanogaster (Deak et al., 1988).

The [^{3}H]-PNA-incubated brains were processed for thaw mount autoradiography as previously described (Bidmon and Koolman, 1989). Hormone localized to a cell or subcellular compartment was quantified by counting silver grains. For the day 3 brain, cells in the pars intercerebralis (Fig. 1A), pars lateralis, ventral tritocerebrum and larval optic center exhibited nuclear binding of PNA. Many small cells around the neuropil, probably interneurons, were also labelled. This suggests that ecdysteroids could influence the integration of motor neuron and neurosecretory cell function. Similarly, neurosecretory cells, motor neurons and small interneurons in the subesophageal ganglion showed nuclear binding of PNA. Non-neural tissues were also screened for PNA binding. Interestingly, prothoracic gland cells and salivary gland cells exhibited specific nuclear binding. In addition, weak nuclear labelling was observed in 55% of the muscle cells and 71% of the fat body cells examined. These results substantiate general observations of developmental effects of 20-HE on larval tissues.

A comparison of the number of neurons labelled with PNA in day 3 and day 4 brains revealed more cells labelled on day 4, an observation which was in agreement with the results of the uptake studies. Also, labelling was found only in the nuclei of cells in the day 4 brain (Fig. 1B), and the labelled small cells were localized to specific regions. Unlike neurons in the day 4 brain, in which only nuclei were labelled, some cells in the day 3 brain exhibited nuclear and cytoplasmic binding, and in a few cells only the cytoplasm was labelled. Specific cytoplasmic binding of PNA has been reported previously for imaginal disc and K_c cells of Drosophila (Beckers et al., 1980). The transition from cytoplasmic and nuclear binding in the day 3 brain to nuclear binding only in the day 4 brain may indicate that ecdysteroid receptors are being synthesized and translocated to the nucleus, as has been shown for the vertebrate glucocorticoid receptor (Van Eekelen et al., 1987).

The fact that binding was specific in the various cellular and subcellular sites in the day 3 and 4 Manduca larval brains, as judged from the results of competition experiments, supports the idea that 20-HE acts on the nervous system to drive development during the commitment period. The pupal commitment peak in the ecdysteroid titer may be involved in inducing the further development of the receptor system, limiting the location of binding sites to the nucleus and evoking binding sites in a new population of cells. The specific binding of PNA to cerebral neurons also supports the observation that the interendocrine effect of 20-HE on CA activity is brain-mediated. A lack of [^{3}H]-PNA binding in CA cells would substantiate the finding that 20-HE does not directly affect the CA (Watson et al., 1987); this is currently being investigated.

This work was supported by NIH grant NS 14816 (NAG) and by a grant from the Deutsche Forschungsgemeinschaft (HJB).

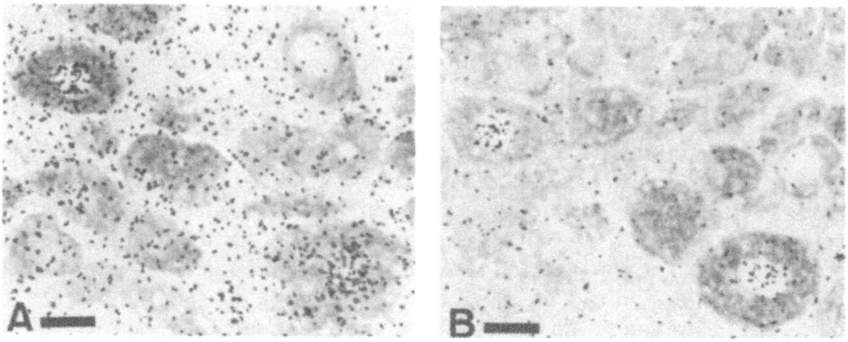

Figure 1. [^{3}H]-PNA binding sites in the mid pars intercerebralis of the fifth stadium Manduca larval brain. A, day 3 plus 12 hr; B, day 4 plus 18 hr. Bar = 10 µm.

REFERENCES

BECKERS Ch., MAROY P., O'CONNOR J.D., and EMMERICH H. (1980) Uptake and release of ^{3}H-ponasterone A by the K_c cell line of *Drosophila melanogaster*. In *Progress in Ecdysone Research* (Ed. by Hoffman J.A.), pp. 335-347, Elsevier, NY.

BIDMON H.-J. and KOOLMAN J. (1989) Ecdysteroid receptors located in the central nervous system of an insect. *Experientia* 45, 106-109.

BOLLENBACHER W.E. (1988) The interendocrine regulation of larval-pupal development in the tobacco hornworm, *Manduca sexta*: a model. *J. Insect Physiol.* 34, 941-948.

DEAK P., ZAVORSKY P., and MAROY P. (1988) Moulting hormone regulates its receptor level in *Drosophila melanogaster*. *Insect Biochem.* 18, 847-852.

GRANGER N.A., WHISENTON, L.R., JANZEN, W.P., and BOLLENBACHER W.E. (1987) Interendocrine control by 20-hydroxyecdysone of the corpora allata of *Manduca sexta* *Insect Biochem.* 17, 949-953.

VAN EEKELEN J.A.M., KISS J.Z., WESTPHAL H.M. and DEKLOET E.R. (1987) Immunocytochemical study on the intracellular localization of the type 2 glucocorticoid receptor in the rat brain. *Brain Res.* 436, 120-128.

WATSON R.D., WHISENTON L.R., BOLLENBACHER W.E. and GRANGER N.A. (1986) Interendocrine regulation of the corpora allata and prothoracic glands of *Manduca sexta*. *Insect Biochem.* 16, 149-155.

From *Insect Neurochemistry and Neurophysiology • 1989* Edited by A. B. Borkovec and E. P. Masler

THE ROLE OF ECDYSONE ON THE CHITINASE ENZYME IN THE DEVELOPMENT OF *SPODOPTERA LITURA*

T. Thangaraj[1,2], T. J. Kelly[2], K. Karaiyan[1], M. Aruchami[1] and A. B. Borkovec[2]

[1]Department of Zoology, Kongunadu Arts & Science College, Coimbatore, India. [2]Insect Reproduction Laboratory, USDA, ARS, Beltsville, MD 20705 USA

One of the early events in the molt of an insect is apolysis, during which detachment of the epidermal cells from the old cuticle occurs. The final event is the shedding of the old cuticle. Between these processes the degraded cuticular materials are absorbed and reutilized. The presence of molting enzymes has been recorded by Plotnikov (1904), and the presence of various lytic enzymes like protease, chitinase and esterases in the molting fluid was later confirmed (Wigglesworth, 1953; Hepburn, 1985). The depolymerization of chitin during different morphogenetic phases by chitinase is a vital step in the life cycle of insects since it allows for successful growth, metamorphosis and reproduction. These physiological events are manifested by the involvement of the hormone, ecdysone, under the master control of the neuropeptide, PTTH. Chitin depolymerization has been reported to occur by the action of two principal enzymes, exochitinase (β-N-acetylglucosaminidase) and endochitinase (chitinase). The endochitinase mediates the breakdown of the larger molecule into simpler units, and the exoenzyme hydrolyses the intermediate oligosaccharides to simpler basic units.

As *Spodoptera litura* undergoes pupation, larvae stop feeding, start wandering, and after a short while the animals burrow into the soil to pupate. Chitinase activity in the cuticle rises from a low level in non-feeding larvae and 0-h pre-pupa to a maximum at the end of the prepupal stage (Fig. 1). The action of molting enzymes is at a maximum around the

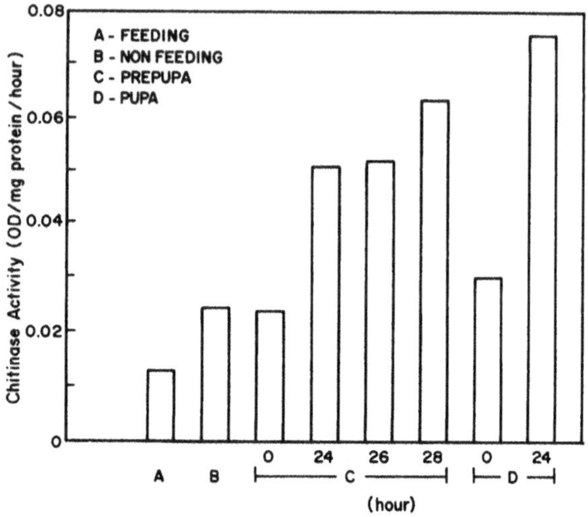

Fig. 1. Chitinase activity in the cuticle of different stages of Spodoptera litura.

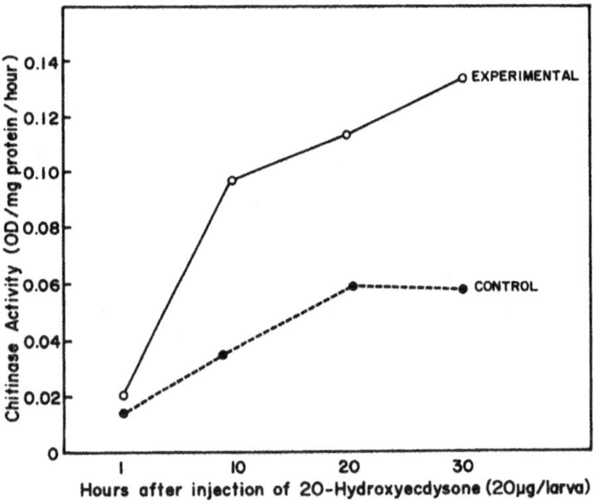

Fig. 2. Chitinase activity in the cuticle of Spodoptera litura after injection of 20-hydroxyecdysone.

time of apolysis in various insects, and this increase in activity has been correlated with the ecdysteroid titer (Kramer et al., 1985).

Administration of different concentrations of 20-hydroxyecdysone to ligated S. litura reveals that the molting hormone activates chitinase in the cuticle when the larvae are injected with 6, 12, 25 or 50 ug of the hormone per larvae. Larvae treated with 25 ug of ecdysone show maximum induction of activity when compared to other concentrations. Induction of chitinase activity is present from 10 h onwards, and maximum activity occurred at 30 h after administration (Fig. 2).

Kimura (1973) suggests that preformed chitinase is released into the endocuticle by ecdysone instead of being newly synthesised and activated in the endocuticle at the right time. Based on in vitro observations, it has been suggested that ecdysone does not act directly on the enzyme but plays a role in the mechanism which transfers the enzyme in the epidermal cells to the endocuticle; this is supported by electron microscopic studies in the silkworm during larval molting (Mori, 1965). It has been observed that the digestion of the endocuticle begins 12 h after the injection of ecdysone in Bombyx mori immediately after ecdysis (Kobayashi et al., 1968).

The ecdysone titer in the hemolymph of S. litura reveals a minor peak at the beginning of the wandering stage and a larger peak at the pupal stage. Though the present study provides evidence for the induction of chitinase, the significance of these two phases of release of ecdysone in relation to the initiation of larval-pupal and adult transformations needs further study. In this context, the report of Fukamizo and Kramer (1987) is of considerable interest. They have suggested a differential response of the individual chitinolytic enzymes to 20-hydroxyecdysone in isolated larval abdomens. Exochitinase activity was stimulated at low ecdysteroid titer, whereas chitinase activity was increased significantly at higher hormonal titer. This finding has been related to the smaller and higher peaks of molting hormone titer in the wandering and pupal periods, respectively. Moreover, mRNA for exochitinase was observed in epidermal tissue during and subsequent to the fifth larval wandering period, whereas mRNA for chitinase did not appear until the middle of the late pupal stage of M. sexta (Fukamizo and

Kramer, 1987) corresponding to the two periods of release of prothoracicotropic hormone for the synthesis of ecdysone (O'Brien et al., 1986).

REFERENCES

Fukamizo T. and Kramer K.J. (1987) Effect of 20-hydroxyecdysone on chitinase and β-N-acetylglucosaminidase during the larval-pupal transformation of Manduca sexta (L). Insect Biochem. 17, 547-550.

Hepburn H.R. (1985) Structure of the integument. In Comprehensive Insect Physiology, Biochemistry and Pharmacology, Vol. 3 (Ed. by Kerkut G.A. and Gilbert L.I.), pp. 1-58. Pergamon Press, Oxford.

Kimura S. (1973) The control of chitinase activity by ecdysterone in larvae of Bombyx mori. J. Insect Physiol. 19, 115-123.

Kobayashi M., Ishitoya Y. and Kimura S. (1968) Autoradiographic studies on chitin and nucleic acid synthesis of the larval integument of Bombyx silkworm under the administration of ecdysone. Ann. meet. Jap. soc. Appl. Ent. Zool. 30.

Kramer K.J., Dziadik-Turner C. and Koga D. (1985) Chitin metabolism in insects. In: Comprehensive Insect Physiology, Biochemistry and Pharmacology, Vol. 3 (Ed. by Kerkut, G.A., and Gilbert, L.I.), pp. 75-115. Pergamon Press, Oxford.

Mori T. (1965). The electron microscopical studies on the fine structure of the integuments in the silkworm Bombyx mori. Kontyu. 33, 385-426.

O'Brien M.A., Granger N.A., Aqui N. and Gilbert L.I. (1986). Prothoracicotropic hormone in the developing brain of the tobacco hornworm, Manduca sexta. Relative amounts of two molecular forms. J. Insect Physiol. 32, 719-725.

Ploynikov W. (1904). Uber die Hauntung and uber einige Elements der Haut bei den Insekten. Z. wiss. Zool. 76, 333-366.

Wigglesworth V.B. (1953) The physiology of the cuticle and of ecdysis in Rhodnius prolixus (Triatomidae, Hemiptera); with special reference to the function of oenocytes and of the dermal glands. Quart. J. Miscrosc. Sci. 76, 269-318.

From *Insect Neurochemistry and Neurophysiology • 1989* Edited by A. B. Borkovec and E. P. Masler

PROSPECTIVE CHEMICAL REGULATORS OF FEMALE REPRODUCTIVE MUSCLE FUNCTION IN THE STABLE FLY *STOMOXYS CALCITRANS*

B. J. Cook and R. M. Wagner

USDA, ARS, VTERL

Route 5, Box 810, College Station, TX 77841

Once Cameron (1953) established that pharmacologically active substances from neuroendocrine glands of insects could affect visceral muscle functions, interest quickly developed in two closely related areas of research: 1) the isolation and identification of prospective regulatory chemicals of the viscera and 2) the determination of the principal neural and hormonal mechanisms that control visceral organ functions. The first efforts of this research were focused on the organs of digestion, and only recently has attention turned to the muscles of the female reproductive system. This group of muscles offers some unique and challenging opportunities for the study of bioregulation because they are not only contract spontaneously but they are also subject to the effects of innervation, circulating hormones, and secretions introduced by the male during mating (Davey, 1985). In light of these facts and with the hope of eventually discovering the chemical messengers in the stable fly that regulate ovulation and oviposition, we began studies on the muscular networks of the ovary and oviduct of that insect.

Ovarian motility in 7-day old flies was recorded by focusing a video camera through a dissecting microscope on the dorsal surface of the exposed reproductive tract *in situ*. Timed sequences of the motility were taped on a video cassette for frame-by-frame analysis. Impedance myographs of ovarian movement were obtained by placing excised ovaries in a 100 μl saline bath and then positioning a pair of 36 gauge silver wires close to the surface of the active tissue. Any small movement was detected by an impedance converter and recorded graphically on a chart recorder. A miniature force transducer system was also developed to detect and quantitate contractile events of the oviduct. Once the oviduct was isolated from the abdomen of the stable fly it was attached to a load cell by a fine strand of hair. Although proctolin was found in whole-body extracts of the stable fly (Holman and Cook, 1979), its tissue distribution was not determined. Consequently, an attempt was made to detect the peptide in extracts of the ovaries and oviduct of 7-day old flies. The dissected tissues were placed in cold 75% methanol/25% 0.1 N HCl (v/v) and sonicated for 3X 15 sec, with intermittant periods of 3 min in ice. The extracts were then centrifuged through a 0.45 μm filter unit and the filtrates were evaporated to dryness and subjected to HPLC fractionation. Initially these tissue extracts were eluted through a sphereisorb C-18 column and a cockroach hindgut bioassay was used to detect proctolin-like activity in the fractions. The active fractions were then further purified on a Microsorb C-1 column. The solvents for both systems consisted of: A) 0.1% TFA by weight in water and B) 50% acetonitrile in 0.1% TFA. The samples were injected in solvent A and a linear gradient was run to 100% B in 60 min.

The ovarian muscles of the stable fly consist of two groups: 1) a network of stellate muscle cells that cover the surface of the ovary and 2) fibers that surround the ovarioles. The spontaneous contraction of the muscles of the ovarian sheath cause oscillations in the diameter of the ovary. Although these events were the most frequent type of activity observed, impedance myographs showed some differences in amplitude. Activation of the muscle fibers that encompass the ovarioles, by comparison, caused a vertical translation of eggs within the tubes. The muscles of the oviduct are composed of an intricate mixture of longitudinal and circular fibers. Myographic recordings of their action revealed both phasic and tonic contractions in the common and lateral oviducts. Three neurochemicals of common occurrence in insects were tested on the contractile properties of the ovaries and oviduct. The presence of both proctolin (Holman and Cook, 1979) and glutamate (Holman and Cook, 1982) have been reported in the stable fly, but octopamine has not. Nevertheless, it is

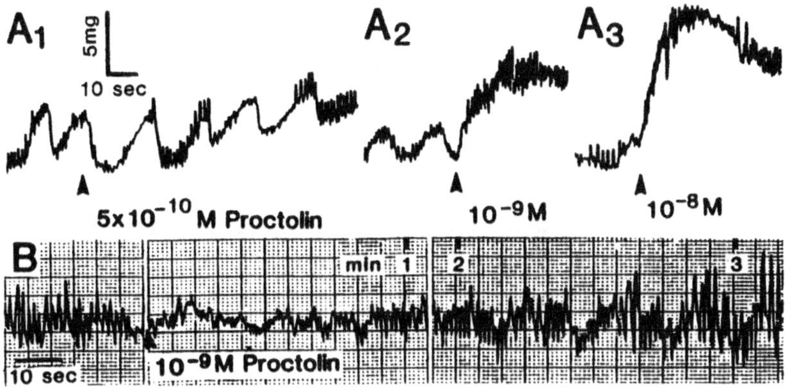

Fig.1. Effects of proctolin on the oviduct and ovarian sheath of the stable fly. (A_{1-3}) Myographic recordings of contractile force of the oviduct in response to increasing concentrations of proctolin. (B) Impedance myograph of ovarian sheath contractions in response to proctolin.

interesting to note that this biogenic amine has been found in the hemolymph of the tsetse fly (Pimley and Stafford, 1985) at a level of 10^{-7}M. Octopamine (10^{-8}M to 10^{-7}M) produced a large increase in the amplitude of ovarian sheath contractions in the stable fly, but the onset of the response often required 2-3 min to develop. The response of the oviduct, by comparison, showed an initial elevation in tonus, followed by a reduction in phasic contractions when 10^{-7}M octopamine was applied. Initially proctolin ($5X10^{-10}$M) caused an increase in the tonus of the oviduct, but little change in the spontaneous phasic events (Fig. $1A_1$). However, at a concentration of 10^{-9}M an evident increase in the frequency and amplitude of phasic contractions was observed (Fig. $1A_2$). At a concentration of 10^{-8}M proctolin the oviduct continued to show an elevation in both its tonic and phasic actions. The ovarian sheath also responded to proctolin, but the response was slower to develop (Fig. 1B). Oviduct and ovarian extracts purified chromatographically by the two HPLC systems mentioned had fractions that contained proctolin-like biological activity. Moreover, the retention time for these fractions corresponded to the proctolin standard in both systems. A quantitative evaluation of the fractions on the hindgut bioassay demonstrated that extracts of the oviduct contained between 1.2-1.7 picomoles of proctolin per organ while ovarian extracts contained only 0.5-0.7 picomoles of proctolin per ovary. Such data provides strong evidence that proctolin is involved in the regulation of reproductive muscle function in the female stable fly. When the oviduct was exposed to glutamate (10^{-4}M) it often caused a brief burst of large phasic contractions that lasted for 20 sec. However, contractions of the ovarian sheath were suppressed by the same concentration of glutamate. These results together with the fact that the thoracic ganglion of the stable fly contains more than twice the amount of glutamate as the brain (Holman and Cook, 1982) suggests that this amino acid is involved in regulating muscle function in the fly as it does in other insects.

REFERENCES

Cameron M. L. (1953) Some pharmacologically active substances found in insects. Ph.D. dissertation, University of Cambridge.

Davey K. G. (1985) The female reproductive tract. In Comprehensive Insect Physiology Biochemistry and Pharmacology Vol. 1 Embryogenesis and Reproduction Ed. G. A. Kerkut and L. I. Gilbert, pp 15-36, Pergamon Press.

Holman G. M. and Cook B. J. (1979) The analytical determination of proctolin by HPLC and its pharmacological action in the stable fly. Comp. Biochem. Physiol. 62C:231-235.

Holman G. M. and Cook B. J. (1982) Physiological amino acids of the nervous system of the stable fly Stomoxys calcitrans. Comp. Biochem. Physiol. 71A:23-27.

Pimley R. W. and Stafford K. A. (1985) Octopamine distribution in the tsetse fly Glossina morsitans. Comp. Biochem. Physiol. 81C(2):297-301.

From *Insect Neurochemistry and Neurophysiology • 1989* Edited by A. B. Borkovec and E. P. Masler

SEROTONERGIC MODULATION OF THE LOCUST MANDIBULAR CLOSER MUSCLE

Richard A. Baines and Roger G.H. Downer

Department of Biology, University of Waterloo, Waterloo, Ontario, Canada N2L 3G1

INTRODUCTION

Within the suboesophageal ganglion of the locust there is a group of 14 anterior-midline cell bodies that stain with antibody to 5-hydroxytryptamine (5HT, serotonin). Six of these cell bodies have previously been termed "the satellite neurones" because their axons run along side the ventrally-directed peripheral nerves of the ganglion forming putative neurohaemal endings in the sheaths (Bråunig, 1987). The other 8 neurones, termed "the link nerve neurones", have similar endings, however, these terminate along the more anterior of the peripheral nerves, in particular the upper reaches of nerve 1D, which contains the motor axons of the mandibular closer muscle. Labelling with anti-5HT also reveals axons among the mandibular closer muscle fibres. The present study demonstrates that 1) 5HT modulates neurally-evoked contractions of the mandibular closer muscle in a similar manner to that of octopamine in the extensor tibiae, 2) drugs that artificially elevate cAMP levels in the mandibular closer muscle mimic the effect of 5HT, and 3) inhibition of adenylate cyclase by adenosine alters the action of 5HT.

Figure 1a shows the effect of 5HT on the contractions produced by stimulation of suboesophageal nerve 1D at a frequency of 1Hz. The contraction amplitude (trace

i) is increased to a peak value of 23%, the contraction rate (trace ii) is increased by 18% while the most marked effect is a large, 50% increase in the relaxation rate (trace iii). The increase in relaxation rate is undoubtedly mediated by a post-synaptic mechanism. Incubation of isolated mandibular closer muscles in saline containing 0.5mM IBMX (a phosphodiesterase inhibitor) and subsequent estimation of cAMP levels by conventional radio-immune assay clearly shows that 5HT elevates the level of this cyclic nucleotide in the muscle (Fig. 1b). The maximal effect was seen with 10^{-6}M 5HT which produced an increase in cAMP levels of 14.1 pmoles/mg protein/min (a 55% increase over control, significant at $P<0.05$).

In this muscle both IBMX (5×10^{-4}M) and forskolin (2×10^{-6}M) elevated the levels of cAMP by 13 and 30% respectively. Addition of IBMX (data not shown) to the muscle preparation increased the contraction amplitude by 18% and the rates of contraction and relaxation by 15 and 38%, respectively. Forskolin (Fig. 2a) also mimicked the effect of 5HT on the muscle increasing contraction amplitude by 20%, contraction rate by 20% and the relaxation rate by 48%. Further confirmation that all three changes in the muscle contraction are mediated by cAMP is that they can be produced by the addition of a phosphodiesterase-resistant cAMP analogue 8-(4-chlorophenylthio) cAMP (Cpt-cAMP) which, at 10^{-3}M produced an increase in contraction amplitude of 19% and increases of 21 and 36% in the rates of contraction and relaxation, respectively (Fig. 2b).

The hypothesis that the 5HT receptors present in the muscle are coupled to an adenylate cyclase is supported by the fact that adenosine blocks the action of 5HT. Adenosine has, in some preparations, been shown to inhibit the adenylate cyclase complex (Morishima, 1980). The presence of adenosine (10^{-4}M) has two effects on the action of 5HT in the muscle preparation. Firstly, it blocks the effect of 5HT on the contraction amplitude and contraction rate, and secondly it substantially prolongs the effect of 5HT on the relaxation rate (Fig. 3). On isolated muscle, adenosine reduces the amount of cAMP produced by a 30 second exposure to 10^{-7}M 5HT by 2.6 pmoles/mg/min (significant $P<0.05$, $n=5$), but has no effect on the amount of cAMP produced by a 2 minute exposure to 5HT. This corroborates with physiological

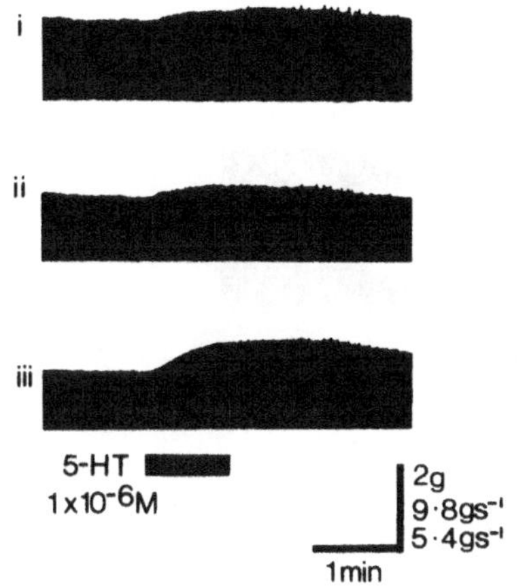

Fig. 1a. Effect of 5HT on the mandibular closer muscles. The contraction amplitude (i), contraction rate (ii) and relaxation rate (iii) are all increased.

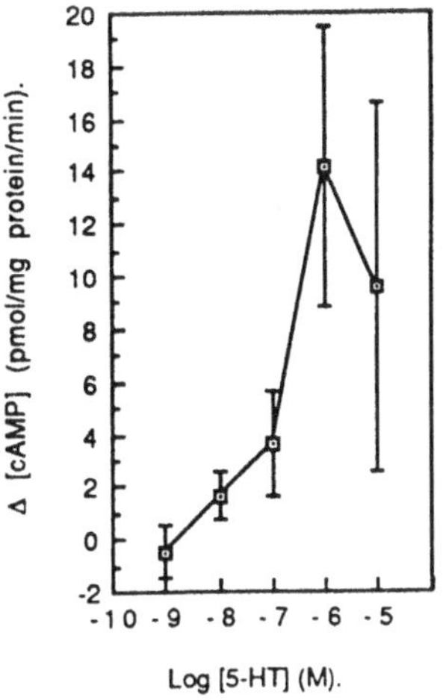

Fig. 1b. 5HT produces a dose-dependent increase in cAMP levels in isolated mandibular closer muscle. (n = 5 $\pm$ SD).

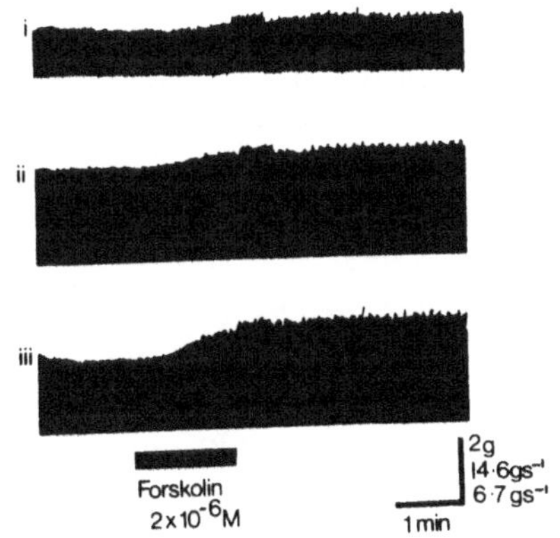

Fig. 2a. Forskolin mimics the effect of 5HT on the closer muscle.

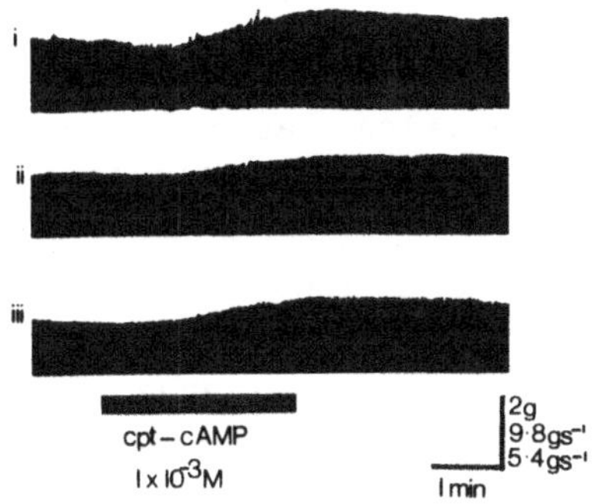

Fig. 2b. 8-(4-chlorophenylthio) cAMP also mimics the effect of 5HT on the closer muscle.

Fig. 3. Adenosine blocks the effect of 5HT on the contraction amplitude (i) and contraction rate (ii), but only delays the effect on the relaxation rate (iii), which is overcome after about two minutes. The effect of adenosine on all three contraction parameters is reversible on wash-out.

data which show that the effect of adenosine on relaxation rate is overcome by 5HT after about two minutes (see Fig. 3).

In summary, the results suggest that the action of 5HT on all three contraction parameters is mediated by cAMP and based on the differential effects seen with adenosine, each parameter may be controlled through a separate 5HT receptor-adenylate cyclase system.

REFERENCES

Bråunig, P. (1987) The satellite nervous system - an extensive neurohaemal network in the locust head. J. Comp. Physiol. A. 160, 69-77.

Morishima, I. (1980) Adenylate cyclase in silkworm. Inhibition by adenosine. Comp. Biochem. Physiol. B. 65, 399-402.

From *Insect Neurochemistry and Neurophysiology • 1989* Edited by A. B. Borkovec and E. P. Masler

OCTOPAMINERGIC MODULATION OF SENSE ORGANS INVOLVED IN LOCUST FLIGHT

J.-M. Ramirez, I. Orchard and K.G. Pearson
Department of Physiology, University of Alberta, Edmonton, Alberta T6G 2H7, and Department of Zoology, University of Toronto, Toronto, Ontario, M5S 1A1, Canada.

INTRODUCTION

The flight behaviour of the locust has been extensively studied and recent attention has been directed towards the possible modulation of flight by Octopamine. Released into the haemolymph, octopamine contributes to the regulation of haemolymph lipid (Orchard, 1987), and modulates neuromuscular transmission and muscle contraction of various flight muscles (Whim and Evans, 1988). Octopamine is also acting centrally and affects the initiation and maintenance of flight (Sombati and Hoyle, 1984). We examined another potential site of action, two proprioceptors involved in the generation of the flight motor pattern: the stretch-receptor (SR) and the tegula. Both sense organs are rhythmically activated during flight and they both play a crucial role in establishing the motor pattern for flight (Wolf and Pearson, 1988). Increase of activity in these sense organs causes an increase in the wingbeat frequency and removal of either one of these sense organs causes a drop in the wingbeat frequency. Thus an octopaminergic modulation of these sense organs might have important functional implications for the generation of the flight rhythm.

OCTOPAMINERGIC MODULATION OF STRETCH-RECEPTOR AND TEGULA

The responses of the SR and the tegula to sinusoidal wing movements were recorded extracellularly and the activity processed and stored on a LSI computer. The action of octopamine was determined in three different preparations. Small volumes (10μl) of octopamine were injected into almost intact animals to demonstrate that octopamine circulating in the haemolymph modulates these sense organs. A similar preparation was also used to study the modulatory effect of certain neurosecretory cells which were selectively stimulated with a second electrode. The direct action of octopamine on the sense organs was demonstrated in perfused isolated thorax preparations (animals without head, abdomen, gut and nervous system) and in perfused "in vitro" preparations (more or less isolated sense-organs). The degree of modulation of the sense organs was consistent for all preparations.

The SR was modulated in a biphasic manner by octopamine. The SR responses were initially decreased (for less than 1 min) and subsequently increased (for approx.15 minutes) (Fig1b). The decrease was not always observed and often activity increased immediately after the onset of octopamine perfusion (Fig.1a,c). The activity increase in the SR occurred in a dose-dependent manner with a threshold at $5x10^{-9}$M and a maximal effect at 10^{-4}M octopamine. The octopaminergic modulation was independent of the frequency at which the receptor was stimulated (examined at 1-18 Hz). After repetitive injection of octopamine into the haemolymph of intact locusts, the SR became often desensitized and was not modulated by further octopamine injection. Desensitization was not seen in perfused thorax preparations in which octopamine was washed out between consecutive octopamine perfusions. Several agonists and antagonists were examined. Serotonin was equally effective, followed by Synephrine, Metanephrine and Tyramine. Dopamine was ineffective. Phentolamine, but not DL-Propranolol antagonized the octopamine effect. Two observations suggest that dorsal

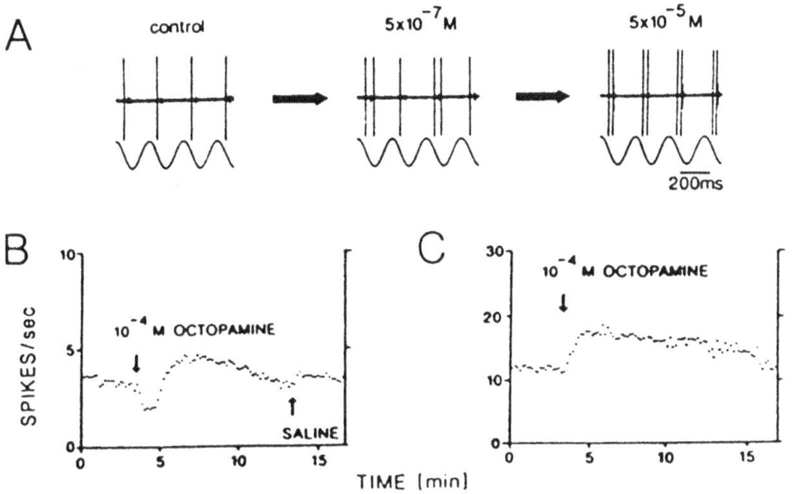

Fig.1. (A) Effect of DL-octopamine on the SR response (upper trace) to sinusoidal wing movements (lower trace). (B,C) 2 Sequential spike histograms demonstrating the effect of DL-octopamine on the SR response. (B) Biphasic response. (C) Increase without initial inhibition. Each point was an average of 10 consecutive cycles. A: obtained in an intact locust, B,C: obtained in isolated thorax preparations.

unpaired median (DUM) neurons contribute to an octopaminergic modulation of the SR during flight. (1) Selective stimulation of these cells modulated the SR response and this effect was blocked by Phentolamine. (2) All DUM cells that have been studied were activated during flight.

Octopamine increased also the activity of the tegula and similar results were obtained in both, intact animals and isolated thorax preparations. In the multi-celled tegula often additional sensory cells were recruited following octopamine injection (Fig.2).

These results suggest that the activity of the SR and the tegula are increased by octopamine following the onset of flight.

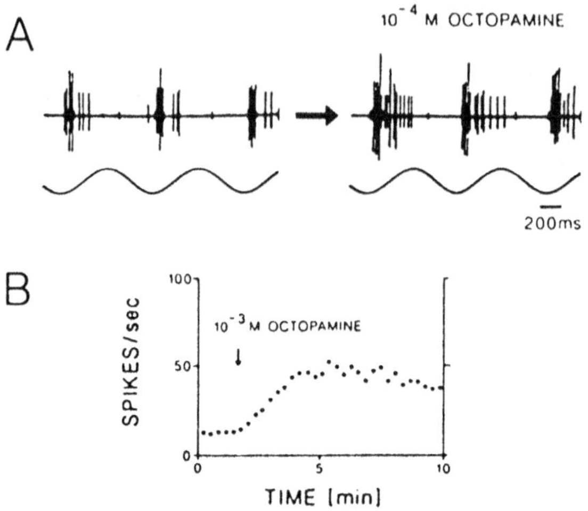

Fig.2. (A) Effect of DL-octopamine on tegula responses (upper traces) to sinusoidal wing movements (lower traces). (B) Sequential spike histogram demonstrating the effect of DL-octopamine on the tegula response. Each point is an average of 20 consecutive cycles. A: obtained in an "in vitro" preparation, B: obtained in an intact perfused locust.

Acknowledgements - This work was supported by grants from MRC Canada to JMR and KGP, AHFMR to JMR, and NSERC Canada to IO.

REFERENCES

Orchard I. (1987) Adipokinetic hormones - an update -. J. Insect Physiol. 33, 451-463.

Sombati S. and Hoyle G. (1984) Generation of specific behaviors in a locust by local release into neuropil of the natural neuromodulator octopamine. J. Neurobiol. 15, 481-506.

Whim M.D. and Evans P.D. (1988) Octopaminergic modulation of flight muscle in the locust. J. exp. Biol. 134, 247-266.

Wolf H. and Pearson K.G. (1988) Proprioceptive input patterns elevator activity in the locust flight system. J. Neurophysiol. 59, 1831-1853.

Neuroanatomy

From *Insect Neurochemistry and Neurophysiology • 1989* Edited by A. B. Borkovec and E. P. Masler

OVERVIEW: PROGRESS IN INSECT NEUROANATOMY AND NEUROGENESIS

S. M. Meola

USDA, ARS, VTERL

Route 5, Box 810, College Station, TX 77840

As in the past, the papers presented at ICINN-3 emphasized the interdependency between neuroanatomy and the disciplines of biochemistry and physiology. To these disciplines is added the contribution from the field of molecular biology with its synergy between biochemistry and genetics. Yet, with all the advances of biotechnology, progress in neuroscience still requires the descriptive analysis of the components of the nervous systems of the large number of species comprising the Class Insecta. In this regard, a new sensory receptor in the ventral body wall of Rhodnius prolixus was described (Chaing et al.). This receptor may monitor pressure resulting from distension of the stomach from a blood meal. Several histochemical stains were used to locate, describe, and quantify several types of neurosecretory cells in the brain of the short horned grasshopper as well as trace their pathway in the nervi corporis cardiaci (Patil). Knowledge of the anatomy of the neuroendocrine system of a species permits studies such as an in vivo comparison of the mode of action of different insecticides on this system. One such study compared the effect of three organophosphate insecticides on the corpora cardiaca of the female roach, Periplaneta americana (Nanda and Banerjee). The authors concluded that the neurotoxic effect of these three insecticides was due to the uncontrolled, accelerated release of neurosecretory products from the perikarya.

In addition to histochemistry and ultrastructure, the availability of synthetic peptides of both invertebrate and vertebrate origin has made possible the more precise localization of specific types of peptides. Antibodies to the molluscan cardioaccelerator, FMRFamide, were used to locate FMRFamide-like cells in the midgut and central nervous system of Heliothis zea, and to determine whether this peptide regulates digestion in this species. FMRFamide-immunoreactive cells have been found to be ubiquitous throughout both invertebrate and vertebrate species, but little is known of the chemical structure or function of the homologues of this peptide. Immunostaining and radioimmunoassay of fed vs. starved larvae indicated this peptide mediates digestion in this lepidopteran. In vitro gut bioassays suggest that the FMRFamide-like peptide may regulate the contractile activity of the gut.
The use of antibodies to a vertebrate peptide permits not only the detection of an immunologically similar compound in insect tissues, but also may indicate a possible function for the insect peptide. A vasopressin-like immunoreactivity was demonstrated in cells of the corpora cardiaca and paired cells in the abdominal ganglia of two lepidopteran species, Lymantria dispar and Manduca sexta (Davis and Lehman). The vasopressin-immunoreactive cells of the ventral nerve cord release their product from perivisceral neurohemal organs. In vitro bioassays on the malpighian tubules of these species indicate that these vasopressin-like cells comprise a diuretic system.

In 1986 the first insect neuropeptide (leucomyosuppressin, LEM-MS) was discovered that inhibited spontaneous contractions of insect muscle fibers. LEM-MS was isolated from head extracts of Leucophaea maderae. Antibodies conjugated to fluorescent and peroxidase probes located the site of synthesis of LEM-MS in the central nervous system of L. maderae (Meola et al.). This study also found that neurons containing LEM-MS immunoreactive material are present in the CNS of the stable fly, and that immunoreactive axons terminated in neurohemal organs as well as at neuroeffector junctions in both species. The neuroeffector junctions containing LEM-MS immunoreactive material were found associated with both motor and sensory effectors. Thus this peptide has both a neuroendocrine and neurotransmitter mode of action.

Serotonin-like immunoreactive neurons were located in the mesothoracic ganglionic mass of Rhodnius prolixus (Orchard et al.). These neurons consist of five unpaired cells that are associated with the abdominal nerves. These nerves appear to release their product into the haemolymph, as well as send processes containing serotonin-like material into the abdominal nerves that extend over the body wall forming synaptoid junctions on the epidermal cells. Biogenic amines have been found to affect tissues directly, act as interneurons, and to be localized in axon terminals containing neuropeptides. Clearly more research needs to be done on the role of these compounds in the integration of physiological processes.

Since ICINN-2 studies involving the use of molecular biology techniques to isolate and characterize genes that code for specific neuropeptides have begun to appear in the literature. These studies have resulted in discovery of a DNA gene from Drosophila that encodes a series of FMRFamide-like peptides, and another that encodes precursors to cholecystokinin-gastrin that have been cloned to Drosophila genes. At ICINN-3, Davis et al. reported the cloning of many new genes from the brain cDNA library of Manduca sexta that are differentially expressed during pupal/adult development. These investigators found four proteins with patterns of differential expression similar to those of the cloned genes. Three of these are putative neurosecretory proteins. In order to prove expression of the peptides synthesized by the cloned DNA in these studies, the natural peptides will have to be isolated and structurally characterized. Another advantage of molecular biology is that it permits investigation of the pathways of neurohormone synthesis from prohormones to the final product.

The use of immunocytochemistry for the localization of homologous neurosecretory products has become an important tool in neuroscience, but must be used with adequate controls. For example one of the most widely used peroxidase techniques can result in nonspecific staining of endogenous peroxidase, which is especially difficult to control in Diptera. Fortunately, improved peroxidase techniques have been developed. In addition, the antigenic determinants required for peptide localization may be as small as two amino acid units. Thus, antibodies for a specific peptide are designated " -like" when localized in a different species or tissue and, as in the case of peptides produced from cloned DNA, will also require isolation and structural and functional characterization in order to be classified as a specific peptide.

An important aspect of the field of neuroscience that has received little attention is the paucity of bioassays for determining the function of the multitude of products produced by the nervous system. The most frequently used assays test myogenic, diuretic, and adipokinetic activity. Since only one of these parameters is usually tested during isolation of a product such as a neuropeptide, it may not be the primary activity of the peptide.

Until 1985 only four neuropeptides had been structurally identified in insects, however with breakthroughs in instrumentation such as ion-pair reverse phase high performance liquid chromatography and fast-atom bombardment mass spectrometry, 41 additional neuropeptides have been identified since then, with 29 identified since ICINN-2. Prior to the development of these technologies, insect endocrinologists used either crude extracts or mammalian peptides to gain some insight into the localization and potential function of insect neuropeptides. With the availability of these techniques coupled with cloned DNA for the production of pure peptides, the field of insect neuroendocrinology is beginning to come into its own.

From *Insect Neurochemistry and Neurophysiology • 1989* Edited by A. B. Borkovec and E. P. Masler

ULTRASTRUCTURE OF A SENSORY RECEPTOR IN THE ABDOMINAL BODY WALL OF **RHODNIUS PROLIXUS**

R. Gary Chiang, Jennifer A. Chiang and Ken G. Davey

Department of Biology, York University
North York, Ontario
Canada, M3J 1P3

The abdominal body wall of the blood-feeding insect, **Rhodnius prolixus**, contains symmetrically arranged sense organs which respond to pressure applied from the haemocoelic side. The organs, which occur on either side of the midline in abdominal segments III to VI, discharge with increasing frequency in response to increasing pressure, and do not accomodate (Chiang and Davey, 1988). Under the light microscope, the receptor appears as a plaque embedded in the body wall. The anterior end of the plaque is joined by a small process to a branch of the segmental nerve in which is located the cell body of the receptor. Extending along the midline of the plaque is a heavily stained filamentous strand which, at its distal end, enlarges into a rod-like structure. The following results describe the appearance of this rod-like structure as revealed by serial thin-section electron microscopy.

The rod-like structure consists of the dendrite surrounded by the densely stained fibrous matrix of a scolopale (see Fig.1). The dendrite enclosed within the scolopale contains a cilium. A small dilation occurs in the mid-region of the cilium, and a spherical dense-staining structure is located in this dilation (Fig.2A). In cross-section, doublets of microtubules are seen in close association with this spherical structure. Along the circumference within the spherical structure are 9 heavily stained bodies, the position of each corresponding

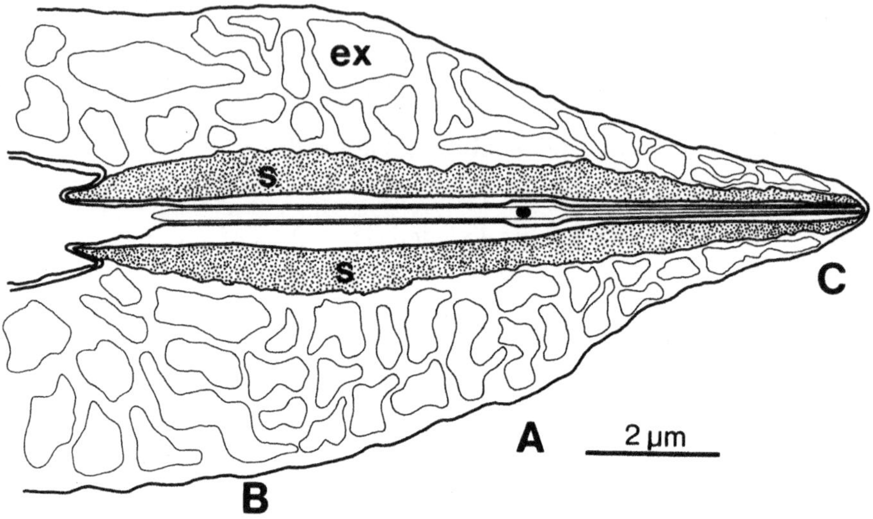

Fig. 1: Line drawing of a longitudinal view of the distal end of the abdominal sensory receptor illustrating the ciliary region of the dendrite enclosed within a scolopale (s) produced by the scolopale cell. Cross-sections from regions marked A, B and C are shown in Fig. 2 A, B and C. ex, extracellular compartments in cyctoplasm of scolopale cell.

to a doublet of microtubules in the cilium. Proximal to the spherical structure, the microtubules are organized in a 9+2+0 arrangment with one microtubule of the pair being filled with dense material and associated with dynein arms (Fig.2B). An additional two microtubules, possibly representing a central pair, are present distal to the dilation where 1. the microtubules appear as less regularly arranged doublets and singlets, 2. amorphous material occurs within the extracellular space between the dendrite and the scolopale, and 3. the circumference of the dendrite is slightly smaller than in the region proximal to the dilation (Fig.2C).

A scolopale cell is folded around the dendrite, and the opposing sides of this cell are joined to produce a "seam" that extends along the long axis of the rod-like structure

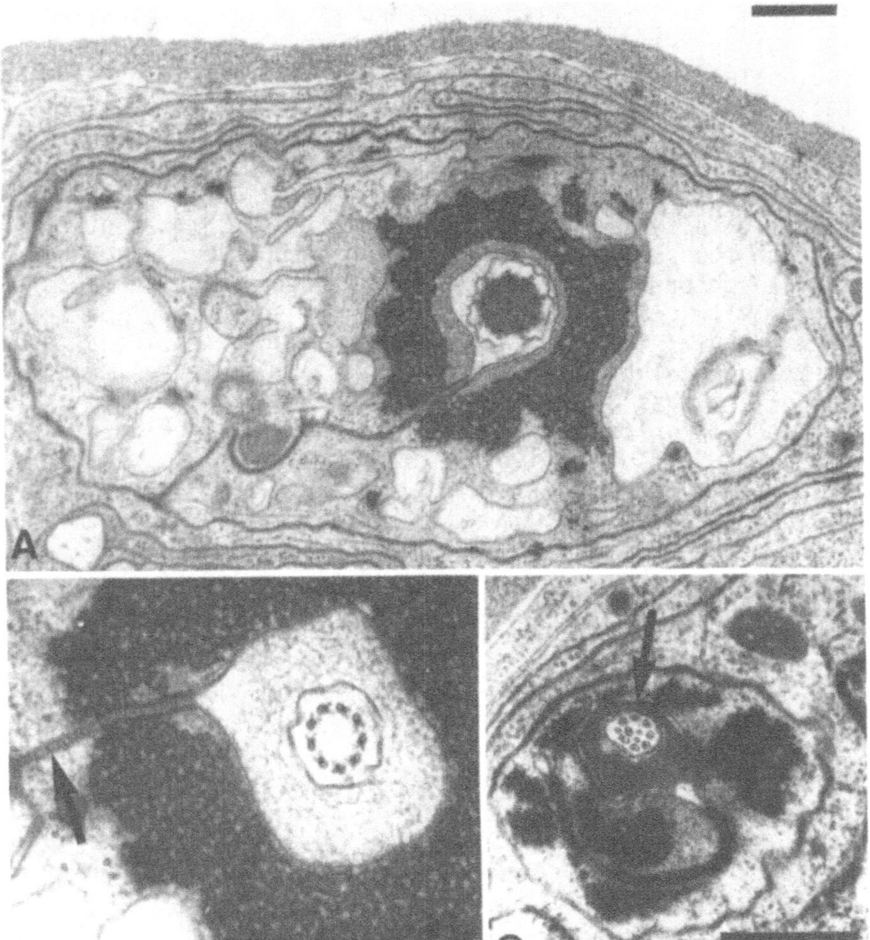

Fig. 2: Electron micrographs of cross-sections through the scolopale of the abdominal sensory receptor illustrating: A. the spherical structure in the dilation of the cilium; B. the regular arrangment of 9 doublets with dynein arms in the proximal region of the scolopale, and the intermediate and septate (arrow) junctions of the "seam"; C. singlets, doublets and the amorphous extracellular material (arrow) in the distal region of the cilium. Sections tilted 50°. Scale mark, 0.5 um.

(Fig.2A,). The scolopale cell gives rise to the scolopale, the tube-like structure completely enclosing the ciliary region of the dendrite. The side of the scolopale facing the cilium is smooth, whereas the side facing the cytoplasm of the scolopale cell is irregular. Proximal to the cilium, the scolopale divides into short columns which fit into invaginations of the dendrite. The seam representing the adhesion of the scolopale cell onto itself displays a strong, intermediate-like junction through the scolopale. Outside the scolopale, this seam is more porous as indicated by the occurrence of a septate junction (Fig.2B). Arising from the septate junction are extracellular invaginations which form an extensive array of extracellular compartments in the scolopale cell (Fig.1). These compartments are continuous with the spaces between the scolopale and the ciliary region of the dendrite.

The presence of a scolopale in this receptor distinguishes it as a chordotonal organ. Chordotonal organs occur throughout the peripheral region of the insect body, and the position and physiology of many of these receptors are well suited for their function as proprioceptors (see Finlayson, 1968). Although this chordotonal organ in **Rhodnius** has structural similarities with other insect chordotonal organs, its position in the mid-ventral body wall and its relative isolation from other body parts, would not support its role as a proprioceptor. On the other hand, these attributes, and the lack of accomodation, are appropriate for a sensory receptor detecting pressure resulting from distension of the stomach by the blood meal.

This work was supported by an operating grant from the Natural Sciences and Engineering Research Council of Canada to K.G.D..

REFERENCES

Chiang R.G. and Davey K.G. (1988) A novel receptor capable of measuring applied pressure in the abdomen of an insect. Science **241**, 1665-1667.

Finlayson L.H. (1968) Proprioceptors in the invertebrates. Symp. zool. Soc. Lond. **23**, 217-249.

From *Insect Neurochemistry and Neurophysiology • 1989* Edited by A. B. Borkovec and E. P. Masler

THE ASSOCIATION OF PROCTOLIN WITH THE FEMALE REPRODUCTIVE TISSUE OF *RHODNIUS PROLIXUS*.

Angela B. Lange

Dept. Zoology, Univ. Toronto,
Erindale College, Mississauga, ON,
Canada, L5L 1C6

Proctolin was first discovered associated with the hindgut of the cockroach *Periplaneta americana* were it was proposed to function as a neurotransmitter (Brown, 1975; Brown and Starratt, 1975). Since that time, proctolin has been shown to be more widely distributed within insects, being present in the central nervous system, and associated with visceral and skeletal muscle (see Orchard *et al.* 1989). Physiological studies subsequently revealed proctolin to be a co-transmitter at skeletal and visceral neuromuscular junctions (Adams and O'Shea, 1983; Lange *et al.*, 1986).

In a recent study Lange *et al.* (1988) used reversed phase-high performance liquid chromatography (RP-HPLC) together with a sensitive bioassay to describe the distribution of proctolin throughout the central nervous system of the blood-feeding bug *Rhodnius prolixus*. The CNS of adult female *Rhodnius* contained the equivalent of about 378 fmol proctolin. The highest content was found in the mesothoracic ganglionic mass, a ganglion consisting of the fused mesothoracic, metathoracic, and abdominal ganglia. The next highest amounts were found respectively in the prothoracic ganglion, sub-oesophageal ganglion and the brain. No proctolin was identified in the corpus cardiacum. Immunohistochemistry of the visceral structures of *Rhodnius* revealed proctolin-like immunoreactivity in terminal arborisations of the reproductive and digestive systems. The present paper will deal with the distribution of proctolin within the reproductive system of *Rhodnius prolixus*.

Tissues were extracted into methanol:water: acetic acid (100:10:1) and subsequently run through Sep-Pak C_{18} cartridges followed by chromatography on RP-HPLC. The fractions from the chromatography were examined for proctolin-like bioactivity (PLB) using the locust oviduct bioassay (Lange *et al.*, 1986). Proctolin produces a dose-dependent tonic contracture of the locust oviducts, and, with the use of this dose response curve, the amount of material co-migrating with ^{3}H proctolin and showing PLB can be quantified. The proctolin content of the various structures of the reproductive system of *Rhodnius* can be seen in Figure 1. The bursa and common oviduct contain larger amounts of proctolin than the spermathecae, lateral oviducts and ovaries. However, if the size of the various structures is taken into account by measuring the protein content, it can be seen that the spermathecae contain the largest amount of proctolin per mg protein, followed by the common oviduct, the bursa and lateral oviducts, with the ovaries containing the lowest amount.

Immunohistochemistry of these structures revealed proctolin-like immunoreactivity in terminal arborisations extending from the lateral oviducts across the common oviduct, spermathecae and along to the bursa. Fine processes were observed branching over the surface of these structures, with intensely staining varicosities and blebs distributed along their length. The spermathecae showed the greatest amount of staining with numerous varicose axons extending across the entire surface. When the primary antiserum was pre-absorbed with authentic proctolin, all staining was abolished. No immunostaining was evident on the cement gland or ovaries. The cement gland was not examined for proctolin content.

The oviducts of *Rhodnius* are sensitive to proctolin, with exposure to proctolin producing a dose-dependent increase in muscle tonus as well as an increase in amplitude and frequency of phasic contractions. The spermathecae are myogenically active displaying a rhythmical pattern (Kuster and Davey, 1986). The addition of proctolin to the spermathecae increased the frequency of the rhythmical contractions of the spermathecae.

In summary, with the use of RP-HPLC, bioassay, and immunohistochemistry , the presence and distribution of material indistinguishable from proctolin has been discovered in the reproductive system of *Rhodnius prolixus*. The physiological relevance of proctolin within this system is currently being investigated.

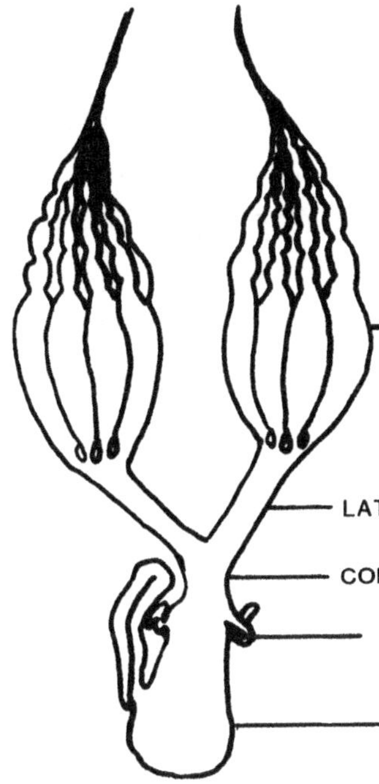

	PROCTOLIN	
	fmol/structure	pmol/mg protein
OVARY	7.6 ± 1.6	0.07 ± 0.01
LATERAL OVIDUCT	6.5 ± 1.1	0.40 ± 0.07
COMMON OVIDUCT	88.6 ± 20.9	2.45 ± 0.58
SPERMATHECA	16.7 ± 1.7	3.11 ± 0.31
BURSA	97.6 ± 17.7	0.46 ± 0.08

Figure 1. Diagrammatic representation of the female reproductive system of *Rhodnius prolixus*. Proctolin content was determined from proctolin-like bioactivity after pooled samples (3-4 tissues/pool) were processed through Sep-Pak and RP-HPLC. Values represent mean ± S.E. (n=3).

REFERENCES

Adams M.E. and O'Shea M. (1983) Peptide co- transmitter at a neuromuscular junction. Science **221**:286-289.

Brown B.E. (1975) Proctolin: A peptide transmitter cnadidate in insects. Life Sci. **17**:1241-1252.

Brown B.E. and Starratt A.N. (1975) Isolation of proctolin, a myotropic peptide, from *Periplaneta americana*. J. Insect Physiol. **21**:1879-1881.

Kuster J.E. and Davey K.G. (1986) Mode of action of cerebral neurosecretory cells on the function of the spermathecae in *Rhodnius prolixus*. Int. J. Invert. Reprod. Devel. **10**:59-69.

Lange A.B., Orchard I. and Adams M.E. (1986) Peptidergic innervation of insect reproductive tissue: The association of proctolin with oviduct visceral musculature. J. comp. Neurol. **254**:279-286.

Lange A.B., Orchard I. and Barrett F.M. (1988) The presence and distribution of proctolin in the blood-feeding bug, *Rhodnius prolixus*. J. Insect Physiol. **34**:379-386.

Orchard I., Belanger J.H. and Lange A.B. (1989) Proctolin: A review with emphasis on insects. J. Neurobiol. (In press).

From *Insect Neurochemistry and Neurophysiology • 1989* Edited by A. B. Borkovec and E. P. Masler

SEROTONIN-LIKE IMMUNOREACTIVE DORSAL UNPAIRED MEDIAN NEURONS IN *RHODNIUS PROLIXUS*

Ian Orchard, Angela B. Lange, Heather Cook and Jan-Marino Ramirez
Department of Zoology, University of Toronto, Toronto, ON, Canada M5S 1A1

Plotnikova (1969) described neurons in the thoracic and abdominal ganglia of locusts which had dorsal cell bodies lying near the midline, and bifurcating axons projecting symmetrically into left and right peripheral nerve roots. Hoyle *et al.* (1974) coined the term dorsal unpaired median (DUM) neurons for these and other neurons of a similar morphology. DUM neurons in a variety of insects have been shown to contain octopamine (Orchard *et al.* 1989) and it has tacitly been assumed that all DUM neurons are octopaminergic. In this paper we provide the first example of serotonin-like immunoreactive DUM neurons in insects.

The basic pattern of serotonin-like immunoreactive staining in the mesothoracic ganglionic mass (MTGM) and abdominal nerves of an unfed fifth instar *Rhodnius prolixus* is shown in Fig. 1A. Of particular interest is the presence of five large dorsal median neurons (labelled DUM 1-5) which display faint immunostaining, and the presence of an extensive neurohaemal complex lying on the abdominal nerves. No serotonin-like immunoreactive processes can be seen entering or leaving the peripheral nerves of this ganglion. Injection of 1 µl of 10^{-3}M 5,7-DHT (5,7-dihydroxytryptamine), a neurotoxin of serotonergic neurons, resulted in enhanced immunostaining of the cell bodies and axons of the five dorsal median neurons (Fig. 1B, Orchard *et al.* 1989) with each of these neurons revealing the characteristic morphology of DUM neurons. The neurons have been labelled DUM 1-5 based upon the abdominal nerve pair along which the axons project. Concomitant with the increase in intensity of immunostaining in the cell bodies and axons was a severe depletion of staining of the neurohaemal areas. We believe 5,7-DHT lesions the terminals of these serotonergic neurons, and results in a build up of serotonin proximal to the site of injury. Support for this suggestion comes from the observation that transection of the abdominal nerves close to the ganglion resulted (after 24 hr) in a similar effect to that seen following 5,7-DHT injection.

Interestingly, cobalt backfills reveal that each of the five abdominal nerves receives axons from two DUM neurons, only one of which is serotonin-like immunoreactive. This was confirmed with double labelling using intracellular injection of Lucifer yellow followed by immunohistochemistry for serotonin (revealed with Texas red-conjugated secondary antibody). The morphology of all ten DUM neurons was similar but only DUM 1-5 were immunoreactive. An example of a serotonin-like immunoreactive DUM neuron injected with Lucifer yellow is shown in Fig. 1C. Each immunoreactive DUM neuron possesses a primary neurite which bifurcates into two neurites projecting laterally, each of which sends an axon along a left and right abdominal nerve. There are four anteriorly projecting neurites and four posteriorly projecting neurites. Extensive branching occurs within the lateral neuropile with numerous varicosities which are serotonin-like immunoreactive.

Intracellular recordings from the cell bodies of the serotonin-like immunoreactive DUM neurons revealed them to have properties similar to other DUM neurons. They produce overshooting action potentials (40-60 mV) with large afterhyperpolarisations (Fig. 1D), and possess different spike types, such as double component spikes (Fig. 1D), which are probably due to multiple spike initiation zones. Double recordings indicated that the DUM neurons are neither electrically nor chemically inter- connected, but do receive similar synaptic input.

In conclusion, we have described a sub-population of five DUM neurons in the MTGM of *Rhodnius* which are serotonin-like immunoreactive. We have enhanced the immunostaining of the cell bodies and axons of these neurons using 5,7-DHT and nerve transection and have further defined their electrophysiology and morphology using intracellular recording and staining. We believe these five DUM neurons produce the extensive serotonin-like immunoreactive peripheral complex in *Rhodnius*. Since peripheral serotonin appears to be integrally involved in feeding activities of this blood-sucking insect (Lange *et al.* 1989), the discovery of these DUM neurons will facilitate studies of the central interactions of serotonergic neurons during feeding.

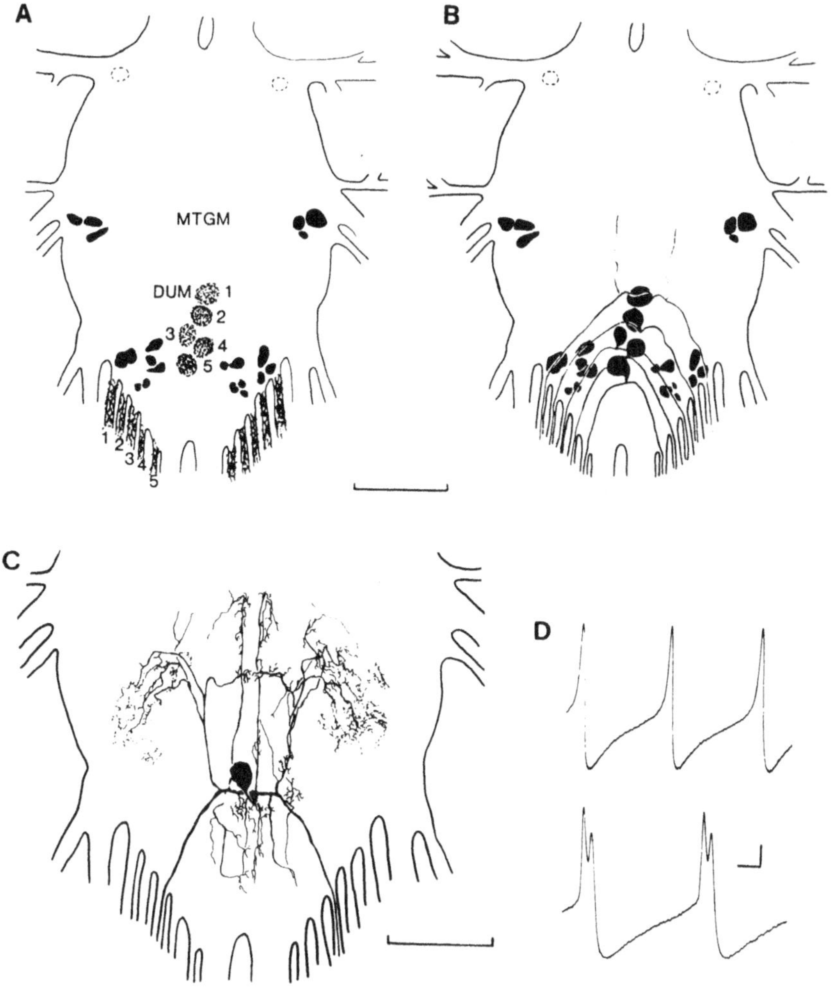

Fig. 1A,B. Serotonin-like immunoreactivity in the MTGM and abdominal nerves (1-5) of *Rhodnius prolixus*. A. Unfed untreated fifth instar. B. Unfed fifth instar injected 24 h previously with 1 µl of 10^{-3}M 5,7-DHT. Note the enhanced staining of the cell bodies and axons of DUM 1-5 and the disappearance of the neurohaemal areas in the abdominal nerves. C. Serotonin-like immunoreactive DUM 3 injected with Lucifer yellow. D. Recordings from the cell body of the DUM 3 filled in C. Scale bar: A,B,C, 200 µm; D, 10 mV, 5msec.(Redrawn from Orchard *et al.* 1989)

REFERENCES

Hoyle G., Dagan D., Moberly B. and Colquhoun W. (1974) Dorsal unpaired median insect neurons make neurosecretory endings on skeletal muscle. J. Exp. Zool. **187**:159-165.

Lange A.B., Orchard I. and Lloyd R.J. (1988) Immunohistochemical and electrochemical detection of serotoninn in the nervous system of the blood-feeding bug, *Rhodnius prolixus.* Arch. Insect Biochem. Physiol. **8**:187-201.

Lange A.B., Orchard I. and Barrett F.M. (1989) Changes in haemolymph serotonin levels associated with feeding in the blood-sucking bug, *Rhodnius prolixus.* J. Insect Physiol. **35**:393-399.

Orchard I., Lange A.B., Cook H. and Ramirez J.-M. (1989) A subpopulation of dorsal unpaired median neurons in the blood-feeding insect, *Rhodnius prolixus*, displays serotonin-like immunoreactivity. J. Comp. Neurol. (in press).

Plotnikova S.N. (1969) Effector neurons with several axons in the ventral nerve cord of the Asian grasshopper *Locusta migratoria.* J. Evol. Biochem. Physiol. **5**:276-277.

From *Insect Neurochemistry and Neurophysiology • 1989* Edited by A. B. Borkovec and E. P. Masler

IMMUNOCYTOCHEMICAL LOCALIZATION OF LEUCOMYOSUPPRESSIN (LEM-MS) IN THE CNS OF THE COCKROACH, LEUCOPHAEA MADERAE AND THE STABLE FLY, STOMOXYS CALCITRANS.

Shirlee Meola, *Mark S. Wright,

G. Mark Holman and J. Mark Thompson

USDA, ARS, VTERL

College Station, Tx 77840,

*Department of Entomology

Texas A&M University

College Station, TX 77843.

Major strides have been made in the past five years in the discovery of new insect neuropeptides due to the development of improved technology and instrumentation. A group of 12 neuropeptides were isolated and identified from extracts of Leucophaea maderae by G. M. Holman and collaborators between 1986 and 1987 (Holman et al., 1988). These neuropeptides consisted of 11 myotropic peptides and one myoinhibitory peptide. Holman et al.(1986) named the latter peptide Leucomyosuppressin (LEM-MS) and determined it to be a decapeptide with the structure, pGlu-Asp-Val-Asp-His-Val-Phe-Leu-Arg-Phe-NH_2. This neuropeptide structurally resembles the molluscan tetrapeptide, FMRFamide (Phe-Met-Arg-Phe-NH_2) which, interestingly, is a cardioaccelerator in molluscs, while LEM-MS inhibits contraction of both visceral and skeletal muscle of insects (Yamamoto et al., 1988). As a prelude to physiological investigations of this peptide, a study was done to locate the cells of origin and the neural pathway of LEM-MS in L. maderae using immunological probes. In order to determine if

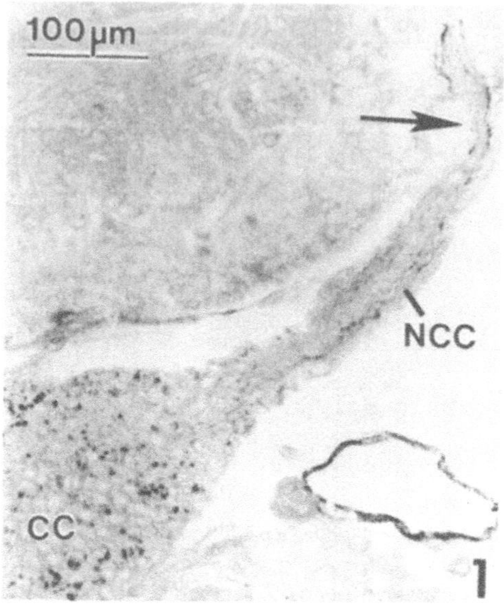

Fig. 1. LEM-MS-reactive axons in one of the nervi corporis cardiaci as it exits (arrow) the brain of the L. maderae. The axons, visualized by the peroxidase probe, terminate at the periphery of the corpus cardiacum (CC).

similar peptides are present in other orders of insects, antibodies to LEM-MS were also tested on tissues of the stable fly, Stomoxys calcitrans. These antibodies were visualized in adult specimens with fluorescent and peroxidase probes.

Neurons synthesizing LEM-MS-immunoreactive material were found throughout the CNS of both species of insects. The major source of LEM-MS-reactive material in the brain of L. maderae was found in two groups of cells. One group is located in the pars intercerebralis, while the second group lies between the protocerebral and deutocerebral lobes of the brain. Axons of the first group enter the nervi corporis cardiaci (NCC) and extend to the corpora cardiaca where many terminate in the neurohemal area at the periphery of this organ (Fig. 1). Some axons of the second group also enter the NCC, but many enter the antennal and circumesophageal nerves. A cluster of 5-6 LEM-MS-reactive

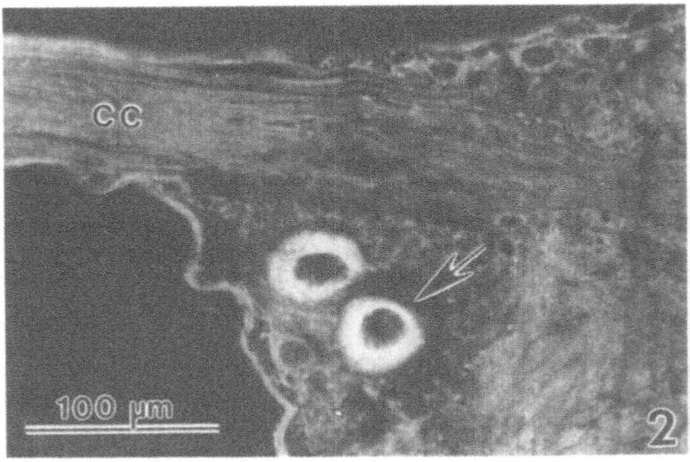

Fig. 2. Two of the 7 pairs of large LEM-MS-reactive neurons (arrow) that lie in the anterior region of the prothoracic neuromere of the stable fly, visualized by Texas red fluorescent probe. CC, cervical connective.

cells are present in the mid-ventral region of the subesophageal ganglion, while a pair of these cells is located in the ventral region of each thoracic ganglion. No intensely reacting cells were found in the abdominal ganglia of L. maderae, but each of these ganglia contain several cells that stain moderately with the LEM-MS probes.

Although the brain and subesophageal ganglion of the stable fly also contained LEM-MS-reactive cells, the majority of the cells reacting to the LEM-MS probes were found in the fused thoracico-abdominal ganglion. This ganglion contains 7 pairs of LEM-MS-reactive cells in the anterior region of the prothoracic neuromere of this ganglion (Fig. 2), and 2 pairs of reactive cells in the ventral region of each thoracic neuromere. Six pairs of immunoreactive cells lie in a chain along the dorsal surface of the abdominal neuromere. A cluster of 4 reactive cells lies at the base of the median abdominal nerve and another group of 6 is located in the mid-ventral region of the abdominal neuromere.

Since no homologous peptide with the structure and myoinhibitory effect of LEM-MS has been found in vertebrates, synthetic analogues of LEM-MS have potential as environmentally safe insecticides.

REFERENCES

Holman, G.M., Wright, M.S. and Nachman, R.J. (1988) Insect neuropeptides: Coming of Age. In ISI Atlas of Sciences: Plants & Animals Vol. 1, pp 129-136.

Holman,G.M., Cook, B.J. and Nachman, R.J. (1986) Isolation, primary structure and synthesis of leucomyosuppressin, an insect neuropeptide that inhibits spontaneous contractions of the cockroach hindgut. Comp.Biochem. Physiol. 85C:329-333.

Yamamoto, D., Ishikawa, S., Holman, G.M., and Nachman R.J. (1988) Leucomyosuppressin, a novel insect neuropeptide inhibits evoked transmitter release at the mealworm neuromuscular junction. Neurosc. Letters 95:137-142.

From *Insect Neurochemistry and Neurophysiology • 1989* Edited by A. B. Borkovec and E. P. Masler

A VASOPRESSIN-LIKE NEUROHEMAL SYSTEM IN LYMANTRIA DISPAR AND MANDUCA SEXTA.

Norman T. Davis and Herman Lehman

Division of Neurobiology, University of Arizona, Tucson, AZ 87521

Using an antiserum to lys-vasopressin (Incstar), a small population of cells is stained which is different from that reported previously in insects. This study was done using an indirect immunofluorescent method (Davis, 1987) applied to the CNS of *Lymantria dispar* (gypsy moth) and *Manduca sexta*. The antiserum was used at working dilutions of 1:2000. Preabsorbtion of the working solution with synthetic arg- and lys-vasopressin (1 uM/ml) completely abolished the staining. The vasopressin-like immunoreceptive (VLI) cells were not stained by an antiserum specific to arg-vasopressin (Peninsula Laboratories) nor by an antiserum to oxytocin (Chemicon). These results suggest that the VLI material has epitopes common to lys- and arg-vasopressin.

In larvae a posterior lateral pair of VLI neurons is found in abdominal ganglia 3-6 and in the first neuromere of the terminal ganglion (figs. 1A,B), normally no other neurons are stained. The principal neurites of the VLI neurons appear to project in the neurophile to a median dendritic arborization, and their axons project by way of the ventral nerve to the perivisceral organ of the next posterior ganglion. Prominent varicose VLI axons are seen in the perivisceral organs associated with abdominal ganglia posterior to the second ganglion (figs. A,B). Since the perivisceral organs have a neurohemal function, these details indicated that a VLI material is probably released as a neurohormone.

The characteristics of the VLI neurons correspond very closely to cell 24 and 25 previously identified in the abdominal ganglia of developing adult M. sexta: these cells, along with two other lateral neurons are responsible for most of the bursicon bioactivity of the abdominal ganglia (Taghert and Truman, 1982). Since only two of the four putative bursiconergic soma are stained, the VLI staining likely indicates colocalization of the VLI material in two of the four bursicon cells.

In early fifth instar larvae kept without food and water for two days, there is a distinct increase in staining of the VLI material in the perivisceral organs (fig. 1C). In addition neurons in the first abdominal ganglion and the intrinsic cells of the CC are stained. Eight hous after these starved larvae have fed, there is a distinct decrease in the staining of the VLI material. Similarly, when 400 ul of water is introduced by canulation of the gut of starved larvae, there is by eight hours after treatment a distinct decrease in the intensity of VLI staining. These changes appear to indicate that lack of food and/or water inhibits the release of the VLI neuropeptide and that its release is stimulated by feeding and forced libation.

In M. sexta the intrinsic cells of the corpora cardiaca show VLI staining in both larvae and pharate adults (fig. 1D). These cells produce an adipokinetic hormone (AKH) and can be stained with an antiserum to AKH. Preabsorption of the lys-vasopressin antiserum with synthetic AKH does not block immunostaining of these intrinsic cells. These results appear to indicate that there is colocalization of LI material and AKH in the intrinsic cells.

The abdominal VLI neurons and their neurohemal projections are present from the first larval instar and persist in postembryonic development into the adult stage. However, a decrease in VLI staining occurs in the prepupal and early pupal stages. In the pharate adult of M. sexta, the staining intensity of the VLI abdominal neurons and of the intrinsic cells of the CC increases, and the volume of the abdominal cells increases about 8 times over that of the fifth instar larva. These observations appear to indicate that the VLI neurohemal system is functional throughout postembryonic development, but that the level of its activity may vary with stages of development.

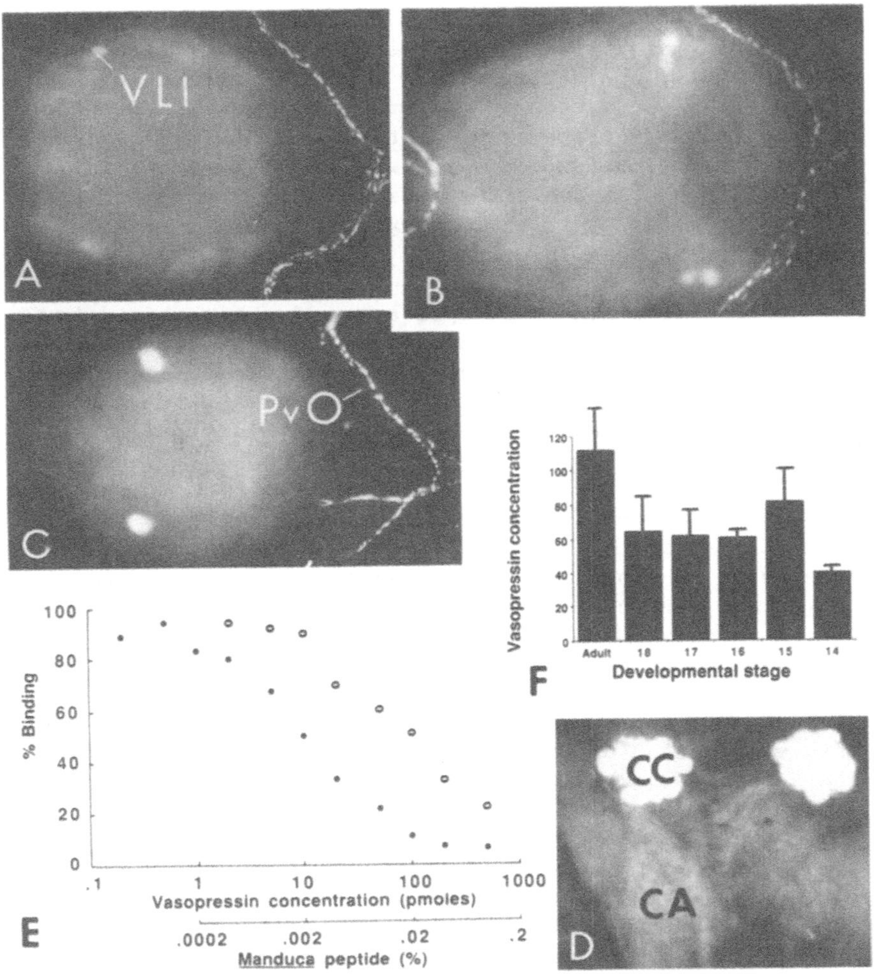

Fig.1A,B - VLI cells and processes in 5th(A) and terminal(B) ganglia of larval L. dispar. (PvO,perivisceral organ). Fig 1C - VLI staining in 5th ganglion of starved larva. Fig.1D VLI staining of cells of corpora cardiaca (CC) of adult M. sexta.(CA, corpus allatum). Fig.1E - Inhibition of binding of vasopressin antiserum to vasopressin thyroglobulin conjugate coated ELISA plates by vasopressin and M. sexta CNS extracts. Concentrations are in pmoles/100ul. Solid circles, vasopressin; open circles,CNS extract. Fig.1F - VLI concentration (pmoles/brain) at stages of development of adult M. sexta. Each bar represents average of 5 brain complexes,±S.D.

Since there appears to be a conspicuous increase in the staining of VLI material during the development of the pharate adult, competitive ELISA's were performed to measure titers. An acetic acid extract was made of 100 CN's of late pharate adults of M. sexta, and an (arg-8) vasopressin-thyroglobulin conjugate was coated on the ELISA plates. Figure 1E shows the similarity of the competitive binding of vasopressin and the extract of CNS's of late pharate adults of M. sexta.

The titer of the VLI peptide was then determined in the brain-retrocerebral complex at successive stages of development of the pharate adult. The results indicate that during this period the titer of VLI peptide about doubles and that there is about 112 pm pmoles of VLI peptide per adult CNS. (fig. 1F).

These various results are consistent with the idea that VLI peptide(s) may be a diuretic hormone. Such a hormone would be expected to be released after feeding and libation and its release would be expected to be inhibited during conditions of thirst, starvation and during the prepupal and pupal stages. The increase in titer of VLI peptide(s) in the pharate adult coincides with the formation of the meconium and diuretic elimination of water at eclosion. This elimination of water serves to reduce the weight of the adult insect, preparatory for flight (Maddrell, 1986)

REFERENCES

Davis, N. T. (1987) Neurosecretory neurons and their projection into the serotonin neurohemal system of the cockroach Periplaneta americana (L.), and identification of motor neurons associated with this system. J. Comp. Neurol. 259:604-621.

Maddrell, S.H.P. (1986) Hormonal control of diuresis in insects. In Insect Neurochemistry and Neurophysiology (eds. A.B. Borkovec and D.B. Gelman). Humana, Clifton NJ.

From *Insect Neurochemistry and Neurophysiology • 1989* Edited by A. B. Borkovec and E. P. Masler

HISTOCYTOCHEMICAL IDENTIFICATION OF NEUROSECRETORY CELLS IN THE BRAIN OF SHORT HORNED GRASSHOPPER

MEENA U. PATIL

Zoology Department, Marathwada University,

AURANGABAD-431 004 INDIA

INTRODUCTION

Neurosecretory cells (NSC) have been investigated histologically and histochemically by several authors (PRENTO, 1972; RAABE, 1983). In most of the papers the A and B cells or their types were described. Presence of C and D cells was first explained in Schistocerca gregaria (HIGHNAM, 1961) and C cells were treated as non neurosecretory. In fact,the brain contains not only A and B cells but also other NSC cell types (RAABE, 1983).

Melanoplus spe. were kept in the laboratory under natural conditions and fed on grass and wheat bran. The brains were dissected out and the histological processing was completed. Sections were stained with Mallory's triple and aldehyde fuchsin (EWEN, 1962) with few modifications.

Four types A, B, C & D NSC were identified. A cells are largest with axons measuring 15-37 µm in diameter(Fig.1). Neurosecretory material (NSM) stained purple with AF and pink with Mallory's triple. A cells are grouped together to form Median neurosecretory cell (MNC) group in the anterior portion of the brain (Fig.2). From MNC axons are descending towards corpora cardiaca forming NCC_1; anterio-medially these axons cross over each other (Fig.3). B cells are 15-20 µm in diameter with bigger nucleus and 2-4 nucleoli. These are very few in number and placed in the peripheral region of the brain (Fig.4).

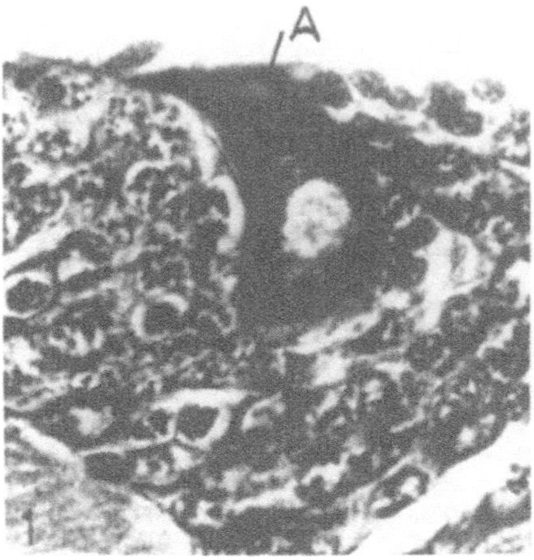

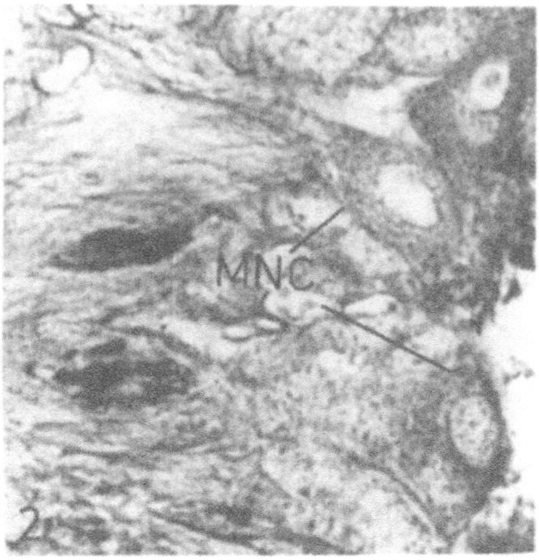

Fig. 1: T.S. through the brain showing 'A' NSC stained with Mallory's triple stain X 800.

Fig. 2: MNC stained with aldehyde fuchsin X 800.

The C & D cells dominate the NSC population and often occur in groups (Fig.5). These are the cells with less diameter (6-12 µm) and 4-6 nucleoli. Because of scanty cytoplasm and bigger diameter of nucleus, they look like rings. The cerebral neuropile contains large amount of neurosecretion and the droplets of it are tracking towards corpus cardiacum (Fig.6).

For obvious reasons, the classification of NSC, which has so far been described, is not conclusive. Depending upon the staining properties of cells in most of the studies NSC have been classifed into either A, B, C, their subtypes. NAYAR (1955) was of the opinion that B cells were A cells with little NSM or might be inactive A cells. JOHANSSON (1958), after various experiments concluded in Oncopeltus that A and B are two distinct cell types. HIGHNAM (1961) described that A & B cells are independent but C & D cells are non-neurosecretory in Locust. The new techniques alcian blue-alcian yellow and paraldehyde fuchsin distinguish between two kinds of A cells (A and A' or A_1 and A_2), which are very alike. But in the present study only type A is concluded. After AF staining, some A cells stained dark and few light in colour. Still I feel that these are only A type, since NSM synthesis/release is different in individual cells. The B cells are altogether different in morpho-

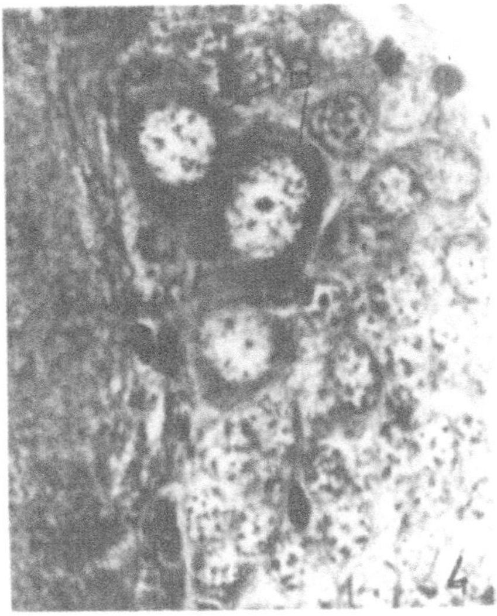

Fig. 3: Section of brain showing crossing over of axonal bundles of NCC_1 indicated by arrow X 400.

Fig. 4: B cells in the brain stained with AF X 800.

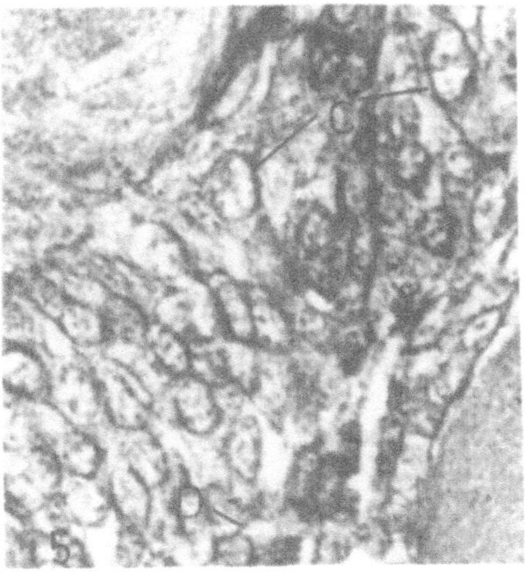

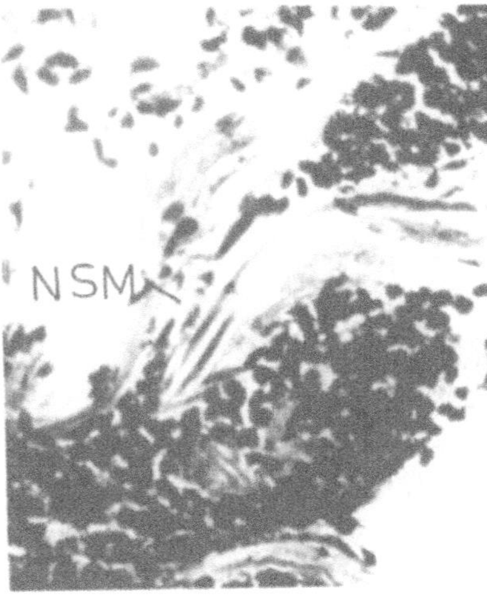

6

Fig. 5: C & D cells stained with aldehyde fuchsin X 800.

Fig. 6: Neurosecretory material (MSM) from neuropile area tracking towards corpora cardiaca X 400.

logy in Melanoplus spe. Thus, there is no point in stating them either inactive A cell or infant A cells. The presence of C & d cells is very clear. Similar cells were also observed in other group of Anthropoda, Crustacea, Macrobrachium kistnensis where their appearance was related to the reproductive cycle of prawn. To conclude their role in Melanoplus requires further detailed study, which is under progress in my laboratory.

REFERENCES

EWEN A.B. (1962) An improved aldehyde fuchsin staining technique for neurosecretory product in insects. Trans. Amer. Micro. Soc. 81, 94-96.

HIGHNAM K.C. (1961) The histology of neurosecretory system of the adult female desert locust Schistocerca gragaria. Quart. J. Microsc. Sci. 100, 27-28.

JOHANSSON A.S. (1958) Relation of nutrition endocrine reproductive functions in the milkweed bug Oncopeltus fasciatus. Nytt Mag. Zool. 7, 1-132.

NAYAR K.K. (1955) Studies of the neurosecretory system of Iphita limbata I. Distribution and structure of the neurosecretory cells of the nerve ring. Biol. Bull. 108, 296-307.

PRENTO P. (1972) Histochemistry of neurosecretion in the pars intercerebralis-corpus cardiacum system of the desert locust Schistocerca gragaria. Gen. Comp. Endocr. 18, 482-

RAABE M. (1983) The neurosecretory-neurochaemal system of insects; Anatomical, structural and physiological data. Adv. Insect Physiol. 17, 205-303.

Participants

TERRY S. ADAMS
USDA-ARS
State University Station
Biosciences Research Laboratory
Department of Biology
Fargo, ND 58105

CHARLES U. ALLEN
University of North Carolina
Department of Cell Biology
and Anatomy
350 Swing Building 217 Box 7090
Chapel Hill, NC 27599

GREG ALLEN
University of North Carolina
Department of Cell Biology
and Anatomy
350 Swing Building CB 7090
Chapel Hill, NC 27599

WILLIAM E. ALLISON
Dow Chemical Company
Box 9002
Walnut Creek, CA 94598-0902

MIRIAM ALTSTEIN
The Volcani Center
Agricultural Research Organization
Department of Entomology
POB 6, Bet Dagan 50 250 Israel

SHALOM APPLEBAUM
The Hebrew University
Faculty of Agriculture
Department of Entomology
PO Box 12
Rehovot 76100 Israel

MARTINE ARPAGAUS
University of Michigan
School of Medicine M 6322
Department of Pharmacology
Ann Arbor, MI 48109-0626

RICHARD A. BAINES
University of Waterloo
Department of Biology
Waterloo, Ontario N2L 3G1
Canada

KIRSTEN BAKKE
North Dakota State University
Stevens Hall
Fargo, ND 58105

ROBERT A. BELL
USDA-ARS
Insect Reproduction Laboratory
Building 306, Room 323, BARC-East
Beltsville, MD 20705

HANS-JURGEN BIDMON
University of North Carolina
Department of Cell Biology
and Anatomy
108 Swing Building 217H
Chapel Hill, NC 27599-7090

GARY BIRD
USDA-ARS
LPSI—Livestock Insects Laboratory
Building 307, Room 120, BARC-East
Beltsville, MD 20705

BRUCE C. BLACK
American Cyanamid Company
Agricultural Research Divison
PO Box 400
Princeton, NJ 08540

ROLF BODMER
UCSF
Howard Hughes Medical Institute
Box 0724
San Francisco, CA 94143

ALEXEJ B. BORKOVEC
USDA-ARS
Insect Reproduction Laboratory
Building 306, Room 323, BARC-East
Beltsville, MD 20705

DOV BOROVSKY
University of Florida–IFAS,
Florida Medical Entomology
Laboratory
200 9th Street SE
Vero Beach, FL 32962

DAVID W. BUSHMAN
University of Maryland
Department of Entomology
Symons Hall
College Park, MD 20742

ALISON CHALMERS
American Cyanamid Company
Agricultural Research Divison
PO Box 400
Princeton, NJ 08540

ANDREW C. CHEN
USDA-ARS
Veterinary Toxicology
and Entomology Research
Laboratory
F&B Road, Route 5, Box 810
College Station, TX 77840

ANN-SHYN CHIANG
Rutgers University
Cook College
Department of Entomology
PO Box 231
New Brunswick, NJ 08903

R. G. CHIANG
York University
Department of Biology
North York, Ontario M3J 1P3
Canada

GEOFFREY M. COAST
University of London
Department of Biology
Birkbeck College-Malet Street
London WCIE 7HX, UK

BENJAMIN J. COOK
USDA-ARS
Veterinary Toxicology
and Entomology Research
Laboratory
F&B Route 5, Box 810
College Station, TX 77840

JOSEPH W. CRIM
University of Georgia
Department of Zoology
Athens, GA 30602

KENNETH G. DAVEY
York University
Department of Biology
4700 Keele Street
North York, Ontario M3J 1P3
Canada

MINH-TAM B. DAVIS
USDA-ARS
Insect Hormone Laboratory
Building 467, BARC-East
Beltsville, MD 20705

NORMAN T. DAVIS
University of Connecticut
Department of Physiology
and Neurobiology
Box U-42
Storrs, CT 06268

ROBIN E. DAVIS
USDA-ARS
Insect Reproduction Laboratory
Building 306, Room 305, BARC-East
Beltsville, MD 20705

CARL V. DEAMICIS
Eli Lilly & Co.
Lilly Research Labs
PO Box 708
Greenfield, IN 46140

ALBERT B. DEMILO
USDA-ARS
Insect Reproduction Laboratory
Building 306, Room 315, BARC-East
Beltsville MD 20705

ELVIRA DOMAN
National Science Foundation
Physiological Process Program
Room 321 NSF
Washington, DC 20550

ROGER G. DOWNER
University of Waterloo
Department of Biology
Waterloo, Ontario N2L 3G1
Canada

STUART J. DUNBAR
ICI Agrochemicals
Jealott's Hill Research Station
Bracknell
Berkshire RG12 6EY, UK

RICHARD A. DYBAS
Merck & Co., Inc.
Hillsborough Road
PO Box 450
Three Bridges, NJ 08887

PETER D. EVANS
University of Cambridge
AFRC Unit—Department of Zoology
Downing Street
Cambridge CB2 3EJ, UK

ROBERT FAUST
USDA-ARS
National Program Staff
Building 005, Room 232, BARC-West
Beltsville, MD 20705

MARK FELDLAUFER
USDA-ARS
Insect Hormone Laboratory
Building 467, BARC-East
Beltsville, MD 20705

HOWARD W. FESCEMYER
USDA-ARS
Insect Reproduction Laboratory
Building 306, Room 304
Beltsville, MD 20705

ELIZABETH FOWLER
Ciba-Geigy
Agricultural Biotechnology
Research Unit
PO Box 12257
Research Triangle Park, NC 27709

MEGUMI FUSE
York University
Department of Biology
North York, Ontario M3J 1P3
Canada

GERD GADE
University of Cape Town
Zoology Department, Private Bag
Rondebosch 7700, South Africa

MICHAL GADOT
Rutgers University
Department of Entomology
Cook College-PO Box 231
New Brunswick, NJ 08903

ROBERT P. GAJEWSKI
Eli Lilly & Company
Lilly Research Laboratories
PO Box 708
Greenfield IN 46140

DALE B. GELMAN
USDA-ARS
Insect Reproduction Laboratory
Building 306, Room 319, BARC-East
Beltsville, MD 20705

JEFFERY W. GERST
North Dakota State University
Department of Zoology
Stevens Hall
Fargo, ND 58105

JADWIGA M. GIEBULTOWICZ
University of Maryland
Department of Zoology
College Park, MD 20742

LAWRENCE I. GILBERT
University of North Carolina
Department of Biology
CB# 3280, Coker Hall
Chapel Hill, NC 27599-3280

JOSIANE GIRARDIE
University of Bordeaux I
DRCNRS Laboratory
of Neuroendocrinology
Avenue des Facultes
33405 Talence Cedex, France

NOELLE A. GRANGER
University of North Carolina
Department of Cell Biology
and Anatomy
217H Swing Building CB#7090
Chapel Hill, NC 27514

JANE A. GREENBLATT
FMC Corporation
PO Box 8
Princeton, NJ 08543

RAZIEL S. HAKIM
Howard University Medical School
Department of Anatomy
520 W Street, NW
Washington, DC 20059

TIMOTHY K. HAYES
Texas A&M University
Biotechnology Support Laboratory
Department of Entomology
College Station, TX 77843

JOHN G. HILDEBRAND
University of Arizona
ARL Division of Neurobiology
611 Gould–Simpson Science Building
Tucson, AZ 85721

CALEB W. HOLYOKE
E. I. DuPont de Nemours
Wilmington
120 Country Club Drive
Newark, DE 19711

FRANK M. HORODYSKI
University of Washington
Department of Zoology
NJ-15
Seattle, WA 98195

R. E. ISAAC
University of Leeds
Department of Pure and Applied
Zoology
Leeds LS2 9JT, UK

GEETA K. JADHAV
Karnatak University
Department of Zoology
Divison of Sericulture
Dharwad 580 003 India

H. JAFFE
Peptide Design
12321 Middlebrook Road
Germantown, MD 20874

STEPHEN P. JENKINS
University of Georgia
Department of Entomology
Athens, GA 30602

JOOS JOOSSE
Vrije Universiteit
Biological Laboratory
De Boelelaan 1087
1081 HV Amsterdam,
The Netherlands

LARRY L. KEELEY
Texas A&M University
Department of Entomology
College Station, TX 77843

THOMAS J. KELLY
USDA-ARS
Insect Reproduction Laboratory
Building 306, Room 320, BARC-East
Beltsville, MD 20705

TOMAS G. KEMPE
University of Maryland
Department of Chemistry
Chemistry Building, Room 2219
College Park, MD 20742

LUKE R. C. KEMPTON
Univeristy of Leeds
Department of Pure and Applied Biology
Leeds, UK

TIMOTHY G. KINGAN
Peptide Design
12321 Middlebrook Road
Germantown, MD 20874

WALDEMAR KLASSEN
USDA-ARS
NPS
Building 005, Room 133, BARC-West
Beltsville, MD 20705

JAN KOCHANSKY
USDA-ARS
Insect Hormone Laboratory
Building 467, BARC-East
Beltsville, MD 20705

CHRISTINE KUKEL
American Cyanamid Company
PO Box 400
Princeton, NJ 08540

ANGELA B. LANGE
University of Toronto
Erindale College
Department of Zoology
3359 Mississauga Road
Mississauga, Ontario L5L 1C6
Canada

HANS LAUFER
University of Connecticut
Genetics and Cell Laboratory
U-125 75N Eagleville Road
Storrs, CT 06268

SARAH Y. LIN
FMC Corporation
Agricultural Chemical Group
Box 8
Princeton, NJ 08543

MARCIA J. LOEB
USDA-ARS
Insect Reproduction Laboratory
Building 306, Room 323, BARC-East
Beltsville, MD 20705

BARRY G. LOUGHTON
York University
Department of Biology
4700 Keele Street
North York, Ontario M3J 1P3
Canada

EDWARD P. MASLER
USDA-ARS
Insect Reproduction Laboratory
Building 306, Room 323, BARC-East
Beltsville, MD 20705

A. R. McQUISTON
University of Toronto
Department of Zoology
25 Harbord Street
Toronto, Ontario M5S 1A1
Canada

JULIUS J. MENN
USDA-ARS
Building 003, Room 201 BARC-West
Beltsville, MD 20705

ROBERT E. MENZER
University of Maryland
Department of Entomology
College Park, MD 20742

SHIRLEE M. MEOLA
USDA-ARS
Veterinary Toxicology
and Entomology Research
Laboratory
Route 5, Box 810
College Station, TX 77840

NITA M. MINNIFIELD
USDA-ARS
Livestock Insects Laboratory
Building 307, Room 120, BARC-East
Beltsville, MD 20705

JUDD O. NELSON
University of Maryland
Department of Entomology
College Park, MD 20742

IAN ORCHARD
University of Toronto
Department of Zoology
25 Harbord Street
Toronto, Ontario M5S 1A1
Canada

LASZLO PAP
CHINOIN
Budapest
Banyaleg, u.2. 1780
Hungary

MENNA U. PATIL
Marathwada University
L-2/2 University Campus
Department of Zoology
Aurangabad 431004 India

M. P. PENER
The Hebrew University
Department of Zoology
91904 Jerusalem, Israel

ADA RAFAELI
The Hebrew University
Faculty of Agriculture
Department of Entomology
Rehovot 76-100 Israel

ASHOK K. RAINA
USDA-ARS
Insect Chemical Ecology Laboratory
Building 402, Room 101, BARC-East
Beltsville, MD 20705

JAN-MARINO RAMIREZ
University of Alberta
Department of Physiology
Facility of Medicine
Edmonton, Alberta T6G 2H7
Canada

R. RIDGWAY
USDA-ARS
Building 402, Room 107 BARC-East
Beltsville, MD 20705

SANDRA ROBB
University of Cambridge, UK
Department of Zoology
Downing Street
Cambridge CB2 3EJ, UK

VINCENT L. SALGADO
Rohm and Haas Company
Research Laboratories
727 Noorestown Road
Spring House, PA 19477

COBY SCHAL
Rutgers University
Department of Entomology
Cook College, PO Box 231
New Brunswick, NJ 08903

MICHAEL E. SCHNEE
E. I. DuPont de Nemours
Agricultural Products E402/RM5233
Experimental Station
Wilmington, DE 19880-0402

DAVID A. SCHOOLEY
University of Nevada, Reno
Biochemistry Department
Reno, NV 89557

V. L. SEVALA
York University-Downs View
Department of Biology
North York, Ontario M3J 1P3
Canada

CAROL A. SHEPPARD
USDA-ARS
Insect Reproduction Laboratory
Building 306, Room 319, BARC-East
Beltsville, MD 20705

KLAUS P. SIEBER
National Institute of Health
NIAID LPD
Building 4, Room 126
Bethesda, MD 20892

ALVIN N. STARRATT
Research Centre Agricultural Canada
1400 Western Road
London, Ontario N6G 2V4
Canada

JOHN G. STOFFALANO, JR.
University of Massachusetts
Department of Entomology
Fernald Hall
Amherst, MA 01003

AKINORI SUZUKI
The University of Tokyo
Bunkyo-ku, Tokyo 113, Japan

JAMES A. SVOBODA
USDA-ARS
Insect Hormone Laboratory
Building 467, BARC-East
Beltsville, MD 20705

CHARLIE S. THOMPSON
University of Toronto
Department of Zoology
25 Harbord Street
Toronto, Ontario M5S 1A1
Canada

DENISE E. THOMSON
American Cyanamid Company
Agricultural Research Division
PO Box 400
Princeton, NJ 08540

B. S. THYAGARAJA
USDA-ARS
Insect Reproduction Laboratory
Building 306, Room 323, BARC-East
Beltsville, MD 20705

STEPHEN S. TOBE
University of Toronto
Department of Zoology
25 Harbord Street
Toronto, Ontario M5S 1A1
Canada

MOSHE TOM
Tulane University, New Orleans
Department of Biology—Stern Hall
2000 Percival
New Orleans, LA 70118

JEAN P. TOUTANT
Case Western Reserve University
School of Medicine
Department of Pharmacology
Cleveland, OH 44106

JOHN E. TREHERNE
University of Cambridge
AFRC—Department of Entomology
Downing Street
Cambridge, CB2 3EJ, UK

JAN A. VEENSTRA
University of Arizona
Department of Entomology
Forbes Building 36
Tucson, AZ 85721

PETER VERHAERT
Zoological Institute KV Leuven
Naamsestraat 59
B-3000
Leuven, Belgium

RENEE M. WAGNER
USDA-ARS
Veterinary Toxicology
and Entomology Research
Laboratory
Route 5, Box 810
College Station, TX 77840

ZHIWEI WANG
York University
Department of Biology
North York, Ontario M3J IP3
Canada

COLIN H. WHEELER
The University of London
Department of Biology
Birkbeck College–Malet Street
London, WC1 7HX , UK

LAVERN R. WHISENTON
Millersville University
Department of Biology
Millersville, PA 27514

CARROLL M. WILLIAMS
Harvard University
Harvard Biological Laboratory
Cambridge, MA 02138

H. ALAN WOOD
Boyce Thompson Institute
Insect Pathology Resource Center
Tower Road
Ithaca, NY 14853

CHIH-MING YIN
University of Massachusetts
Department of Entomology
Amherst, MA 01003

SHINJI YOKOI
Sankyo Company, Ltd.
Agricultural Chemical Research
Laboratory
1041, YASU, YASU-CHO,
YASU-GUN
Shiga-Ken 520-23, Japan

Author Index (Contributors)

Adams, T. S., 381
Altaratz, M., 333
Altstein, M., 303
Applebaum, S. W., 333
Arpagaus, M., 239
Aruchami, M., 409

Baines, R. A., 417
Bell, R. A., 255, 337, 341, 345, 353, 357, 361, 377
Bidmon, H.-J., 405
Bird, T. G., 271, 299
Bishoff, S. T., 365
Bodmer, R., 63
Borkovec, A. B., 337, 341, 345, 353, 357, 361, 377, 409
Borovsky, D., 251
Bradfield, J. Y., 163
Brown, M. R., 401
Burns, E., 321
Bushman, D. W., 393

Carlson, D. A., 251
Chen, A. C., 389
Chiang, A.-S., 317, 321
Chiang, J. A., 431
Chiang, R. G., 431
Coast, G. M., 227
Cook, B. J., 413
Cook, H., 439
Cournil, I., 397
Crim, J. W., 401

Davey, K. G., 325, 431
Davis, M.-T. B., 295
Davis, N. T., 447
Davis, R. E., 377
DeMilo, A. B., 255, 259, 341
Downer, R. G. H., 275, 417
Drake, A. F., 235
Dunkelblum, E., 303

Evans, P. D., 397

Fescemyer, H. W., 349, 377
Ford, M. M., 247
Fugo, H., 211

Gäde, G., 231, 235, 243
Gadot, M., 317, 321
Gazit, Y., 303
Gelman, D. B., 353, 357, 361
Geraerts, W. P. M., 3
Gerst, J. W., 381
Giebultowicz, J. M., 299, 313
Gilbert, L. I., 333, 365
Girardie, A., 219
Girardie, J., 219
Goldsworthy, G. J., 227, 235
Granger, N. A., 405

Hakim, R. S., 295
Hamblin, M., 285
Hayes, D. K., 263
Hayes, T. K., 163, 243, 247

Herzog, G. A., 401
Holman, G. M., 443
Horodyski, F. M., 215
Howes, E. A., 105
Huet, J.-C., 219
Hughes, P. R., 285

Imberski, R. B., 337, 345, 353, 357
Isaac, R. E., 267, 281
Ishizaki, H., 211

Jadhav, G. K., 369
Jenkins, A. C., 401
Joosse, J., 3

Kallapur, V. L., 369
Kamito, T., 211
Karaiyan, K., 409
Kataoka, H., 39
Keeley, L. L., 163, 247
Kelly, T. J., 255, 337, 341, 345, 349, 353, 357, 361, 377, 409
Kempe, T. G., 127
Kempton, L. R. C., 281
Klun, J. A., 309
Kochansky, J., 271
Kono, T., 211
Kramer, S. J., 39

Lange, A. B., 435, 439
Lea, A. O., 401
Lehman, H., 447
Loeb, M. J., 259
Loughton, B. G., 385

Masler, E. P., 255, 337, 341, 345, 349, 353, 357, 361, 377
McQuiston, A. R., 329
Meola, S. M., 427, 443
Mizoguchi, A., 211

Nagasawa, H., 211
Nelson, J. O., 393

Orchard, I., 421, 439

Patil, M. U., 451
Pearson, K. G., 421
Pener, M. P., 365
Pernollet, J.-C., 219
Philp, R. J., 227
Pillmoor, J. B., 281
Priestly, R. M., 267

Rafaeli, A., 309
Raina, A. K., 207, 299, 309, 313
Ramirez, J.-M., 421, 439
Richard, D. S., 333
Riddiford, L. M., 215
Robb, S., 397
Rountree, D. B., 365

Sakagami, Y., 211
Sato, B., 211
Schal, C., 317, 321
Schofield, P. K., 105
Schooley, D. A., 39
Segal, D., 333
Sevala, V. L., 325
Sheppard, C. A., 259
Smith, P. J. S., 105
Soroker, V., 309
Stoffolano, J. G., Jr., 373
Stumpf, W. E., 405
Suzuki, A., 211
Svoboda, J. A., 295

Tanaka, H., 211
Thangaraj, T., 409
Thompson, C. S., 329
Thompson, J. M., 443
Thornton, J. M., 235
Thyagaraja, B. S., 337, 341, 345, 353, 357, 361, 377
Tobe, S. S., 329
Toschi, A., 39
Totty, N.F., 227
Toutant, J.-P., 239
Treherne, J. E., 105
Truman, J. W., 215

Uebel, E. C., 313

van Beek, N., 285
Veenstra, J. A., 223
Verhaert, P. D., 275

Wagner, R. M., 271, 413
Walgraeve, H. R., 275
Wheeler, C. H., 227, 235
Willis, R. J., 281
Wilmot, C. M., 235
Wood, H. A., 285
Wright, M. S., 443

Yin, C.-M., 291, 373

Zou, B.-X., 373

Subject Index

Abd-A, 89
Abdominal, 67
Abdominal ganglia, 22, 435
Accessory glands, 299
Accommodate, 431
Ace mutants, 240
Acetate, 54
Acetonitrile, 53, 57
Acetyl, 132
Acetylcholinesterase, 239–242
  amphiphilic, 239, 242
  dimer, 239, 242
  disulfide linked
    in *Drosophila*, 239
Achaete-scute complex, 75, 80
Acheta, 167
Acheta domesticus, 227
Acid
  cleavage, 50
  activitation ratio, 339, 350, 366, 440
  acylation, 132
  address, 244, 246
  adenylate cyclase, 179, 386, 417
Adaptive changes, 8
Adipokinetic hormone, 28, 41, 43, 165–184, 235, 263, 385, 448
  AKH-I, 167, 174, 235–237, 268
  AKH-II, 167, 174, 235–237
  AKH-like peptides, 238
  bioanalogs of, 166
  effect on lipid, 167, 168, 235
  precursor-related peptides, 174
Adipokinetic/hypertrehalosemic hormone family, 166, 169, 176–178, 180
Adipokinetic/red pigment concentrating hormone family, 224, 231
Adsorption chromatography, 40
Adsorptive losses, 47
Aedes
  aegypti, 169, 251, 252, 402
  atropalpus, 251, 255
Affinity
  chromatography, 47
  column, 264
African rosechafers, 231
After-discharges, 23
Agonist, 129, 235, 271, 405
AKH, *see* adipokinetic hormone
AKH/HrGH, *see* adipokinetic/hypertrehalosemic hormone family
AKH/RPCH, *see* adipokinetic/red pigment concentrating hormone family
Alkyl-bonded silica, 46
Alkyl groups, 52
Allatostatins, 165, 185
Allatotropin, 43, 55, 56, 61, 165, 170, 171, 185
Amastatin, 268
American cockroach, 275
Amidated C-terminal, 235
Amidation, 132, 174, 233
Amide bond, 132, 53
Amino acid
  analysis, 228, 232
  replacement analogs, 244, 248
  sequence, 212, 216
D-amino acids, 131, 132

L-amino acids, 131, 132
Amino peptidase, 132, 140, 175
Amino protecting groups, 132
4-Aminopyridine, 330–332
Ammonium acetate, 50
Amphipathic molecule, 238
Amphihilic, 239
Analgesic, 128, 136
Analogs, 131, 190–193, 252
Angiotensin, 133
Angiotensinogen, 133
Anion exchange chromatography, 54, 219
Anodonta cygnae, 10
Antagonists, 129, 271
Antennal nerves, 444
Antheraea polyphemus, 179
Antidiuresis, 169, 179, 185
Antidiuretic, 221, 181
  factor, 394
  hormone, 165, 189
  gene, 188
Antifeedant, 365
Antigonadotropins, 165, 251
Antihormones, 189
Anti-juvenile, 221
Antioxidant, 49
Antiparallel dimer, 17
Antipeptides, 189
Aplysia californica, 17, 22, 24–27, 30
Apolysis, 409, 411
Apterous
  gene, 335
  mutation, 333
Arborization, 447
Arginine, 136
Arginine vasopressin, 181
Aromatic
  regions, 53
  residues, 249
Aromaticity, 249
Ascorbate, 282
Aspartic acid, 49
Atrial natriuretic peptide, 389
Autoexcitatory mechanism, 24
Autographa californica nuclear polyhedrosis virus, 286
Autoimmune disorder, 128
Autoradiography, 406
Azadirachtin, 365–368

Baculovirus, 187, 285–287
  cloning-expression vectors, 186
  expression vector, 188, 286
  vectors, 188
Bag cell system, 22–25
Basal
  lamina, 319
  metabolism, 169
Basic peptides, 45
Binding, 271
  pockets, 132
  sites, 406
Bioanalog, 109
Bioassay, 301, 393, 414
  for proctolin, 414–416
Biocontrol agent, 187
Bioengineering, 191
Biogenic amine, 179, 180, 191, 381, 415
  effect on neurosecretion, 179, 180
  in invertebrates, 179
Blaberus discoidalis, 167, 184, 247
  heartbeat rate, 169
  hypertrehalosemic hormone, 166, 168, 175–178, 243, 244, 247–250
Blatella germanica, 317, 321, 323
Blocked *N*-terminals, 223
Blood, 393
  brain barrier, 31, 105–123
  octopamine and, 110, 113
  potassium permeability of, 111
  meal, 390
Bombyx mori, 171, 217, 369–372
  diapause in, 185
  eclosion hormone in, 29, 170, 171
  eclosion hormone gene in embryo, 211
  EH gene in, 211–213
  MRCH in, 170, 171
  PTTH in, 10, 42, 44, 61, 170, 190
Bombyxin II, 10
Boophilus microplus, 281
Bovine
  blood, 390
  plasma, 391
  serum albumin, 48
Bradykinins, 21

Brain, 259, 275, 361, 444, 445, 451–453
  cDNA library from, 295
  extracts of, 309, 346, 365
  polypeptides in, 295
  PTTH in, 337, 340, 347
  removal of, 345, 347
Brain-retrocerebral complex, 450
BrdU, *see* 5-Bromo-2-deoxyuridine
  5-Bromo-2-deoxyuridine, 69, 70, 117–120
B-turn, 236, 237, 244, 248, 249
  prediction of, 236
Buffer salts, 54
Bursa copulatrix, 314
Bursicon, 165, 448

C_{18},C_8,C_4,C_3 Functionality, 52
Calcitonin, 131
Calcium, 325–332
  channels, 329
  channel blocker, 326, 330
  ions, 329
Calfluxin, 24
Calling behavior, 309
Calmodulin, 386
Capillary gas chromatography, 303, 309
Captopril, 130
Carausius, 167, 177
  HrTH II, 167
α-Carbons, 131
Carboxymethyl column, 54
Carboxypeptidase, 132, 133, 136, 140, 142, 233
Carboxyterminal fragments, 385
Cardioaccelatory factor, 43
  hormone, 165, 166, 169, 179
  peptide, 223
Cardioexcitatory peptide, 401
Cardioinhibitory peptide, 167
Catalytic subunit, 239
Catecholamine, 284
Cation exchange chromatography, 46, 54
Cattle tick, *see Boophilus microplus*
Caudo-dorsal cells, 22, 26
CD, *see* circular dichroism spectroscopy
CDCH gene family, 25, 27
cDNA clones, 380
cDNA library, 216, 295, 297
Cell
  cultures, 320
  membranes, 130
  number, 317
  size, 318
Cell-cell
  communication, 63
  interaction, 63
Cellular identities, 64
Central nervous system, 68, 138
Cepaea, 18
Channels, *see* ion channels
Chemical synthesis, 128
Chiral, 131
Chitin, 409
Chitinase enzyme, 409, 411
Chlordimeform, 180
Chloride-transport factor, 165
Cholecystokinin, 130, 178
Chorodontal organ, 67, 434
Chromatography, *see* affinity, anion exchange, gel-filtration, ion exchange, reverse phase, size-exclusion
Chrysodeixis chalcites, 303
Chymotrypsin, 133
Ciliary region, 432
Cilium, 431
Circular dichroism spectroscopy, 236, 237, 249
Circumesophageal nerves, 444
Classical transmitters, 4
Clearance, 130, 131, 134
Cleavage sites, 174
Clones, 216
Cloning, 295
CNS, *see* central nervous system
Cobalt backfills, 440
Cockroach, *see Blaberus discoidalis, Blatella germanica, Leucophaea maderae, Diploptera punctata, Nauphoeta cinerea, Periplaneta americana*
Cocoon shell, 372
Coding region, 286
Codon, 213

Coelenterates, 7, 20
Coleoptera, 231
Collagenase, 318
Collateral system, 22
Color change, 164
Column packing, 52
Commitment, 406
Competitive inhibitor, 283
Confirmation, 244
Conopressins, 16
Consensus sequence, 213
Conserved amino acids, 83
Contraction, 417
Conus
  geographus, 16
  striatis, 16
Corazonin, 223
Corn earworm, *see Heliothis zea*
Corpora allata, 164, 317, 321, 329, 332, 394, 395, 405, 407, 448, 449
  volume of, 323
Corpora cardiaca, 41, 163, 172, 174, 179, 180, 185, 186, 217, 223–227, 272, 276, 278, 279, 370, 371, 394, 402, 444, 448, 449, 451
  nervous connection, 219
  neuropeptides in, 223–226, 272
Corpora cardiaca/corpora allata complex, 56, 163, 276, 277, 340
Cotton bollworm, 285
Counter-current distribution, 40
Cross-linkages, 131
C-terminal
  amidated, 235
  cyclization 132, 239
Cut gene, 80
Cut gene product, 81
Cuticle, 67, 409, 410
  bristles in, 67
  plasticity factor, 165
Cuticular structures, 84
Cyanogen bromide cleavage, 149
Cyclic AMP, 179, 182, 309, 417
  levels in fat body, 167
Cyclic GMP, 179
Cyclic nucleotides, *see* cyclic AMP, GMP
Cytochromes, 168
  synthesis of, 169
Cytoplasm, 407

Dale's concept, 5
Daughterless gene, 75
Deamidase, 175
Decapeptide, 443
Defatting, 44
Degradation, 130, 131, 134, 192, 271
  by proteases, 42
Degradative, enzymes, 174
  proteases, 187, 193
Dendrite, 85, 431, 432
Desalted, 50
Design of peptides, 127
Deutocerebral lobe, 444
Developmental
  fate, 79
  genes, 93
Diapause, 341–343
  hormone, 165
  neuroendocrine involvement in, 164
Diethylaminoethyl column, 54
Differential packaging, 24
Differentiated tissues, 64
Digestive processes, 401
5, 7 Dihydroxytryptamine, 439, 441
Dimer precursors, 29
Diploptera punctata, 329
Disparlure, 313, 315
Disulfide bonds, 131, 240
Diuresis, 169, 393–396
Diuretic
  factors, 394
  hormone, 43, 56, 165, 170, 171, 179, 181, 185, 188, 189, 392, 450
  peptide, 227–230
Divide symetrically, 71
Division patterns, 66, 71
DNA replication pattern, 69
Dopamine, 180, 283
Dorsal unpaired median (DUM) neurons, 339–341
Dorso-ventral genes, 74
Dose-response, 244, 248, 261, 340
Dot-blotted RNA, 297
Douglas fir Tussock moth, 285

Downstream, 93
Drosophila embryo, 68, 76, 77
Drosophila melanogaster, 63–103, 171, 217, 218, 239–242, 267–269, 333–335, 406, 407
  acetylcholinesterase in, 239–242
Drosulfakinin, 171, 172
Drug design, 129
Dynein arms, 433

Ecdysiotropin, 361–364 (*see also* EDNH, PTTH, Testi Ecdysiotropin)
Ecdysis, 215
Ecdysone, 359, 361–364, 409–412
  3-dehydro, 358, 361–364
  20-hydroxy, 379, 405
Ecdysteroid, 170, 183, 341–343, 349–352, 357, 358, 373–376, 381, 405
  in hemolymph, 252
  ketoreductase and, 340, 357–359
  receptors, 407
  synthesis, 165
  titers, 343, 351, 374, 379, 406, 411
Eclosion, 211, 350, 393
Eclosion hormone, 29–31, 43, 179, 185, 189, 211–213, 215–218
  precursor of, 30
Ectodermal cells, 66
Edman degradation, 224, 228
ED_{50} values, 340
EDNH, *see* egg development neurosecretory hormone
Egg development neurosecretory hormone, 165, 185, 251
Egg(s), 357–359
  development of, 251
  diapause in, 185
  laying control in snails, 22–27
  maturation, 381
EH, *see* eclosion hormone
Electron microscopy, 431, 433
Electrophoresis, 264, 296, 377, 379
Electroretinogram, 87
ELH gene family, 26
ELISA, *see* enzyme-linked immunosorbant assay
Embryogenesis, 341
Embryonic PTTH, 354, 357
Embryonic peripheral nervous system, 67
Enalapril, 130
Endocrinology, 341
Endopeptidase, 141, 175, 211, 267–269
  β-Endorphin, 136, 137
Endosulfan, 369–372
Energy metabolism, 169
Engrailed gene, 88
Enkephalin, 128, 129, 135–137, 268–269
Enzyme
  amidation, 174
  degradation by, 129, 138
  inhibitors, 130
  membrane-bound, 267
  metabolism, 127
  processing by, 186, 191
  linked immunosorbent assay, 373, 449, 450
Epitopes, 447
Erythropoietin, 131
es
  organ development, 84
  to ch transformations, 84
Ethanol, 53
European cornborer, *see Ostrinia nubilalis*
Even-skipped, 92
Exopeptidases, 132
Expression vectors, 184
External sensory organs, 67
Extraction, 44

FAB-MS, 229
Face fly, *see Musca autumnalis*
Fat body, 235, 377, 407
  of *Blaberus*, 168
  cyclic AMP in, 167, 179, 235
  lipid stores in, 168, 179
  metabolism in, 169
  mitochondriogenesis in, 168
  proline synthesis in, 168
Field trial, 287
Final purification step, 52
Flanking residues, 136
Flexible linear molecules, 131

Flight, 421–424
Flight behavior, 421
Fluid secretion, 169, 170
Fluorescent probes, 444
FMRF amide, 167, 171, 401–404, 443
family, 17–19, 400
gene, 17
-like peptides, 397–400
precursor, 17
-related peptides, 167, 171
Follicles, 382
Follicle cells, 325–327
Foreign gene, 286
Formamidine insecticides, 180, 191
Formamyl, 132
Forskolin, 179, 386, 419
Frontal ganglion, 394, 401
Fushi tarazu gene, 92
Fusion protein, 149

Ganglionectomy, 394
Ganglion mother cells, 79
Gap junction, 329
Gap junctional conductance, 330
Gastrin, 172, 178, 181, 182
Gel electrophoresis, 239, 296
Gel-filtration chromatography, 46, 268
Gene
expression, 27, 295–298
families, 25–27
network, 63
promotor, 286
therapy, 127, 129
transfer, 128
Genetic
analysis, 65
engineering, 185–187, 189, 190
manipulation, 64, 65
Genome, 7
Genomic
DNA, 213
library, 213, 216
Glia, 105, 115, 116
Glial cells, 67
Gliosis, 117, 118, 120, 121
pGlu-aminopeptidase, 175
Glucagoncagon, 131, 181, 182
Glucocorticoid receptor, 407
Glutamate, 414
Glycine-phenylalanine bond, 140
Glycogen, 168, 369
Glycolipid anchors, 240
Gonadotropic factor, 381
Gonotrophic cycle, 317, 329
Grace's medium, 406
Gradient, 53
Granule-containing cells, 116, 117
Grasshopper, 66
central nervous system of, 78
Growth
hormone, 131
rate, 345
regulator, 365
Gryllus bimaculatus
AKH bioanalogs in, 166
Gypsy moth, *see Lymantria dispar*

hANP, *see* Human atrial natriuretic peptide
Haemocytes, 116–122
Half-life, 42, 130
Head
activator, 20, 21
critical period, 347
factor, 345
inhibitor, 20
litigation, 347
Heartbeat, 124, 125
Heartrate, 164, 165, 169
Heliothis armigera, 309–312
H. peltigera, 303–306
H. viresecens, 188, 189
H. zea, 166, 171, 172, 226, 299–302, 310–313, 393–395, 401–404
AKH in, 166
Heme synthesis, 168, 169
Hemimetabolous, 295
Hemolymph, 272, 390, 391, 393
ecdystero1d, 345, 352, 367, 406
lipid, 221, 421
proteins, 379
trehalose, 247
vitellogenin, 377, 378, 381–383
Heptafluorobutyric acid, 50, 223
Heteroallelic, 242
Hierarchy of genes, 66
High performance liquid chroma-

tography, 223, 227, 232, 243, 247, 252, 255–257, 259–261, 268, 272, 282, 309, 319, 362, 394, 397, 398, 414
High stringency hybridization, 216
Hindgut, 390, 391
contractility of, 165, 168, 226
Holometabolous, 295
Homeobox, 83
Homeodomains, 83
Homeostasis, 164
Homologous peptide, 445
Homology, 83
Hormone address, 244, 246
Hormone message, 244, 246
House fly, *see Musca domestica*
HPLC, *see* high performance liquid chromatography
5HT receptors, *see* 5-Hydroxytrypt-amine
HTH, *see* Hypertrehalosemic hormone
Human atrial natriuretic peptide, 389–392
Hyalophora cecropia, 299
Hydra, 6, 20, 21
Hydrocarbon chains, 52
Hydrogen bonding, 141
Hydrolysis
acid, 224
of neuropeptides, 267
Hydrophilic peptides, 52
Hydrophobic
domain, 240
interactions, 46, 240
peptides, 47, 52
signal sequence, 147
Hydrophobicity, 249
6-Hydroxydopamine, 381–383
Hydroxylamine, 240, 379, 405, 407, 410, 411
5-Hydroxytryptamine, 180, 417–420
receptors, 418
Hyperglycemic
condition, 370
hormone, 165, 184
Hyperlipaemia, 221, 232
Hyperplasia, 80
Hypertrehalosemic effects, 221
Hypertrehalosemic hormone, 180, 226, 243–246, 247–251
in *Blaberus*, 166–168, 175–177
in *Carausius*, 167, 177
Hypoglycemic hormone, 165
Hypolipaemic hormone, 385–387
Hypothalamic releasing peptides, 179
Hypotrehalosemic hormone, 264

IBMX, *see* Isobutyl-methylxanthine
Immunoaffinity support, 47
Immunocytochemistry, 397, 401, 443–445
staining, 283, 439
Immunoelectrophoresis, 381
Immunofluorescent, 447
Immunogenicity, 135
Immunohistochemistry, 213, 436, 440
Immunological probes, 443
Immunoprecipitation, 283
Immunoreactivity
insulin-like, 387
proctolin-like, 436
In situ hybridization, 217, 297
In vitro
assay for PTTH, 349
metabolism, 39
radiochemical assay, 333
In vivo bioassay, 345–347, 393
Inhibitors, 282
of protein kinase, 327
Inositol phosphates, 275–279
Insect
control, 183, 185, 186
neuropeptides, 130, 443
viruses, 128, 130
Insecticides, 180, 187, 369–372, 446
Insulin, 9–15, 128, 147–150, 181, 182, 385
gene, 14
human, 129, 131, 147–150, 170
-like growth factor II, 170
-like peptide, 10
-like precursor molecule, 387
receptors, 10
-related peptides, 9–15
Interactions, 63

Intercellular junctions, 319
Interendocrine mechanisms, 405
Interferon, 131, 132
Internal stretch receptor, 67
Intrinsic
  activity, 127
  cells, 448
Intron, 213
Invaginations, 434
Ion channels, 329–332
  blockers of, 330
  calcium and, 329, 332
  potassium and, 330
  sodium and, 332
Ion exchange chromatography, 40, 42, 46, 51, 54, 55, 59, 227
Ion-pairing and suppressing reagents, 50, 53
Ion pairs, 54
Ionic balance, 393
Isobutyl-methylxanthine, 179, 418
Isocratic, 53

JH, *see* juvenile hormone
Junctions
  intermediate, 433
  intermediate-like, 434
  septate, 433, 434
Juvenile hormone, 165, 169–183, 185, 221, 251, 325–327, 329, 333–335, 373–377, 383, 405
  analog of, 374, 378, 379
  biosynthesis of, 165, 333, 374, 375
  inhibition of, 165
  synthesis of, 165, 333
Juvenile hormone III, 317–319, 321, 322
  biosynthesis, 317–319

Ketoecdysteroid reductase, 355, 361–364

Lambda EMBL-3 vector, 213
Lambda Zap phage, 297
Lateral inhibition, 78
Lateral neurosecretory cells, 347, 387
Leader sequence, 147
Lepidoptera: Noctuidea, 259–261, 303–306
Lepidopterous insects, 285
Leptinotarsa decemelineata, 231
Leucokinins, 167–172
Leucomyosuppressin, 167, 168, 404, 443–446
  synthetic analogs of, 446
Leucophaea maderae, 163, 167–171, 323, 400, 443–446
  myokinins of, 167–168
Leucosulfakinin, 167, 223–226, 404
Ligation, 345
Lineage relationships, 71
Locust, 421–424, (*see also Locusta* and *Schistocerca*)
  adipokinetic hormone, 167, 176, 177
  diuretic hormone of, 171
Locusta migratoria, 11, 41, 166, 171, 219–221, 227, 400
  ACTH-like peptides in, 227
  AVP-like peptides in, 227
  diuretic peptides in, 227
  suboesophageal ganglion of, 400
Lucifer yellow, 440
Luteinizing hormone releasing hormone, 131
Lymantria dispar, 259–261, 310, 313–315, 337–340, 341–343, 345–347, 349–352, 353–355, 357–359, 361–364, 377–380
Lymnaea stagnalis, 10–12, 17, 19, 22–27
Lyophilizing, 49
Lysine, 136

Macrocallista nimbosa, 17
Malpighian tubules, 169, 175, 179, 188, 227, 391
Mandibular closer muscle, 417–420
Manduca, 170–181
Manduca sexta, 29, 30, 215–218, 295–298, 365–368, 393, 405–408, 447–450
  adipkinetic hormone in, 166, 184
  allatotropin in, 43, 55, 56, 171
  brain polypeptides of, 296
  eclosion hormone in, 29, 170, 171
  FMRFamide in, 402
  gene expression in, 295–298

prepro adipkinetic hormone in, 173
Mass spectrometry, 252
Mating, 299
May beetles, 231
md neurons, 67
Mealworms, 231
Median abdominal nerve, 445
Median neurosecretory cells, 347, 387, 451
Melanization and reddish coloration hormone, 170, 171
Melanoplus, 452–454
Melolontha, 231, 232
hippocastan, 231
Membrane-bound enzymes, 267
Mesothoracic ganglia, 435
Metabolic
rate, 372
stabilization, 132
Metabolite-mobilizing peptides, 183
Metallo-endopeptidase, 175
Metamorphosis, 185, 215, 295–296, 353, 378
Metathoracic ganglia, 435
Methanol, 53
Methionine, 49, 319, 380
Methoprene, 374, 378
Methyl farnesoate, 333
2-(Methylthio) ethanol, 49
Microbore columns, 51
Microtubules, 432
Midgut, 251, 390, 391, 401–404
Mobile phase, 53
Mode of action, 179, 192
Modulation, 417, 421
Molecular mechanisms, 63
Molecular size, 42
Molluscan
cardio-acceleratory peptide, 167
insulin-related peptide, 11–12
tetrapeptide, 443
Molting, 164, 345–347, 409
and metamorphosis, 164, 185, 345
inhibition, 365–367
Monoclonal antibodies, 189
Monolayer, 317
Morphine, 129
Morphogenic gradients, 65
Mosaic flies, 86
Mosquito oostatic hormone, 251
Mother cell, 94
MRCH, *see* melanization and reddish coloration hormone
Multigene families, 26
Multivoltine, 372
Musca autumnalis, 263–265, 271–273
Musca domestica, 251, 255–257, 299, 381–383
Muscle
skeletal, 443
visceral, 443
Mutations, 71, 240, 335
Myc-gene family, 78
Myoactive peptides, 406
Myographic recordings, 414
Myoinhibitory peptides, 443
Myokinins, 167, 168, 183, 193
Mystropic peptides, 176, 443
Mytilis galloprovincialis, 10

N-terminal
cyclization, 132
methylation, 142
Na^+,K^+-ATPase, 325
Natural sequence, 128
Nauphoeta cinerae, 166, 167
Nerves
antennal, 444
circumesophageal, 444
Nervous corpora cardiaca, 219, 443
Network of genes, 63,
Neural lamella, 107
Neural precursor, 63
Neurite, 440
Neuroblasts, 66, 78
Neurodevelopment, 63–96
Neuroendocrine cascade, 373
Neurogenic
genes, 79
model, 72
mutants, 80
Neurohaemal area, 440, 444
Neurohaemal system, 447–450
Neurohormone, 53, *see* review (163–203) and specific hormone
D, 43

genes 182, 185, 187, 190
mode of action, 179, 192
polytropic, 221
Neuromere, 445
Neuromodulators, 5, 135
Neuronal
connectivity, 64
messengers, 6–8
microenvironment, 105
precursor, 66
Neurons, 67, 407
Neuroparsins, 219–221
Neuropeptide, 3–31, 39–61, 135, 167, 180–182, 215, 223–226, 259, 263–265, 267–269, 271–274, 303, 443, (*see also* specific neuropeptide)
analogs, 184
applications to insect population management, 263
isolation of, 39–62
Neuropile, 440, 452
Neurosecretion, 4, 180, 185, (*see also* neurohormone, neuropeptide)
Neurosecretory activity, 179
Neurosecretory cells, 163, 213, 217, 221, 387, 451
lateral, 347
median, 347, 451
Neurosecretory material, 451
Neurotransmitters, 5, 43, 135, 281
Noctuidae, 188, 303
Nonsynaptic release, 22
Northern blots, 217
Nuclear binding, 406
Numb gene, 93

Octopamine, 110–113, 180, 281, 414, 421, 422, 439
perfusion of, 422
Octopaminergic, 422, 439
Oligonucleotide probes, 213, 216
Ommatidia, 87
Oocytes, 326
development of, 321
length of, 322
Oogenesis, 373
Ooplasm, 382
Oostatic
factors, 185, 256
hormone, 255–257
Ootheca, 321
Opioid, 129
antagonism, 18
peptide, 18, 139
Organic modifier, 53
Ostrinia nubilalis, 361–364
Ouabain, 325
Ovarian
cycle, 321
development, 381–383
Ovarian maturation assay, 256
Ovariectomized, 321–324
Ovaries, 252, 255, 414
Ovarioles, 325
Oviduct, 414, 436
Ovulation, 22
hormone, 165
Oxidation, 49
Oxytocin, 16, 131
vasopressin family, 181

Pandalus, 167
Parenchymal cells, 329
Pars intercerebralis, 406, 408, 444
Pars lateralis, 406
Patency, 325
Pathfinding, 64
Pathways, 64
PBAN, *see* pheromone biosynthesis activating neuropeptide
Pepstatin A, 45, 56
Peptides, (*see also* neuropeptide, neurohormone)
AKH-like, 238
bioengineering of, 190
blocked N-and C-termini of, 267
C, 10
degrading enzymes and, 131
drugs, 134
drug design, 127
FMRFamide-like, 397–400
FMRFamide-related, 167
homologous, 445
immunogenicity of, 135
losses during isolation, 47–50
mapping of, 228
mechanisms of separation, 51

metabolism of, 132
mimic of, 129–130
myoactive, 404
myoinhibitory, 443
myotropic, 443
secondary structures of, 237
Peptidases, 133
Peptidergic factors, 361
Peripheral expression genes, 27
Peripheral nervous system, 66–71
Perineurium, 107, 110, 116–121
Periplaneta americana, 42, 43, 223–226, 275–279, 435
cardioaccelerator of, 169
hyperglycemic response in, 178
proctolin in, 42, 175, 435
Periplanetin, 43, 224
Perisulfakinin, 223
Perivisceral organ, 447, 449
Permissive temperature, 241
Peroxidase probes, 444
Pest control, 183
Pesticides, 285–287
pH gradient, 55
Pharate adult, 56, 211, 296, 448
Phenotypic changes, 65
Phentolamine, 423
Phenyl-bonded silicas, 224
Phenylisothiocyanate, 224
Phenylmethylsulfonyl fluoride, 45, 56
Pheromone, 299–302, 303–306
biosynthesis activating neuropeptide, 140, 171, 299–302, 303–306, 312
factor, 309
glands, 309
synthesis regulating hormone, 170, 185
synthesis stimulating hormone, 165
synthetic PBAN, 301
Phormia regina, 373–376
Phosphatidylinositol, 239
Phosphodiesterase inhibitor, 418
Phosphoric acid, 53
Phosphoinositide metabolism, 275
Phosphoramidon, 269
Phosphorylase activity, 370
Photophase, 305
Photoreceptors, 87
Pine sawfly, 285
Plaque, 431
Plasma half-life, 130
Plasma membrane, 332
PMSF, *see* phenylmethylsulfonyl-fluoride
PNS, 71
Polyclonal antisera, 378
Polyfunctional, 137
Polyhedrin gene promotor, 286
Polymer-based columns, 53
Polypropylene, 47
Polytropic neurohormone, 221
Ponasterone A, 405–408
Pore size, 52
Positional information, 75
Postembryonic
development, 405
neurogenesis, 295
Posterior compartment, 89
Potassium channel, *see* ion channels
Potato beetle, 231
Precipitation, 42
Precocene II, 373–376
Precursors, 271
Prehatch embryonic eggs, 337
Preparative columns, 51
Prepattern genes, 74
Pressure, 431, 434
Probe, 216, 443
Proboscis, 84
Processing enzymes, 174, 186, 191
Proctodaeum, 362
Proctolin, 39–41, 165, 166, 178, 180, 183, 414, 416, 436–437
bioassay for, 229
degradation by aminopeptidase, 78
hindgut muscle and, 74–78
-like immunoreactivity, 436
Proenzymes, 132
Prohormones, 24, 132, 136, 137, 151, 172, 174, 186, 253
Prolidase activity, 229
Proline, 168
Proneural
clusters, 78

genes, 74–78
Propanol, 47
Proprioceptors, 421, 434
Protease digestion, 354
Proteases, 240, 251–253
Protein
engineering, 185
kinase C, 325–327
metabolism, 132
sequence similarity, 78
sequencer, 220
sequencing, 272
synthesis, 169, 185, 296
synthesis stimulating hormone, 165
Proteinase K, 240
Prothoracic glands, 337–340, 349–355, 357–359, 361–364
activation, 347
bioassay, 347, 354
Prothoracicotropic hormone, 10, 12, 165, 170, 181, 183, 185, 189, 190, 337–340, 341–343, 345–347, 359, 361, 365–368, 387, 396
embryonic, 354
isolation from *B. mori*, 42
post-embryonic, 354
titer of, 337
Protocerebral lobe, 444
PTH-amino acid, 224
PTTH, *see* prothoracicotropic hormone
pUC18 vectors, 213
Pupae, 349, 350
Pupal commitment, 347
Pupariation hormone, 165, 185
Pupation, 345
Pyroglutamate aminopeptidase, 223, 228, 233
Pyroglutamyl, 132

Radiochemical assay, 319, 321, 333, 373
Radioimmunoassay, 259, 338, 342, 357, 361–363, 365, 373, 381, 389, 397, 401, 406
RCA, *see* radiochemical assay
Receptor, 137, 138, 247, 253, 263, 271, 325
binding, 272
interactors with, 238
neuroendocrine, 184
Receptivity termination factor, 299
Recognition, 244, 247
Recombinant DNA technology, 127, 130, 149, 151, 186
Red pigment concentrating hormone, 28, 166
(*see also* adipokinetic/red pigment concentrating hormone family)
Reductive alkylation, 211
Reproduction, 164
Renin, 133
Resolving power, 46
Restriction map, 212, 216
Restrictive temperature, 241
Reverse-phase, 42, 46, 51
Rhodnius prolixus, 251, 325–327, 435–437, 439–441
RIA, *see* radioimmunoassay
Ring glands, 263, 333
Rocket immunoelectrophoresis, 381
Romalea microptera, 166

Salivary gland, 391
Salt gradient, 55
Salt regulating hormones, 193
Samia cynthia ricini, 170, 190
Sarcophga bullata, 381
Scarabaeidae, 231
Schistocerca gregaria, 28, 41, 166, 167, 175, 267, 268, 397–400
S. migratoria, 166
S. nitans, 166, 174
Schisto-FLRF-amide, 397
Scolopale cell, 432–434
Scolopales, 81
Scotophase, 299, 305, 310
Second messengers, 179, 192
Secondary precursor, 94
Secondary structures of peptides, 237
Segment polarity, 74
Segmental
arrangements, 65
identities, 80
nerve, 431

Selective glial disruption, 115
Selective glial lesioning, 117
Selectivity, 52–54
Selector genes, 80–93
Semi-preparative
  columns, 51
  reverse-phase liquid chronotography, 59
Semi-synthesis, 149
Sense organs, 421–424, 431
Sensilla, 66
Sensory organs, 71
Sensory receptor, 431–434
Sequence analysis, 227
Sequencing, 219–221
Serotonin, 440, (*see also* 5-hydroxytryptamine)
Serotonergic neurons, 439
Sex pheromone, 313–315
  biosynthesis, 303–306
Sexual
  behavior, 334
  receptivity, 321
Signal, 321
  peptide, 216
  sequence, 136, 147
Silkworm, *see Bombyx mori, Samia cynthia ricini*
Size-exclusion chromatography, 46, 227, 255
Skeletal muscle, 443
Solid-phase extraction, 50
Somatostatin, 131
Southern blot, 216
Species selectivity, 245
Specific antibody, 82
Specification of neurons, 66
Spermatheca, 314, 315, 436
Spermathophore, 314
Spodoptera litura, 409–412
Stable analogs, 131
Stable fly, *see Stomoxys calcitrans*
Stem cell, 66
  division pattern, 71
Step elution, 56
Step gradients, 47, 57
Stomoxys calcitrans, 389–392, 413–416, 443–449
Stretch receptor, 67, 421
Structure-activity relationships, 235
Structure-activity studies, 175–179, 243–246, 247–250
Subesophageal ganglion, 387, 400, 406, 445
Sulfopropyl, 46, 54
Support cell, 82
Surgical manipulation, 64
Switching valve, 49
Synaptic membrane, 267, 268
Synaptogenesis, 64
Synganglion, 283
Synthetic
  analogs, 446
  peptide, 211
  pesticides, 286

Tabanidae, 166, 167
Tabanus atratus, 264
Tanning hormone, 189
Target cell receptor, 247
Tarsi, 84
Tegula, 421
Tenebrio molitor, 231–233
Terminal abdominal ganglia, 309
Terminal codon, 213
Termination codon, 216
Testis
  ecdysiotropin, 259–261
  ecdysteroid production, 259
Tetraethylammonium, 330
Tetrodotoxin, 330
Thiodiglycol, 49
Thoracic, 67
Thoracic ganglion, 391, 445
Thoracico-abdominal ganglion, 445
Three-dimensional structure, 131, 132, 151
Transduction, 244, 247
Transperineurial resistance, 110
Trehalose, 221
  synthesis of, 182
Triethylammonium
  acetate, 50, 54
  phosphate, 54
Trifluoroacetic acid, 50, 53, 57, 223
Trypsin, 133, 253
Tryptic digest, 52
Turn frequency coefficients, 248

Tyramine-hydroxylase, 281–284
Tyrosine-glycine bond, 140

Ultrabithorax, 89
Unio Pictorum, 10

V8 protease, 211
Vasopressin, 16, 17, 400, 449
Vasopressin-like immunoreactive cells, 447
Vasotocin, 16
Vertebrates, 163, 179, 181, 184, 446
  Arg-vasopressin, 181
  insulin, 170
Viral genome, 286
Viral pesticides, 285
Visceral muscle, 443
Vitellin, 373
Vitellogenesis, 333, 334, 377–381
Vitellogenin, 251, 325, 373–376, 381–383

Water
  balance, 185
  excretion, 165
  reabsorption, 165
  regulating hormones, 193
Western blots, 377
Wild-type, 76

Xba I - Bam HI, 213

Yolk granules, 382

Z-11-16: ald, 310
Z(11)-hexadecenal, 311
Zophobas rugipes, 231–233

GPSR Compliance

The European Union's (EU) General Product Safety Regulation (GPSR) is a set of rules that requires consumer products to be safe and our obligations to ensure this.

If you have any concerns about our products, you can contact us on ProductSafety@springernature.com

In case Publisher is established outside the EU, the EU authorized representative is:

Springer Nature Customer Service Center GmbH
Europaplatz 3
69115 Heidelberg, Germany

Batch number: 10761588

Printed by Printforce, the Netherlands